STUDENT SOLUTIONS MANUAL

VOLUMES TWO AND THREE: CHAPTERS 21–44

SEARS & ZEMANSKY'S

UNIVERSITY PHYSICS

13TH EDITION

LEWIS FORD
WAYNE ANDERSON

PEARSON

Boston Columbus Indianapolis New York San Francisco Upper Saddle River
Amsterdam Cape Town Dubai London Madrid Milan Munich Paris Montreal Toronto
Delhi Mexico City São Paulo Sydney Hong Kong Seoul Singapore Taipei Tokyo

Publisher:	Jim Smith
Executive Editor:	Nancy Whilton
Project Editor:	Chandrika Madhavan
Editorial Manager:	Laura Kenney
Managing Editor:	Corinne Benson
Production Project Manager:	Beth Collins
Production Management and Compositor:	PreMediaGlobal
Manufacturing Buyer:	Jeffrey Sargent
Executive Marketing Manager:	Kerry Chapman
Cover Design:	Derek Bacchus and Seventeenth Street Design
Cover Photo Credits:	Getty Images/Mirko Cassanelli
Cover and Text Printer:	Bind-Rite Graphics

ISBN 10: 0-321-69667-0

ISBN 13: 978-0-321-69667-0

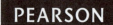

1 2 3 4 5 6 7 8 9 10—BR—14 13 12 11

CONTENTS

PREFACE

This Student Solutions Manual, Volume 2, contains detailed solutions for approximately one-third of the Exercises and Problems in Chapters 21 through 44 of the Thirteenth Edition of University Physics by Hugh Young and Roger Freedman. The Exercises and Problems included in this manual are selected solely from the odd-numbered Exercises and Problems in the text (for which the answers are tabulated at the back of the textbook). The Exercises and Problems included were not selected at random but rather were carefully chosen to include at least one representative example of each problem type. The remaining Exercises and Problems, for which solutions are not given here, constitute an ample set of problems for you to tackle on your own. In addition, there are the Challenge Problems in the text for which no solutions are given here.

This manual greatly expands the set of worked-out examples that accompanies the presentation of physics laws and concepts in the text. This manual was written to provide you with models to follow in working physics problems. The problems are worked out in the manner and style in which you should carry out your own problem solutions.

The Student Solutions Manual Volume 1 companion volume is also available from your college bookstore.

Lewis Ford
Wayne Anderson
Sacramento, CA

ELECTRIC CHARGE AND ELECTRIC FIELD

21

21.3. **IDENTIFY:** From your mass estimate the number of protons in your body. You have an equal number of electrons.

SET UP: Assume a body mass of 70 kg. The charge of one electron is -1.60×10^{-19} C.

EXECUTE: The mass is primarily protons and neutrons of $m = 1.67 \times 10^{-27}$ kg. The total number of

protons and neutrons is $n_{\text{p and n}} = \dfrac{70 \text{ kg}}{1.67 \times 10^{-27} \text{ kg}} = 4.2 \times 10^{28}$. About one-half are protons, so

$n_p = 2.1 \times 10^{28} = n_e$. The number of electrons is about 2.1×10^{28}. The total charge of these electrons is

$Q = (-1.60 \times 10^{-19} \text{ C/electron})(2.10 \times 10^{28} \text{ electrons}) = -3.35 \times 10^9$ C.

EVALUATE: This is a huge amount of negative charge. But your body contains an equal number of protons and your net charge is zero. If you carry a net charge, the number of excess or missing electrons is a very small fraction of the total number of electrons in your body.

21.9. **IDENTIFY:** Apply Coulomb's law.

SET UP: Consider the force on one of the spheres.

(a) EXECUTE: $q_1 = q_2 = q$

$F = \dfrac{1}{4\pi\epsilon_0} \dfrac{|q_1 q_2|}{r^2} = \dfrac{q^2}{4\pi\epsilon_0 r^2}$ so $q = r\sqrt{\dfrac{F}{(1/4\pi\epsilon_0)}} = 0.150 \text{ m} \sqrt{\dfrac{0.220 \text{ N}}{8.988 \times 10^9 \text{ N} \cdot \text{m}^2/\text{C}^2}} = 7.42 \times 10^{-7}$ C (on each)

(b) $q_2 = 4q_1$

$F = \dfrac{1}{4\pi\epsilon_0} \dfrac{|q_1 q_2|}{r^2} = \dfrac{4q_1^2}{4\pi\epsilon_0 r^2}$ so $q_1 = r\sqrt{\dfrac{F}{4(1/4\pi\epsilon_0)}} = \dfrac{1}{2} r\sqrt{\dfrac{F}{(1/4\pi\epsilon_0)}} = \dfrac{1}{2}(7.42 \times 10^{-7} \text{ C}) = 3.71 \times 10^{-7}$ C.

And then $q_2 = 4q_1 = 1.48 \times 10^{-6}$ C.

EVALUATE: The force on one sphere is the same magnitude as the force on the other sphere, whether the spheres have equal charges or not.

21.11. **IDENTIFY:** Apply $F = ma$, with $F = k\dfrac{|q_1 q_2|}{r^2}$.

SET UP: $a = 25.0g = 245 \text{ m/s}^2$. An electron has charge $-e = -1.60 \times 10^{-19}$ C.

EXECUTE: $F = ma = (8.55 \times 10^{-3} \text{ kg})(245 \text{ m/s}^2) = 2.09$ N. The spheres have equal charges q, so

$F = k\dfrac{q^2}{r^2}$ and $|q| = r\sqrt{\dfrac{F}{k}} = (0.150 \text{ m}) \sqrt{\dfrac{2.09 \text{ N}}{8.99 \times 10^9 \text{ N} \cdot \text{m}^2/\text{C}^2}} = 2.29 \times 10^{-6}$ C.

$N = \dfrac{|q|}{e} = \dfrac{2.29 \times 10^{-6} \text{ C}}{1.60 \times 10^{-19} \text{ C}} = 1.43 \times 10^{13}$ electrons. The charges on the spheres have the same sign so the

electrical force is repulsive and the spheres accelerate away from each other.

EVALUATE: As the spheres move apart the repulsive force they exert on each other decreases and their acceleration decreases.

21.15.   **IDENTIFY:** Apply Coulomb's law. The two forces on q_3 must have equal magnitudes and opposite directions.

SET UP: Like charges repel and unlike charges attract.

EXECUTE: The force $\vec{F}_2$ that q_2 exerts on q_3 has magnitude $F_2 = k\dfrac{|q_2 q_3|}{r_2^2}$ and is in the $+x$-direction.

$\vec{F}_1$ must be in the $-x$-direction, so q_1 must be positive. $F_1 = F_2$ gives $k\dfrac{|q_1||q_3|}{r_1^2} = k\dfrac{|q_2||q_3|}{r_2^2}$.

$|q_1| = |q_2|\left(\dfrac{r_1}{r_2}\right)^2 = (3.00 \text{ nC})\left(\dfrac{2.00 \text{ cm}}{4.00 \text{ cm}}\right)^2 = 0.750 \text{ nC}.$

EVALUATE: The result for the magnitude of q_1 doesn't depend on the magnitude of q_2.

21.19.   **IDENTIFY and SET UP:** Apply Coulomb's law to calculate the force exerted by q_2 and q_3 on q_1. Add these forces as vectors to get the net force. The target variable is the x-coordinate of q_3.

EXECUTE: $\vec{F}_2$ is in the x-direction.

$F_2 = k\dfrac{|q_1 q_2|}{r_{12}^2} = 3.37 \text{ N}$, so $F_{2x} = +3.37 \text{ N}$

$F_x = F_{2x} + F_{3x}$ and $F_x = -7.00 \text{ N}$

$F_{3x} = F_x - F_{2x} = -7.00 \text{ N} - 3.37 \text{ N} = -10.37 \text{ N}$

For F_{3x} to be negative, q_3 must be on the $-x$-axis.

$F_3 = k\dfrac{|q_1 q_3|}{x^2}$, so $|x| = \sqrt{\dfrac{k|q_1 q_3|}{F_3}} = 0.144 \text{ m}$, so $x = -0.144 \text{ m}$

EVALUATE: q_2 attracts q_1 in the $+x$-direction so q_3 must attract q_1 in the $-x$-direction, and q_3 is at negative x.

21.23.   **IDENTIFY:** We use Coulomb's law to find each electrical force and combine these forces to find the net force.

SET UP: In the O-H-N combination the O^- is 0.170 nm from the H^+ and 0.280 nm from the N^-. In the N-H-N combination the N^- is 0.190 nm from the H^+ and 0.300 nm from the other N^-. Like charges repel and unlike charges attract. The net force is the vector sum of the individual forces. The force due to each pair of charges is $F = k\dfrac{|q_1 q_2|}{r^2} = k\dfrac{e^2}{r^2}$.

EXECUTE: **(a)** $F = k\dfrac{|q_1 q_2|}{r^2} = k\dfrac{e^2}{r^2}$.

O-H-N:

$O^- - H^+$: $F = (8.99 \times 10^9 \text{ N} \cdot \text{m}^2/\text{C}^2)\dfrac{(1.60 \times 10^{-19} \text{ C})^2}{(0.170 \times 10^{-9} \text{ m})^2} = 7.96 \times 10^{-9} \text{ N}$, attractive

$O^- - N^-$: $F = (8.99 \times 10^9 \text{ N} \cdot \text{m}^2/\text{C}^2)\dfrac{(1.60 \times 10^{-19} \text{ C})^2}{(0.280 \times 10^{-9} \text{ m})^2} = 2.94 \times 10^{-9} \text{ N}$, repulsive

N-H-N:

$N^- - H^+$: $F = (8.99 \times 10^9 \text{ N} \cdot \text{m}^2/\text{C}^2)\dfrac{(1.60 \times 10^{-19} \text{ C})^2}{(0.190 \times 10^{-9} \text{ m})^2} = 6.38 \times 10^{-9} \text{ N}$, attractive

$N^- - N^-$: $F = (8.99 \times 10^9 \text{ N} \cdot \text{m}^2/\text{C}^2)\dfrac{(1.60 \times 10^{-19} \text{ C})^2}{(0.300 \times 10^{-9} \text{ m})^2} = 2.56 \times 10^{-9} \text{ N}$, repulsive

The total attractive force is $1.43 \times 10^{-8} \text{ N}$ and the total repulsive force is $5.50 \times 10^{-9} \text{ N}$. The net force is attractive and has magnitude $1.43 \times 10^{-8} \text{ N} - 5.50 \times 10^{-9} \text{ N} = 8.80 \times 10^{-9} \text{ N}$.

(b) $F = k\dfrac{e^2}{r^2} = (8.99 \times 10^9 \text{ N} \cdot \text{m}^2/\text{C}^2)\dfrac{(1.60 \times 10^{-19} \text{ C})^2}{(0.0529 \times 10^{-9} \text{ m})^2} = 8.22 \times 10^{-8} \text{ N}.$

EVALUATE: The bonding force of the electron in the hydrogen atom is a factor of 10 larger than the bonding force of the adenine-thymine molecules.

21.27. **IDENTIFY:** The acceleration that stops the charge is produced by the force that the electric field exerts on it. Since the field and the acceleration are constant, we can use the standard kinematics formulas to find acceleration and time.

(a) SET UP: First use kinematics to find the proton's acceleration. $v_x = 0$ when it stops. Then find the electric field needed to cause this acceleration using the fact that $F = qE$.

EXECUTE: $v_x^2 = v_{0x}^2 + 2a_x(x - x_0)$. $0 = (4.50 \times 10^6 \text{ m/s})^2 + 2a(0.0320 \text{ m})$ and $a = 3.16 \times 10^{14} \text{ m/s}^2$. Now find the electric field, with $q = e$. $eE = ma$ and

$E = ma/e = (1.67 \times 10^{-27} \text{ kg})(3.16 \times 10^{14} \text{ m/s}^2)/(1.60 \times 10^{-19} \text{ C}) = 3.30 \times 10^6 \text{ N/C}$, to the left.

(b) SET UP: Kinematics gives $v = v_0 + at$, and $v = 0$ when the electron stops, so $t = v_0/a$.

EXECUTE: $t = v_0/a = (4.50 \times 10^6 \text{ m/s})/(3.16 \times 10^{14} \text{ m/s}^2) = 1.42 \times 10^{-8} \text{ s} = 14.2 \text{ ns}$

(c) SET UP: In part (a) we saw that the electric field is proportional to m, so we can use the ratio of the electric fields. $E_e/E_p = m_e/m_p$ and $E_e = (m_e/m_p)E_p$.

EXECUTE: $E_e = [(9.11 \times 10^{-31} \text{ kg})/(1.67 \times 10^{-27} \text{ kg})](3.30 \times 10^6 \text{ N/C}) = 1.80 \times 10^3 \text{ N/C}$, to the right

EVALUATE: Even a modest electric field, such as the ones in this situation, can produce enormous accelerations for electrons and protons.

21.31. **IDENTIFY:** For a point charge, $E = k\dfrac{|q|}{r^2}$. The net field is the vector sum of the fields produced by each charge. A charge q in an electric field $\vec{E}$ experiences a force $\vec{F} = q\vec{E}$.

SET UP: The electric field of a negative charge is directed toward the charge. Point A is 0.100 m from q_2 and 0.150 m from q_1. Point B is 0.100 m from q_1 and 0.350 m from q_2.

EXECUTE: **(a)** The electric fields at point A due to the charges are shown in Figure 21.31a.

$E_1 = k\dfrac{|q_1|}{r_{A1}^2} = (8.99 \times 10^9 \text{ N} \cdot \text{m}^2/\text{C}^2)\dfrac{6.25 \times 10^{-9} \text{ C}}{(0.150 \text{ m})^2} = 2.50 \times 10^3 \text{ N/C}$

$E_2 = k\dfrac{|q_2|}{r_{A2}^2} = (8.99 \times 10^9 \text{ N} \cdot \text{m}^2/\text{C}^2)\dfrac{12.5 \times 10^{-9} \text{ C}}{(0.100 \text{ m})^2} = 1.124 \times 10^4 \text{ N/C}$

Since the two fields are in opposite directions, we subtract their magnitudes to find the net field.

$E = E_2 - E_1 = 8.74 \times 10^3 \text{ N/C}$, to the right.

(b) The electric fields at point B are shown in Figure 21.31b.

$E_1 = k\dfrac{|q_1|}{r_{B1}^2} = (8.99 \times 10^9 \text{ N} \cdot \text{m}^2/\text{C}^2)\dfrac{6.25 \times 10^{-9} \text{ C}}{(0.100 \text{ m})^2} = 5.619 \times 10^3 \text{ N/C}$

$E_2 = k\dfrac{|q_2|}{r_{B2}^2} = (8.99 \times 10^9 \text{ N} \cdot \text{m}^2/\text{C}^2)\dfrac{12.5 \times 10^{-9} \text{ C}}{(0.350 \text{ m})^2} = 9.17 \times 10^2 \text{ N/C}$

Since the fields are in the same direction, we add their magnitudes to find the net field.

$E = E_1 + E_2 = 6.54 \times 10^3 \text{ N/C}$, to the right.

(c) At A, $E = 8.74 \times 10^3 \text{ N/C}$, to the right. The force on a proton placed at this point would be

$F = qE = (1.60 \times 10^{-19} \text{ C})(8.74 \times 10^3 \text{ N/C}) = 1.40 \times 10^{-15} \text{ N}$, to the right.

EVALUATE: A proton has positive charge so the force that an electric field exerts on it is in the same direction as the field.

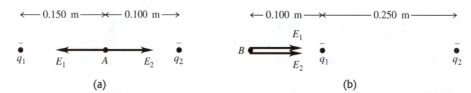

Figure 21.31

21.33.   **IDENTIFY:** Eq. (21.3) gives the force on the particle in terms of its charge and the electric field between the plates. The force is constant and produces a constant acceleration. The motion is similar to projectile motion; use constant acceleration equations for the horizontal and vertical components of the motion.
(a) SET UP: The motion is sketched in Figure 21.33a.

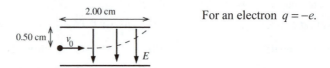

For an electron $q = -e$.

Figure 21.33a

$\vec{F} = q\vec{E}$ and q negative gives that $\vec{F}$ and $\vec{E}$ are in opposite directions, so $\vec{F}$ is upward. The free-body diagram for the electron is given in Figure 21.33b.

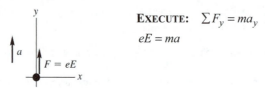

EXECUTE: $\sum F_y = ma_y$

$eE = ma$

Figure 21.33b

Solve the kinematics to find the acceleration of the electron: Just misses upper plate says that $x - x_0 = 2.00$ cm when $y - y_0 = +0.500$ cm.

x-component

$v_{0x} = v_0 = 1.60 \times 10^6$ m/s, $a_x = 0$, $x - x_0 = 0.0200$ m, $t = ?$

$x - x_0 = v_{0x}t + \frac{1}{2}a_x t^2$

$t = \dfrac{x - x_0}{v_{0x}} = \dfrac{0.0200 \text{ m}}{1.60 \times 10^6 \text{ m/s}} = 1.25 \times 10^{-8}$ s

In this same time t the electron travels 0.0050 m vertically:

y-component

$t = 1.25 \times 10^{-8}$ s, $v_{0y} = 0$, $y - y_0 = +0.0050$ m, $a_y = ?$

$y - y_0 = v_{0y}t + \frac{1}{2}a_y t^2$

$a_y = \dfrac{2(y - y_0)}{t^2} = \dfrac{2(0.0050 \text{ m})}{(1.25 \times 10^{-8} \text{ s})^2} = 6.40 \times 10^{13} \text{ m/s}^2$

(This analysis is very similar to that used in Chapter 3 for projectile motion, except that here the acceleration is upward rather than downward.) This acceleration must be produced by the electric-field force: $eE = ma$

$$E = \frac{ma}{e} = \frac{(9.109 \times 10^{-31} \text{ kg})(6.40 \times 10^{13} \text{ m/s}^2)}{1.602 \times 10^{-19} \text{ C}} = 364 \text{ N/C}$$

Note that the acceleration produced by the electric field is <u>much</u> larger than g, the acceleration produced by gravity, so it is perfectly ok to neglect the gravity force on the elctron in this problem.

(b) $a = \dfrac{eE}{m_p} = \dfrac{(1.602 \times 10^{-19} \text{ C})(364 \text{ N/C})}{1.673 \times 10^{-27} \text{ kg}} = 3.49 \times 10^{10} \text{ m/s}^2$

This is much less than the acceleration of the electron in part (a) so the vertical deflection is less and the proton won't hit the plates. The proton has the same initial speed, so the proton takes the same time $t = 1.25 \times 10^{-8}$ s to travel horizontally the length of the plates. The force on the proton is downward (in the same direction as $\vec{E}$, since q is positive), so the acceleration is downward and $a_y = -3.49 \times 10^{10}$ m/s^2. $y - y_0 = v_{0y}t + \frac{1}{2}a_y t^2 = \frac{1}{2}(-3.49 \times 10^{10}$ m/s$^2)(1.25 \times 10^{-8}$ s$)^2 = -2.73 \times 10^{-6}$ m. The displacement is 2.73×10^{-6} m, downward.

(c) EVALUATE: The displacements are in opposite directions because the electron has negative charge and the proton has positive charge. The electron and proton have the same magnitude of charge, so the force the electric field exerts has the same magnitude for each charge. But the proton has a mass larger by a factor of 1836 so its acceleration and its vertical displacement are smaller by this factor.
(d) In each case $a \gg g$ and it is reasonable to ignore the effects of gravity.

21.37. **IDENTIFY:** The forces the charges exert on each other are given by Coulomb's law. The net force on the proton is the vector sum of the forces due to the electrons.
SET UP: $q_e = -1.60 \times 10^{-19}$ C. $q_p = +1.60 \times 10^{-19}$ C. The net force is the vector sum of the forces exerted by each electron. Each force has magnitude $F = k\dfrac{|q_1 q_2|}{r^2} = k\dfrac{e^2}{r^2}$ and is attractive so is directed toward the electron that exerts it.
EXECUTE: Each force has magnitude

$$F_1 = F_2 = k\frac{|q_1 q_2|}{r^2} = k\frac{e^2}{r^2} = \frac{(8.988 \times 10^9 \text{ N} \cdot \text{m}^2/\text{C}^2)(1.60 \times 10^{-19} \text{ C})^2}{(1.50 \times 10^{-10} \text{ m})^2} = 1.023 \times 10^{-8} \text{ N}.$$ The vector force

diagram is shown in Figure 21.37.

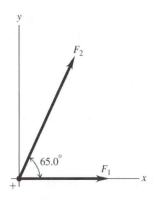

Figure 21. 37

Taking components, we get $F_{1x} = 1.023 \times 10^{-8}$ N; $F_{1y} = 0$. $F_{2x} = F_2 \cos 65.0° = 4.32 \times 10^{-9}$ N;

$F_{2y} = F_2 \sin 65.0° = 9.27 \times 10^{-9}$ N. $F_x = F_{1x} + F_{2x} = 1.46 \times 10^{-8}$ N; $F_y = F_{1y} + F_{2y} = 9.27 \times 10^{-9}$ N.

$F = \sqrt{F_x^2 + F_y^2} = 1.73 \times 10^{-8}$ N. $\tan\theta = \dfrac{F_y}{F_x} = \dfrac{9.27 \times 10^{-9} \text{ N}}{1.46 \times 10^{-8} \text{ N}} = 0.6349$ which gives

$\theta = 32.4°$. The net force is 1.73×10^{-8} N and is directed toward a point midway between the two electrons.
EVALUATE: Note that the net force is less than the algebraic sum of the individual forces.

21.39. **IDENTIFY:** Find the angle θ that $\hat{r}$ makes with the $+x$-axis. Then $\hat{r} = (\cos\theta)\hat{i} + (\sin\theta)\hat{j}.$.
SET UP: $\tan\theta = y/x$

EXECUTE: **(a)** $\tan^{-1}\left(\dfrac{-1.35}{0}\right) = -\dfrac{\pi}{2}$ rad. $\hat{r} = -\hat{j}.$

(b) $\tan^{-1}\left(\dfrac{12}{12}\right) = \dfrac{\pi}{4}$ rad. $\hat{r} = \dfrac{\sqrt{2}}{2}\hat{i} + \dfrac{\sqrt{2}}{2}\hat{j}.$

(c) $\tan^{-1}\left(\dfrac{2.6}{+1.10}\right) = 1.97 \text{ rad} = 112.9°.$ $\hat{r} = -0.39\hat{i} + 0.92\hat{j}$ (Second quadrant).

EVALUATE: In each case we can verify that $\hat{r}$ is a unit vector, because $\hat{r} \cdot \hat{r} = 1$.

21.41.   **IDENTIFY** and **SET UP:** Use $\vec{E}$ in Eq. (21.3) to calculate $\vec{F}$, $\vec{F} = m\vec{a}$ to calculate $\vec{a}$, and a constant acceleration equation to calculate the final velocity. Let $+x$ be east.

(a) EXECUTE: $F_x = |q|E = (1.602 \times 10^{-19} \text{ C})(1.50 \text{ N/C}) = 2.403 \times 10^{-19} \text{ N}$

$a_x = F_x/m = (2.403 \times 10^{-19} \text{ N})/(9.109 \times 10^{-31} \text{ kg}) = +2.638 \times 10^{11} \text{ m/s}^2$

$v_{0x} = +4.50 \times 10^5 \text{ m/s}, a_x = +2.638 \times 10^{11} \text{ m/s}^2, x - x_0 = 0.375 \text{ m}, v_x = ?$

$v_x^2 = v_{0x}^2 + 2a_x(x - x_0)$ gives $v_x = 6.33 \times 10^5 \text{ m/s}$

EVALUATE: $\vec{E}$ is west and q is negative, so $\vec{F}$ is east and the electron speeds up.

(b) EXECUTE: $F_x = -|q|E = -(1.602 \times 10^{-19} \text{ C})(1.50 \text{ N/C}) = -2.403 \times 10^{-19} \text{ N}$

$a_x = F_x/m = (-2.403 \times 10^{-19} \text{ N})/(1.673 \times 10^{-27} \text{ kg}) = -1.436 \times 10^8 \text{ m/s}^2$

$v_{0x} = +1.90 \times 10^4 \text{ m/s}, a_x = -1.436 \times 10^8 \text{ m/s}^2, x - x_0 = 0.375 \text{ m}, v_x = ?$

$v_x^2 = v_{0x}^2 + 2a_x(x - x_0)$ gives $v_x = 1.59 \times 10^4 \text{ m/s}$

EVALUATE: $q > 0$ so $\vec{F}$ is west and the proton slows down.

21.47.   **IDENTIFY:** $E = k\dfrac{|q|}{r^2}$. The net field is the vector sum of the fields due to each charge.

SET UP: The electric field of a negative charge is directed toward the charge. Label the charges q_1, q_2 and q_3, as shown in Figure 21.47a. This figure also shows additional distances and angles. The electric fields at point P are shown in Figure 21.47b. This figure also shows the xy coordinates we will use and the x and y components of the fields $\vec{E}_1$, $\vec{E}_2$ and $\vec{E}_3$.

EXECUTE: $E_1 = E_3 = (8.99 \times 10^9 \text{ N} \cdot \text{m}^2/\text{C}^2)\dfrac{5.00 \times 10^{-6} \text{ C}}{(0.100 \text{ m})^2} = 4.49 \times 10^6 \text{ N/C}$

$E_2 = (8.99 \times 10^9 \text{ N} \cdot \text{m}^2/\text{C}^2)\dfrac{2.00 \times 10^{-6} \text{ C}}{(0.0600 \text{ m})^2} = 4.99 \times 10^6 \text{ N/C}$

$E_y = E_{1y} + E_{2y} + E_{3y} = 0$ and $E_x = E_{1x} + E_{2x} + E_{3x} = E_2 + 2E_1 \cos 53.1° = 1.04 \times 10^7 \text{ N/C}$

$E = 1.04 \times 10^7 \text{ N/C}$, toward the $-2.00 \ \mu\text{C}$ charge.

EVALUATE: The x-components of the fields of all three charges are in the same direction.

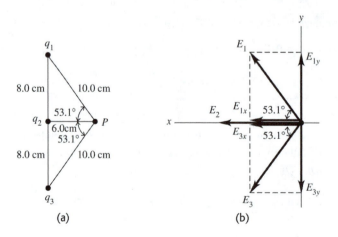

Figure 21.47

21.53. **IDENTIFY:** For a ring of charge, the electric field is given by Eq. (21.8). $\vec{F} = q\vec{E}$. In part (b) use Newton's third law to relate the force on the ring to the force exerted by the ring.

SET UP: $Q = 0.125 \times 10^{-9}$ C, $a = 0.025$ m and $x = 0.400$ m.

EXECUTE: (a) $\vec{E} = \dfrac{1}{4\pi\epsilon_0} \dfrac{Qx}{(x^2 + a^2)^{3/2}} \hat{\imath} = (7.0 \text{ N/C})\hat{\imath}$.

(b) $\vec{F}_{\text{on ring}} = -\vec{F}_{\text{on q}} = -q\vec{E} = -(-2.50 \times 10^{-6}\text{C})(7.0 \text{ N/C})\hat{\imath} = (1.75 \times 10^{-5} \text{ N})\hat{\imath}$

EVALUATE: Charges q and Q have opposite sign, so the force that q exerts on the ring is attractive.

21.55. **IDENTIFY:** We must use the appropriate electric field formula: a uniform disk in (a), a ring in (b) because all the charge is along the rim of the disk, and a point-charge in (c).

(a) **SET UP:** First find the surface charge density (Q/A), then use the formula for the field due to a disk of charge, $E_x = \dfrac{\sigma}{2\epsilon_0}\left[1 - \dfrac{1}{\sqrt{(R/x)^2 + 1}}\right]$.

EXECUTE: The surface charge density is $\sigma = \dfrac{Q}{A} = \dfrac{Q}{\pi r^2} = \dfrac{6.50 \times 10^{-9}\text{C}}{\pi(0.0125 \text{ m})^2} = 1.324 \times \text{C/m}^2$.

The electric field is

$$E_x = \dfrac{\sigma}{2\epsilon_0}\left[1 - \dfrac{1}{\sqrt{(R/x)^2 + 1}}\right] = \dfrac{1.324 \times 10^{-5} \text{ C/m}^2}{2(8.85 \times 10^{-12} \text{ C}^2/\text{N}\cdot\text{m}^2)}\left[1 - \dfrac{1}{\sqrt{\left(\dfrac{1.25 \text{ cm}}{2.00 \text{ cm}}\right)^2 + 1}}\right]$$

$E_x = 1.14 \times 10^5$ N/C, toward the center of the disk.

(b) **SET UP:** For a ring of charge, the field is $E = \dfrac{1}{4\pi\epsilon_0} \dfrac{Qx}{(x^2 + a^2)^{3/2}}$.

EXECUTE: Substituting into the electric field formula gives

$$E = \dfrac{1}{4\pi\epsilon_0} \dfrac{Qx}{(x^2 + a^2)^{3/2}} = \dfrac{(9.00 \times 10^9 \text{ N}\cdot\text{m}^2/\text{C}^2)(6.50 \times 10^{-9}\text{C})(0.0200 \text{ m})}{\left[(0.0200 \text{ m})^2 + (0.0125 \text{ m})^2\right]^{3/2}}$$

$E = 8.92 \times 10^4$ N/C, toward the center of the disk.

(c) **SET UP:** For a point charge, $E = (1/4\pi\epsilon_0)q/r^2$.

EXECUTE: $E = (9.00 \times 10^9 \text{ N}\cdot\text{m}^2/\text{C}^2)(6.50 \times 10^{-9} \text{ C})/(0.0200 \text{ m})^2 = 1.46 \times 10^5$ N/C

(d) **EVALUATE:** With the ring, more of the charge is farther from P than with the disk. Also with the ring the component of the electric field parallel to the plane of the ring is greater than with the disk, and this component cancels. With the point charge in (c), all the field vectors add with no cancellation, and all the charge is closer to point P than in the other two cases.

21.61. (a) **IDENTIFY:** Use Coulomb's law to calculate each force and then add them as vectors to obtain the net force. Torque is force times moment arm.

SET UP: The two forces on each charge in the dipole are shown in Figure 21.61a.

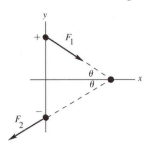

$\sin\theta = 1.50/2.00$ so $\theta = 48.6°$

Opposite charges attract and like charges repel.

$F_x = F_{1x} + F_{2x} = 0$

Figure 21. 61a

EXECUTE: $F_1 = k\dfrac{|qq'|}{r^2} = k\dfrac{(5.00 \times 10^{-6}\ \text{C})(10.0 \times 10^{-6}\ \text{C})}{(0.0200\ \text{m})^2} = 1.124 \times 10^3\ \text{N}$

$F_{1y} = -F_1 \sin\theta = -842.6\ \text{N}$

$F_{2y} = -842.6\ \text{N}$ so $F_y = F_{1y} + F_{2y} = -1680\ \text{N}$ (in the direction from the $+5.00\text{-}\mu\text{C}$ charge toward the $-5.00\text{-}\mu\text{C}$ charge).

EVALUATE: The x-components cancel and the y-components add.

(b) SET UP: Refer to Figure 21.61b.

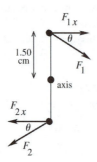

The y-components have zero moment arm and therefore zero torque.

F_{1x} and F_{2x} both produce clockwise torques.

Figure 21. 61b

EXECUTE: $F_{1x} = F_1 \cos\theta = 743.1\ \text{N}$

$\tau = 2(F_{1x})(0.0150\ \text{m}) = 22.3\ \text{N} \cdot \text{m}$, clockwise

EVALUATE: The electric field produced by the $-10.00\ \mu\text{C}$ charge is not uniform so Eq. (21.15) does not apply.

21.67.   **(a) IDENTIFY:**   Use Coulomb's law to calculate the force exerted by each Q on q and add these forces as vectors to find the resultant force. Make the approximation $x \gg a$ and compare the net force to $F = -kx$ to deduce k and then $f = (1/2\pi)\sqrt{k/m}$.

SET UP: The placement of the charges is shown in Figure 21.67a.

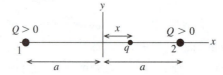

Figure 21. 67a

EXECUTE: Find the net force on q.

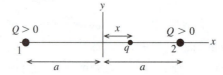 $F_x = F_{1x} + F_{2x}$ and $F_{1x} = +F_1, F_{2x} = -F_2$

Figure 21. 67b

$F_1 = \dfrac{1}{4\pi\epsilon_0} \dfrac{qQ}{(a+x)^2}, F_2 = \dfrac{1}{4\pi\epsilon_0} \dfrac{qQ}{(a-x)^2}$

$F_x = F_1 - F_2 = \dfrac{qQ}{4\pi\epsilon_0} \left[\dfrac{1}{(a+x)^2} - \dfrac{1}{(a-x)^2} \right]$

$F_x = \dfrac{qQ}{4\pi\epsilon_0 a^2} \left[+\left(1 + \dfrac{x}{a}\right)^{-2} - \left(1 - \dfrac{x}{a}\right)^{-2} \right]$

Since $x \ll a$ we can use the binomial expansion for $(1 - x/a)^{-2}$ and $(1 + x/a)^{-2}$ and keep only the first two terms: $(1 + z)^n \approx 1 + nz$. For $(1 - x/a)^{-2}$, $z = -x/a$ and $n = -2$ so $(1 - x/a)^{-2} \approx 1 + 2x/a$. For $(1 + x/a)^{-2}$, $z = +x/a$ and $n = -2$ so $(1 + x/a)^{-2} \approx 1 - 2x/a$. Then $F \approx \dfrac{qQ}{4\pi\epsilon_0 a^2}\left[\left(1 - \dfrac{2x}{a}\right) - \left(1 + \dfrac{2x}{a}\right)\right] = -\left(\dfrac{qQ}{\pi\epsilon_0 a^3}\right)x$.

For simple harmonic motion $F = -kx$ and the frequency of oscillation is $f = (1/2\pi)\sqrt{k/m}$. The net force here is of this form, with $k = qQ/\pi\epsilon_0 a^3$. Thus $f = \dfrac{1}{2\pi}\sqrt{\dfrac{qQ}{\pi\epsilon_0 ma^3}}$.

(b) The forces and their components are shown in Figure 21.67c.

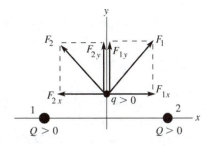

Figure 21.67c

The x-components of the forces exerted by the two charges cancel, the y-components add, and the net force is in the $+y$-direction when $y > 0$ and in the $-y$-direction when $y < 0$. The charge moves away from the origin on the y-axis and never returns.

EVALUATE: The directions of the forces and of the net force depend on where q is located relative to the other two charges. In part (a), $F = 0$ at $x = 0$ and when the charge q is displaced in the $+x$- or $-x$-direction the net force is a restoring force, directed to return q to $x = 0$. The charge oscillates back and forth, similar to a mass on a spring.

21.69.   **IDENTIFY:**   Use Coulomb's law for the force that one sphere exerts on the other and apply the 1st condition of equilibrium to one of the spheres.

(a) SET UP: The placement of the spheres is sketched in Figure 21.69a.

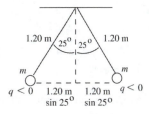

Figure 21.69a

The free-body diagrams for each sphere are given in Figure 21.69b.

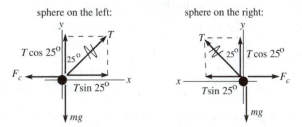

Figure 21.69b

F_c is the repulsive Coulomb force exerted by one sphere on the other.

(b) EXECUTE: From either force diagram in part (a): $\sum F_y = ma_y$

$$T\cos 25.0° - mg = 0 \text{ and } T = \frac{mg}{\cos 25.0°}$$

$$\sum F_x = ma_x$$

$$T\sin 25.0° - F_c = 0 \text{ and } F_c = T\sin 25.0°$$

Use the first equation to eliminate T in the second: $F_c = (mg/\cos 25.0°)(\sin 25.0°) = mg\tan 25.0°$

$$F_c = \frac{1}{4\pi\epsilon_0}\frac{|q_1 q_2|}{r^2} = \frac{1}{4\pi\epsilon_0}\frac{q^2}{r^2} = \frac{1}{4\pi\epsilon_0}\frac{q^2}{\left[2(1.20 \text{ m})\sin 25.0°\right]^2}$$

Combine this with $F_c = mg\tan 25.0°$ and get $mg\tan 25.0° = \dfrac{1}{4\pi\epsilon_0}\dfrac{q^2}{\left[2(1.20 \text{ m})\sin 25.0°\right]^2}$

$$q = (2.40 \text{ m})\sin 25.0°\sqrt{\frac{mg\tan 25.0°}{(1/4\pi\epsilon_0)}}$$

$$q = (2.40 \text{ m})\sin 25.0°\sqrt{\frac{(15.0\times 10^{-3} \text{ kg})(9.80 \text{ m/s}^2)\tan 25.0°}{8.988\times 10^9 \text{ N}\cdot\text{m}^2/\text{C}^2}} = 2.80\times 10^{-6} \text{ C}$$

(c) The separation between the two spheres is given by $2L\sin\theta$. $q = 2.80\,\mu\text{C}$ as found in part (b).

$F_c = (1/4\pi\epsilon_0)q^2/(2L\sin\theta)^2$ and $F_c = mg\tan\theta$. Thus $(1/4\pi\epsilon_0)q^2/(2L\sin\theta)^2 = mg\tan\theta$.

$$(\sin\theta)^2\tan\theta = \frac{1}{4\pi\epsilon_0}\frac{q^2}{4L^2 mg} = (8.988\times 10^9 \text{ N}\cdot\text{m}^2/\text{C}^2)\frac{(2.80\times 10^{-6} \text{ C})^2}{4(0.600 \text{ m})^2(15.0\times 10^{-3} \text{ kg})(9.80 \text{ m/s}^2)} = 0.3328.$$

Solve this equation by trial and error. This will go quicker if we can make a good estimate of the value of θ that solves the equation. For θ small, $\tan\theta \approx \sin\theta$. With this approximation the equation becomes $\sin^3\theta = 0.3328$ and $\sin\theta = 0.6930$, so $\theta = 43.9°$. Now refine this guess:

θ	$\sin^2\theta\tan\theta$	
45.0°	0.5000	
40.0°	0.3467	
39.6°	0.3361	
39.5°	0.3335	
39.4°	0.3309	so $\theta = 39.5°$

EVALUATE: The expression in part (c) says $\theta \to 0$ as $L \to \infty$ and $\theta \to 90°$ as $L \to 0$. When L is decreased from the value in part (a), θ increases.

21.71.   **IDENTIFY** and **SET UP:** Use Avogadro's number to find the number of Na^+ and Cl^- ions and the total positive and negative charge. Use Coulomb's law to calculate the electric force and $\vec{F} = m\vec{a}$ to calculate the acceleration.

(a) EXECUTE: The number of Na^+ ions in 0.100 mol of NaCl is $N = nN_A$. The charge of one ion is $+e$, so the total charge is $q_1 = nN_A e = (0.100 \text{ mol})(6.022\times 10^{23} \text{ ions/mol})(1.602\times 10^{-19} \text{ C/ion}) = 9.647\times 10^3 \text{ C}$.

There are the same number of Cl^- ions and each has charge $-e$, so $q_2 = -9.647\times 10^3 \text{ C}$.

$$F = \frac{1}{4\pi\epsilon_0}\frac{|q_1 q_2|}{r^2} = (8.988\times 10^9 \text{ N}\cdot\text{m}^2/\text{C}^2)\frac{(9.647\times 10^3 \text{ C})^2}{(0.0200 \text{ m})^2} = 2.09\times 10^{21} \text{ N}$$

(b) $a = F/m$. Need the mass of 0.100 mol of Cl^- ions. For Cl, $M = 35.453\times 10^{-3} \text{ kg/mol}$, so

$$m = (0.100 \text{ mol})(35.453\times 10^{-3} \text{ kg/mol}) = 35.45\times 10^{-4} \text{ kg. Then } a = \frac{F}{m} = \frac{2.09\times 10^{21} \text{ N}}{35.45\times 10^{-4} \text{ kg}} = 5.90\times 10^{23} \text{ m/s}^2.$$

(c) EVALUATE: Is is not reasonable to have such a huge force. The net charges of objects are rarely larger than $1\,\mu C$; a charge of 10^4 C is immense. A small amount of material contains huge amounts of positive and negative charges.

21.73. **IDENTIFY:** The electric field exerts a horizontal force away from the wall on the ball. When the ball hangs at rest, the forces on it (gravity, the tension in the string, and the electric force due to the field) add to zero.
SET UP: The ball is in equilibrium, so for it $\sum F_x = 0$ and $\sum F_y = 0$. The force diagram for the ball is given in Figure 21.73. F_E is the force exerted by the electric field. $\vec{F} = q\vec{E}$. Since the electric field is horizontal, $\vec{F}_E$ is horizontal. Use the coordinates shown in the figure. The tension in the string has been replaced by its x- and y-components.

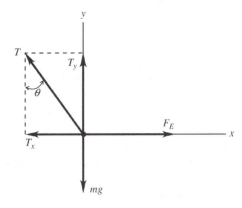

Figure 21.73

EXECUTE: $\sum F_y = 0$ gives $T_y - mg = 0$. $T\cos\theta - mg = 0$ and $T = \dfrac{mg}{\cos\theta}$. $\sum F_x = 0$ gives $F_E - T_x = 0$.

$F_E - T\sin\theta = 0$. Combing the equations and solving for F_E gives

$$F_E = \left(\frac{mg}{\cos\theta}\right)\sin\theta = mg\tan\theta = (12.3\times10^{-3}\text{ kg})(9.80\text{ m/s}^2)(\tan 17.4°) = 3.78\times10^{-2}\text{ N}. \quad F_E = |q|E \text{ so}$$

$$E = \frac{F_E}{|q|} = \frac{3.78\times10^{-2}\text{ N}}{1.11\times10^{-6}\text{ C}} = 3.41\times10^4\text{ N/C}. \text{ Since } q \text{ is negative and } \vec{F}_E \text{ is to the right, } \vec{E} \text{ is to the left in the figure.}$$

EVALUATE: The larger the electric field E the greater the angle the string makes with the wall.

21.77. **IDENTIFY:** Use Coulomb's law to calculate the forces between pairs of charges and sum these forces as vectors to find the net charge.
(a) SET UP: The forces are sketched in Figure 21.77a.

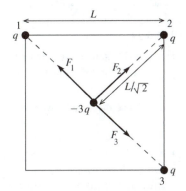

EXECUTE: $\vec{F}_1 + \vec{F}_3 = 0$, so the net force is $\vec{F} = \vec{F}_2$.

$$F = \frac{1}{4\pi\epsilon_0}\frac{q(3q)}{(L/\sqrt{2})^2} = \frac{6q^2}{4\pi\epsilon_0 L^2}, \text{ away from the vacant corner.}$$

Figure 21. 77a

(b) SET UP: The forces are sketched in Figure 21.77b.

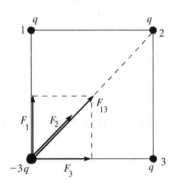

Figure 21. 77b

EXECUTE: $F_2 = \dfrac{1}{4\pi\epsilon_0}\dfrac{q(3q)}{(\sqrt{2}L)^2} = \dfrac{3q^2}{4\pi\epsilon_0(2L^2)}$

$F_1 = F_3 = \dfrac{1}{4\pi\epsilon_0}\dfrac{q(3q)}{L^2} = \dfrac{3q^2}{4\pi\epsilon_0 L^2}$

The vector sum of F_1 and F_3 is $F_{13} = \sqrt{F_1^2 + F_3^2}$.

$F_{13} = \sqrt{2}F_1 = \dfrac{3\sqrt{2}q^2}{4\pi\epsilon_0 L^2};\;\; \vec{F}_{13}$ and $\vec{F}_2$ are in the same direction.

$F = F_{13} + F_2 = \dfrac{3q^2}{4\pi\epsilon_0 L^2}\left(\sqrt{2} + \dfrac{1}{2}\right),$ and is directed toward the center of the square.

EVALUATE: By symmetry the net force is along the diagonal of the square. The net force is only slightly larger when the $-3q$ charge is at the center. Here it is closer to the charge at point 2 but the other two forces cancel.

21.81. **IDENTIFY:** Estimate the number of protons in the textbook and from this find the net charge of the textbook. Apply Coulomb's law to find the force and use $F_{\text{net}} = ma$ to find the acceleration.

SET UP: With the mass of the book about 1.0 kg, most of which is protons and neutrons, we find that the number of protons is $\frac{1}{2}(1.0\text{ kg})/(1.67\times10^{-27}\text{ kg}) = 3.0\times10^{26}$.

EXECUTE: **(a)** The charge difference present if the electron's charge was 99.999% of the proton's is

$\Delta q = (3.0\times10^{26})(0.00001)(1.6\times10^{-19}\text{ C}) = 480\text{ C}.$

(b) $F = k(\Delta q)^2/r^2 = k(480\text{ C})^2/(5.0\text{ m})^2 = 8.3\times10^{13}\text{ N},$ and is repulsive.

$a = F/m = (8.3\times10^{13}\text{ N})/(1\text{ kg}) = 8.3\times10^{13}\text{ m/s}^2.$

EXECUTE: **(c)** Even the slightest charge imbalance in matter would lead to explosive repulsion!

21.83. **IDENTIFY:** The only external force acting on the electron is the electrical attraction of the proton, and its acceleration is toward the center of its circular path (that is, toward the proton). Newton's second law applies to the proton and Coulomb's law gives the electrical force on it due to the proton.

SET UP: Newton's second law gives $F_C = m\dfrac{v^2}{r}.$ Using the electrical force for F_C gives $k\dfrac{e^2}{r^2} = m\dfrac{v^2}{r}.$

EXECUTE: Solving for v gives $v = \sqrt{\dfrac{ke^2}{mr}} = \sqrt{\dfrac{(8.99\times10^9\text{ N}\cdot\text{m}^2/\text{C}^2)(1.60\times10^{-19}\text{ C})^2}{(9.109\times10^{-31}\text{ kg})(5.29\times10^{-11}\text{ m})}} = 2.19\times10^6\text{ m/s}.$

EVALUATE: This speed is less than 1% the speed of light, so it is reasonably safe to use Newtonian physics.

21.89. **IDENTIFY:** Divide the charge distribution into infinitesimal segments of length dx. Calculate E_x and E_y due to a segment and integrate to find the total field.

SET UP: The charge dQ of a segment of length dx is $dQ = (Q/a)dx$. The distance between a segment at x and the charge q is $a + r - x$. $(1-y)^{-1} \approx 1 + y$ when $|y| \ll 1.$

EXECUTE: **(a)** $dE_x = \dfrac{1}{4\pi\epsilon_0}\dfrac{dQ}{(a+r-x)^2}$ so $E_x = \dfrac{1}{4\pi\epsilon_0}\displaystyle\int_0^a \dfrac{Q\,dx}{a(a+r-x)^2} = \dfrac{1}{4\pi\epsilon_0}\dfrac{Q}{a}\left(\dfrac{1}{r} - \dfrac{1}{a+r}\right).$

$a + r = x,$ so $E_x = \dfrac{1}{4\pi\epsilon_0}\dfrac{Q}{a}\left(\dfrac{1}{x-a} - \dfrac{1}{x}\right).$ $E_y = 0.$

(b) $\vec{F} = q\vec{E} = \dfrac{1}{4\pi\epsilon_0}\dfrac{qQ}{a}\left(\dfrac{1}{r} - \dfrac{1}{a+r}\right)\hat{i}$.

EVALUATE: **(c)** For $x \gg a$, $F = \dfrac{kqQ}{ax}((1-a/x)^{-1} - 1) = \dfrac{kqQ}{ax}(1 + a/x + \cdots - 1) \approx \dfrac{kqQ}{x^2} \approx \dfrac{1}{4\pi\epsilon_0}\dfrac{qQ}{r^2}$. (Note

that for $x \gg a$, $r = x - a \approx x$.) The charge distribution looks like a point charge from far away, so the force

takes the form of the force between a pair of point charges.

21.91. **IDENTIFY:** Apply Eq. (21.9) from Example 21.10.

SET UP: $a = 2.50$ cm. Replace Q by $|Q|$. Since Q is negative, $\vec{E}$ is toward the line of charge and

$$\vec{E} = -\dfrac{1}{4\pi\epsilon_0}\dfrac{|Q|}{x\sqrt{x^2+a^2}}\hat{i}.$$

EXECUTE: $\vec{E} = -\dfrac{1}{4\pi\epsilon_0}\dfrac{|Q|}{x\sqrt{x^2+a^2}}\hat{i} = -\dfrac{1}{4\pi\epsilon_0}\dfrac{7.00\times10^{-9}\ \text{C}}{(0.100\ \text{m})\sqrt{(0.100\ \text{m})^2+(0.025\ \text{m})^2}}\hat{i} = (-6110\ \text{N/C})\hat{i}$.

(b) The electric field is less than that at the same distance from a point charge (6300 N/C). For large x,

$(x+a)^{-1/2} = \dfrac{1}{x}(1+a^2/x^2)^{-1/2} \approx \dfrac{1}{x}\left(1 - \dfrac{a^2}{2x^2}\right)$, which gives $E_{x\to\infty} = \dfrac{1}{4\pi\epsilon_0}\dfrac{Q}{x^2}\left(1 - \dfrac{a^2}{2x^2} + \cdots\right)$. The first

correction term to the point charge result is negative.

(c) For a 1% difference, we need the first term in the expansion beyond the point charge result to be less

than 0.010: $\dfrac{a^2}{2x^2} \approx 0.010 \Rightarrow x \approx a\sqrt{1/(2(0.010))} = (0.025\ \text{m})\sqrt{1/0.020} \Rightarrow x \approx 0.177$ m.

EVALUATE: At $x = 10.0$ cm (part b), the exact result for the line of charge is 3.1% smaller than for a point

charge. It is sensible, therefore, that the difference is 1.0% at a somewhat larger distance, 17.7 cm.

21.93. **IDENTIFY:** Apply Eq. (21.11).

SET UP: $\sigma = Q/A = Q/\pi R^2$. $(1+y^2)^{-1/2} \approx 1 - y^2/2$, when $y^2 \ll 1$.

EXECUTE: **(a)** $E = \dfrac{\sigma}{2\epsilon_0}[1 - (R^2/x^2 + 1)^{-1/2}]$.

$E = \dfrac{7.00\ \text{pC}/\pi(0.025\ \text{m})^2}{2\epsilon_0}\left[1 - \left(\dfrac{(0.025\ \text{m})^2}{(0.200\ \text{m})^2} + 1\right)^{-1/2}\right] = 1.56$ N/C, in the $+x$-direction.

(b) For $x \gg R$, $E = \dfrac{\sigma}{2\epsilon_0}[1 - (1 - R^2/2x^2 + \cdots)] \approx \dfrac{\sigma}{2\epsilon_0}\dfrac{R^2}{2x^2} = \dfrac{\sigma\pi R^2}{4\pi\epsilon_0 x^2} = \dfrac{Q}{4\pi\epsilon_0 x^2}$.

(c) The electric field of (a) is less than that of the point charge (0.90 N/C) since the first correction term to

the point charge result is negative.

(d) For $x = 0.200$ m, the percent difference is $\dfrac{(1.58-1.56)}{1.56} = 0.01 = 1\%$. For $x = 0.100$ m,

$E_{\text{disk}} = 6.00$ N/C and $E_{\text{point}} = 6.30$ N/C, so the percent difference is $\dfrac{(6.30-6.00)}{6.30} = 0.047 \approx 5\%$.

EVALUATE: The field of a disk becomes closer to the field of a point charge as the distance from the disk

increases. At $x = 10.0$ cm, $R/x = 25\%$ and the percent difference between the field of the disk and the field

of a point charge is 5%.

21.95. **IDENTIFY:** Find the resultant electric field due to the two point charges. Then use $\vec{F} = q\vec{E}$ to calculate the

force on the point charge.

SET UP: Use the results of Problems 21.90 and 21.89.

EXECUTE: (a) The y-components of the electric field cancel, and the x-component from both charges, as given in Problem 21.90, is $E_x = \dfrac{1}{4\pi\epsilon_0} \dfrac{-2Q}{a} \left(\dfrac{1}{y} - \dfrac{1}{(y^2 + a^2)^{1/2}} \right)$. Therefore,

$\vec{F} = \dfrac{1}{4\pi\epsilon_0} \dfrac{-2Qq}{a} \left(\dfrac{1}{y} - \dfrac{1}{(y^2 + a^2)^{1/2}} \right) \hat{i}$. If $y \gg a$, $\vec{F} \approx \dfrac{1}{4\pi\epsilon_0} \dfrac{-2Qq}{ay} (1 - (1 - a^2/2y^2 + \cdots)) \hat{i} = -\dfrac{1}{4\pi\epsilon_0} \dfrac{Qqa}{y^3} \hat{i}$.

(b) If the point charge is now on the x-axis the two halves of the charge distribution provide different forces, though still along the x-axis, as given in Problem 21.89: $\vec{F}_+ = q\vec{E}_+ = \dfrac{1}{4\pi\epsilon_0} \dfrac{Qq}{a} \left(\dfrac{1}{x-a} - \dfrac{1}{x} \right) \hat{i}$

and $\vec{F}_- = q\vec{E}_- = -\dfrac{1}{4\pi\epsilon_0} \dfrac{Qq}{a} \left(\dfrac{1}{x} - \dfrac{1}{x+a} \right) \hat{i}$. Therefore, $\vec{F} = \vec{F}_+ + \vec{F}_- = \dfrac{1}{4\pi\epsilon_0} \dfrac{Qq}{a} \left(\dfrac{1}{x-a} - \dfrac{2}{x} + \dfrac{1}{x+a} \right) \hat{i}$. For

$x \gg a$, $\vec{F} \approx \dfrac{1}{4\pi\epsilon_0} \dfrac{Qq}{ax} \left(\left(1 + \dfrac{a}{x} + \dfrac{a^2}{x^2} + \cdots \right) - 2 + \left(1 - \dfrac{a}{x} + \dfrac{a^2}{x^2} - \cdots \right) \right) \hat{i} = \dfrac{1}{4\pi\epsilon_0} \dfrac{2Qqa}{x^3} \hat{i}$.

EVALUATE: If the charge distributed along the x-axis were all positive or all negative, the force would be proportional to $1/y^2$ in part (a) and to $1/x^2$ in part (b), when y or x is very large.

21.97. **IDENTIFY:** Divide the charge distribution into small segments, use the point charge formula for the electric field due to each small segment and integrate over the charge distribution to find the x and y components of the total field.

SET UP: Consider the small segment shown in Figure 21.97a.

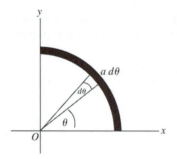

EXECUTE: A small segment that subtends angle $d\theta$ has length $a\,d\theta$ and contains charge

$dQ = \left(\dfrac{a\,d\theta}{\frac{1}{2}\pi a} \right) Q = \dfrac{2Q}{\pi} d\theta$. ($\frac{1}{2}\pi a$ is the total

length of the charge distribution.)

Figure 21.97a

The charge is negative, so the field at the origin is directed toward the small segment. The small segment is located at angle θ as shown in the sketch. The electric field due to dQ is shown in Figure 21.97b, along with its components.

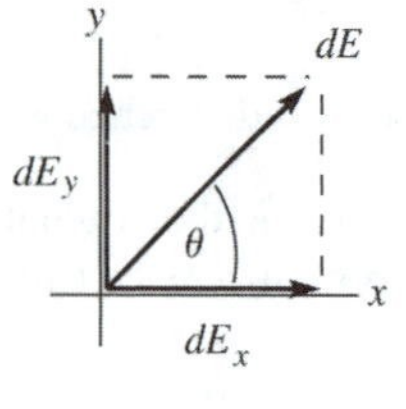

$$dE = \dfrac{1}{4\pi\epsilon_0} \dfrac{|dQ|}{a^2}$$

$$dE = \dfrac{Q}{2\pi^2\epsilon_0 a^2} d\theta$$

Figure 21.97b

$dE_x = dE\cos\theta = (Q/2\pi^2\epsilon_0 a^2)\cos\theta\,d\theta$

$E_x = \int dE_x = \dfrac{Q}{2\pi^2\epsilon_0 a^2} \int_0^{\pi/2} \cos\theta\,d\theta = \dfrac{Q}{2\pi^2\epsilon_0 a^2} (\sin\theta \,\big|_0^{\pi/2}) = \dfrac{Q}{2\pi^2\epsilon_0 a^2}$

$$dE_y = dE\sin\theta = (Q/2\pi^2\epsilon_0 a^2)\sin\theta\, d\theta$$

$$E_y = \int dE_y = \frac{Q}{2\pi^2\epsilon_0 a^2}\int_0^{\pi/2}\sin\theta\, d\theta = \frac{Q}{2\pi^2\epsilon_0 a^2}(-\cos\theta\Big|_0^{\pi/2}) = \frac{Q}{2\pi^2\epsilon_0 a^2}$$

EVALUATE: Note that $E_x = E_y$, as expected from symmetry.

21.99.   **IDENTIFY:** Each wire produces an electric field at P due to a finite wire. These fields add by vector addition.

SET UP: Each field has magnitude $\dfrac{1}{4\pi\epsilon_0}\dfrac{Q}{x\sqrt{x^2+a^2}}$. The field due to the negative wire points to the left,

while the field due to the positive wire points downward, making the two fields perpendicular to each other and of equal magnitude. The net field is the vector sum of these two, which is

$E_{net} = 2E_1\cos 45° = 2\dfrac{1}{4\pi\epsilon_0}\dfrac{Q}{x\sqrt{x^2+a^2}}\cos 45°$. In part (b), the electrical force on an electron at P is eE.

EXECUTE: **(a)** The net field is $E_{net} = 2\dfrac{1}{4\pi\epsilon_0}\dfrac{Q}{x\sqrt{x^2+a^2}}\cos 45°$.

$$E_{net} = \frac{2(9.00\times10^9\,\text{N}\cdot\text{m}^2/\text{C}^2)(2.50\times10^{-6}\,\text{C})\cos 45°}{(0.600\,\text{m})\sqrt{(0.600\,\text{m})^2+(0.600\,\text{m})^2}} = 6.25\times10^4\ \text{N/C}.$$

The direction is 225° counterclockwise from an axis pointing to the right at point P.

(b) $F = eE = (1.60\times10^{-19}\ \text{C})(6.25\times10^4\ \text{N/C}) = 1.00\times10^{-14}\ \text{N},$ opposite to the direction of the electric field, since the electron has negative charge.

EVALUATE: Since the electric fields due to the two wires have equal magnitudes and are perpendicular to each other, we only have to calculate one of them in the solution.

21.101.   **IDENTIFY:** Each sheet produces an electric field that is independent of the distance from the sheet. The net field is the vector sum of the two fields.

SET UP: The formula for each field is $E = \sigma/2\epsilon_0$, and the net field is the vector sum of these,

$E_{net} = \dfrac{\sigma_B}{2\epsilon_0} \pm \dfrac{\sigma_A}{2\epsilon_0} = \dfrac{\sigma_B \pm \sigma_A}{2\epsilon_0}$, where we use the + or − sign depending on whether the fields are in the

same or opposite directions and σ_B and σ_A are the magnitudes of the surface charges.

EXECUTE: **(a)** The fields add and point to the left, giving $E_{net} = 1.19\times10^6$ N/C.

(b) The fields oppose and point to the left, so $E_{net} = 1.19\times10^5$ N/C.

(c) The fields oppose but now point to the right, giving $E_{net} = 1.19\times10^5$ N/C.

EVALUATE: We can simplify the calculations by sketching the fields and doing an algebraic solution first.

GAUSS'S LAW

22.5. **IDENTIFY:** The flux through the curved upper half of the hemisphere is the same as the flux through the flat circle defined by the bottom of the hemisphere because every electric field line that passes through the flat circle also must pass through the curved surface of the hemisphere.

SET UP: The electric field is perpendicular to the flat circle, so the flux is simply the product of E and the area of the flat circle of radius r.

EXECUTE: $\Phi_E = EA = E(\pi r^2) = \pi r^2 E$

EVALUATE: The flux would be the same if the hemisphere were replaced by any other surface bounded by the flat circle.

22.9. **IDENTIFY:** Apply the results in Example 22.5 for the field of a spherical shell of charge.

SET UP: Example 22.5 shows that $E = 0$ inside a uniform spherical shell and that $E = k\dfrac{|q|}{r^2}$ outside the shell.

EXECUTE: **(a)** $E = 0$.

(b) $r = 0.060$ m and $E = (8.99 \times 10^9 \ \text{N} \cdot \text{m}^2/\text{C}^2)\dfrac{35.0 \times 10^{-6} \ \text{C}}{(0.060 \ \text{m})^2} = 8.74 \times 10^7$ N/C.

(c) $r = 0.110$ m and $E = (8.99 \times 10^9 \ \text{N} \cdot \text{m}^2/\text{C}^2)\dfrac{35.0 \times 10^{-6} \ \text{C}}{(0.110 \ \text{m})^2} = 2.60 \times 10^7$ N/C.

EVALUATE: Outside the shell the electric field is the same as if all the charge were concentrated at the center of the shell. But inside the shell the field is not the same as for a point charge at the center of the shell, inside the shell the electric field is zero.

22.11. **(a) IDENTIFY and SET UP:** It is rather difficult to calculate the flux directly from $\Phi_E = \int \vec{E} \cdot d\vec{A}$ since the magnitude of $\vec{E}$ and its angle with $d\vec{A}$ varies over the surface of the cube. A much easier approach is to use Gauss's law to calculate the total flux through the cube. Let the cube be the Gaussian surface. The charge enclosed is the point charge.

EXECUTE: $\Phi_E = Q_{\text{encl}}/\epsilon_0 = \dfrac{6.20 \times 10^{-6} \ \text{C}}{8.854 \times 10^{-12} \ \text{C}^2/\text{N} \cdot \text{m}^2} = 7.002 \times 10^5 \ \text{N} \cdot \text{m}^2/\text{C}.$ By symmetry the flux is the same through each of the six faces, so the flux through one face is $\frac{1}{6}(7.002 \times 10^5 \ \text{N} \cdot \text{m}^2/\text{C}) = 1.17 \times 10^5 \ \text{N} \cdot \text{m}^2/\text{C}.$

(b) EVALUATE: In part (a) the size of the cube did not enter into the calculations. The flux through one face depends only on the amount of charge at the center of the cube. So the answer to (a) would not change if the size of the cube were changed.

22.15. **IDENTIFY:** Each line lies in the electric field of the other line, and therefore each line experiences a force due to the other line.

SET UP: The field of one line at the location of the other is $E = \dfrac{\lambda}{2\pi\epsilon_0 r}$. For charge $dq = \lambda dx$ on one line, the force on it due to the other line is $dF = Edq$. The total force is $F = \int Edq = E \int dq = Eq$.

EXECUTE: $E = \dfrac{\lambda}{2\pi\epsilon_0 r} = \dfrac{5.20\times10^{-6}\ \text{C/m}}{2\pi(8.854\times10^{-12}\ \text{C}^2/(\text{N}\cdot\text{m}^2))(0.300\ \text{m})} = 3.116\times10^5$ N/C. The force on one line

due to the other is $F = Eq$, where $q = \lambda(0.0500\ \text{m}) = 2.60\times10^{-7}$ C. The net force is

$F = Eq = (3.116\times10^5\ \text{N/C})(2.60\times10^{-7}\ \text{C}) = 0.0810$ N.

EVALUATE: Since the electric field at each line due to the other line is uniform, each segment of line experiences the same force, so all we need to use is $F = Eq$, even though the line is *not* a point charge.

22.19.   **IDENTIFY:**   Add the vector electric fields due to each line of charge. $E(r)$ for a line of charge is given by Example 22.6 and is directed toward a negative line of charge and away from a positive line.
   SET UP: The two lines of charge are shown in Figure 22.19.

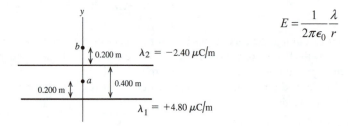

$$E = \frac{1}{2\pi\epsilon_0}\frac{\lambda}{r}$$

Figure 22.19

EXECUTE: **(a)** At point a, $\vec{E}_1$ and $\vec{E}_2$ are in the $+y$-direction (toward negative charge, away from positive charge).

$E_1 = (1/2\pi\epsilon_0)[(4.80\times10^{-6}\ \text{C/m})/(0.200\ \text{m})] = 4.314\times10^5$ N/C

$E_2 = (1/2\pi\epsilon_0)[(2.40\times10^{-6}\ \text{C/m})/(0.200\ \text{m})] = 2.157\times10^5$ N/C

$E = E_1 + E_2 = 6.47\times10^5$ N/C, in the y-direction.

(b) At point b, $\vec{E}_1$ is in the $+y$-direction and $\vec{E}_2$ is in the $-y$-direction.

$E_1 = (1/2\pi\epsilon_0)[(4.80\times10^{-6}\ \text{C/m})/(0.600\ \text{m})] = 1.438\times10^5$ N/C

$E_2 = (1/2\pi\epsilon_0)[(2.40\times10^{-6}\ \text{C/m})/(0.200\ \text{m})] = 2.157\times10^5$ N/C

$E = E_2 - E_1 = 7.2\times10^4$ N/C, in the $-y$-direction.

EVALUATION: At point a the two fields are in the same direction and the magnitudes add. At point b the two fields are in opposite directions and the magnitudes subtract.

22.21.   **IDENTIFY:**   The electric field inside the conductor is zero, and all of its initial charge lies on its outer surface. The introduction of charge into the cavity induces charge onto the surface of the cavity, which induces an equal but opposite charge on the outer surface of the conductor. The net charge on the outer surface of the conductor is the sum of the positive charge initially there and the additional negative charge due to the introduction of the negative charge into the cavity.

(a) SET UP: First find the initial positive charge on the outer surface of the conductor using $q_i = \sigma A$, where A is the area of its outer surface. Then find the net charge on the surface after the negative charge has been introduced into the cavity. Finally, use the definition of surface charge density.
   EXECUTE: The original positive charge on the outer surface is

$$q_i = \sigma A = \sigma(4\pi r^2) = (6.37\times10^{-6}\ \text{C/m}^2)4\pi(0.250\ \text{m})^2 = 5.00\times10^{-6}\ \text{C}$$

After the introduction of $-0.500\ \mu$C into the cavity, the outer charge is now

$$5.00\ \mu\text{C} - 0.500\ \mu\text{C} = 4.50\ \mu\text{C}$$

The surface charge density is now $\sigma = \dfrac{q}{A} = \dfrac{q}{4\pi r^2} = \dfrac{4.50\times10^{-6}\ \text{C}}{4\pi(0.250\ \text{m})^2} = 5.73\times10^{-6}\ \text{C/m}^2$

(b) SET UP: Using Gauss's law, the electric field is $E = \dfrac{\Phi_E}{A} = \dfrac{q}{\epsilon_0 A} = \dfrac{q}{\epsilon_0 4\pi r^2}$.

EXECUTE: Substituting numbers gives

$$E = \frac{4.50 \times 10^{-6} \text{ C}}{(8.85 \times 10^{-12} \text{ C}^2/\text{N} \cdot \text{m}^2)(4\pi)(0.250 \text{ m})^2} = 6.47 \times 10^5 \text{ N/C}.$$

(c) SET UP: We use Gauss's law again to find the flux. $\Phi_E = \frac{q}{\epsilon_0}$.

EXECUTE: Substituting numbers gives

$$\Phi_E = \frac{-0.500 \times 10^{-6} \text{ C}}{8.85 \times 10^{-12} \text{ C}^2/\text{N} \cdot \text{m}^2} = -5.65 \times 10^4 \text{ N} \cdot \text{m}^2/\text{C}.$$

EVALUATE: The excess charge on the conductor is still $+5.00 \ \mu\text{C}$, as it originally was. The introduction of the $-0.500 \ \mu\text{C}$ inside the cavity merely induced equal but opposite charges (for a net of zero) on the surfaces of the conductor.

22.23.   **IDENTIFY:**   The magnitude of the electric field is constant at any given distance from the center because the charge density is uniform inside the sphere. We can use Gauss's law to relate the field to the charge causing it.

(a) SET UP: Gauss's law tells us that $EA = \frac{q}{\epsilon_0}$, and the charge density is given by $\rho = \frac{q}{V} = \frac{q}{(4/3)\pi R^3}$.

EXECUTE: Solving for q and substituting numbers gives

$q = EA\epsilon_0 = E(4\pi r^2)\epsilon_0 = (1750 \text{ N/C})(4\pi)(0.500 \text{ m})^2(8.85 \times 10^{-12} \text{ C}^2/\text{N} \cdot \text{m}^2) = 4.866 \times 10^{-8} \text{ C}.$ Using the

formula for charge density we get $\rho = \frac{q}{V} = \frac{q}{(4/3)\pi R^3} = \frac{4.866 \times 10^{-8} \text{ C}}{(4/3)\pi(0.355 \text{ m})^3} = 2.60 \times 10^{-7} \text{ C/m}^3.$

(b) SET UP: Take a Gaussian surface of radius $r = 0.200 \text{ m}$, concentric with the insulating sphere. The

charge enclosed within this surface is $q_{\text{encl}} = \rho V = \rho \left(\frac{4}{3}\pi r^3 \right)$, and we can treat this charge as a point-

charge, using Coulomb's law $E = \frac{1}{4\pi\epsilon_0} \frac{q_{\text{encl}}}{r^2}$. The charge beyond $r = 0.200 \text{ m}$ makes no contribution to

the electric field.

EXECUTE: First find the enclosed charge:

$$q_{\text{encl}} = \rho \left(\frac{4}{3}\pi r^3 \right) = (2.60 \times 10^{-7} \text{ C/m}^3) \left[\frac{4}{3}\pi(0.200 \text{ m})^3 \right] = 8.70 \times 10^{-9} \text{ C}$$

Now treat this charge as a point-charge and use Coulomb's law to find the field:

$$E = (9.00 \times 10^9 \text{ N} \cdot \text{m}^2/\text{C}^2) \frac{8.70 \times 10^{-9} \text{ C}}{(0.200 \text{ m})^2} = 1.96 \times 10^3 \text{ N/C}$$

EVALUATE: Outside this sphere, it behaves like a point-charge located at its center. Inside of it, at a distance r from the center, the field is due only to the charge between the center and r.

22.25.   **IDENTIFY:**   The uniform electric field of the sheet exerts a constant force on the proton perpendicular to the sheet, and therefore does not change the parallel component of its velocity. Newton's second law allows us to calculate the proton's acceleration perpendicular to the sheet, and uniform-acceleration kinematics allows us to determine its perpendicular velocity component.

SET UP: Let $+x$ be the direction of the initial velocity and let $+y$ be the direction perpendicular to the

sheet and pointing away from it. $a_x = 0$ so $v_x = v_{0x} = 9.70 \times 10^2 \text{ m/s}$. The electric field due to the sheet is

$E = \frac{\sigma}{2\epsilon_0}$ and the magnitude of the force the sheet exerts on the proton is $F = eE$.

EXECUTE: $E = \frac{\sigma}{2\epsilon_0} = \frac{2.34 \times 10^{-9} \text{ C/m}^2}{2(8.854 \times 10^{-12} \text{ C}^2/(\text{N} \cdot \text{m}^2))} = 132.1 \text{ N/C}.$ Newton's second law gives

$a_y = \frac{Eq}{m} = \frac{(132.1 \text{ N/C})(1.602 \times 10^{-19} \text{ C})}{1.673 \times 10^{-27} \text{ kg}} = 1.265 \times 10^{10} \text{ m/s}^2.$ Kinematics gives

$v_y = v_{0y} + a_y y = (1.265 \times 10^{10}$ m/s$^2)(5.00 \times 10^{-8}$ s$) = 632.7$ m/s. The speed of the proton is the magnitude of its velocity, so $v = \sqrt{v_x^2 + v_y^2} = \sqrt{(9.70 \times 10^2$ m/s$)^2 + (632.7$ m/s$)^2} = 1.16 \times 10^3$ m/s.

EVALUATE: We can use the constant-acceleration kinematics equations because the uniform electric field of the sheet exerts a constant force on the proton, giving it a constant acceleration. We could *not* use this approach if the sheet were replaced with a sphere, for example.

22.27. **IDENTIFY:** The field of the sphere exerts a force on the object as it accelerates away from the sphere, and therefore does work on it. Coulomb's law gives the force that the sphere exerts on the object.

SET UP: The sphere carries charge $Q = \rho V = \rho \left(\dfrac{4}{3} \pi R^3 \right)$ and produces an electric field $E = \dfrac{kQ}{r^2}$ for points outside its surface. The work done on the object is $W = \int_R^\infty F(r)\, dr$.

EXECUTE: $Q = \rho V = \rho \left(\dfrac{4}{3} \pi R^3 \right) = (7.20 \times 10^{-9}$ C/m$^3) \left(\dfrac{4}{3} \pi \right)(0.160$ m$)^3 = 1.235 \times 10^{-10}$ C. Outside the sphere, $E = \dfrac{kQ}{r^2}$. The work done on the object is

$$W = \int_R^\infty F(r)\, dr = kQq \int_R^\infty \frac{dr}{r^2} = \frac{kQq}{R} = \frac{(8.988 \times 10^9 \text{ N} \cdot \text{m}^2/\text{C}^2)(1.235 \times 10^{-10} \text{ C})(3.40 \times 10^{-6} \text{ C})}{0.160 \text{ m}}.$$

$W = 2.36 \times 10^{-5}$ J.

EVALUATE: Even though the force on the sphere extends to infinity, it does finite work because it gets weaker and weaker as the distance from the sphere increases.

22.31. **IDENTIFY:** Apply Gauss's law to a Gaussian surface and calculate E.

(a) SET UP: Consider the charge on a length l of the cylinder. This can be expressed as $q = \lambda l$. But since the surface area is $2\pi R l$ it can also be expressed as $q = \sigma 2\pi R l$. These two expressions must be equal, so $\lambda l = \sigma 2\pi R l$ and $\lambda = 2\pi R \sigma$.

(b) Apply Gauss's law to a Gaussian surface that is a cylinder of length l, radius r, and whose axis coincides with the axis of the charge distribution, as shown in Figure 22.31.

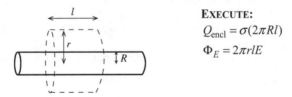

EXECUTE:
$Q_{encl} = \sigma(2\pi R l)$
$\Phi_E = 2\pi r l E$

Figure 22.31

$\Phi_E = \dfrac{Q_{encl}}{\epsilon_0}$ gives $2\pi r l E = \dfrac{\sigma(2\pi R l)}{\epsilon_0}$

$E = \dfrac{\sigma R}{\epsilon_0 r}$

(c) EVALUATE: Example 22.6 shows that the electric field of an infinite line of charge is $E = \lambda / 2\pi\epsilon_0 r$.

$\sigma = \dfrac{\lambda}{2\pi R}$, so $E = \dfrac{\sigma R}{\epsilon_0 r} = \dfrac{R}{\epsilon_0 r} \left(\dfrac{\lambda}{2\pi R} \right) = \dfrac{\lambda}{2\pi\epsilon_0 r}$, the same as for an infinite line of charge that is along the axis of the cylinder.

22.37. **(a) IDENTIFY:** Find the net flux through the parallelepiped surface and then use that in Gauss's law to find the net charge within. Flux out of the surface is positive and flux into the surface is negative.

SET UP: $\vec{E}_1$ gives flux out of the surface. See Figure 22.37a.

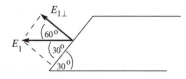

Figure 22.37a

EXECUTE: $\Phi_1 = +E_{1\perp}A$

$A = (0.0600 \text{ m})(0.0500 \text{ m}) = 3.00 \times 10^{-3} \text{ m}^2$

$E_{1\perp} = E_1 \cos 60° = (2.50 \times 10^4 \text{ N/C}) \cos 60°$

$E_{1\perp} = 1.25 \times 10^4 \text{ N/C}$

$\Phi_{E_1} = +E_{1\perp}A = +(1.25 \times 10^4 \text{ N/C})(3.00 \times 10^{-3} \text{ m}^2) = 37.5 \text{ N} \cdot \text{m}^2/\text{C}$

SET UP: $\vec{E}_2$ gives flux into the surface. See Figure 22.37b.

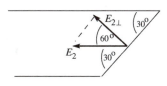

Figure 22.37b

EXECUTE: $\Phi_2 = -E_{2\perp}A$

$A = (0.0600 \text{ m})(0.0500 \text{ m}) = 3.00 \times 10^{-3} \text{ m}^2$

$E_{2\perp} = E_2 \cos 60° = (7.00 \times 10^4 \text{ N/C}) \cos 60°$

$E_{2\perp} = 3.50 \times 10^4 \text{ N/C}$

$\Phi_{E_2} = -E_{2\perp}A = -(3.50 \times 10^4 \text{ N/C})(3.00 \times 10^{-3} \text{ m}^2) = -105.0 \text{ N} \cdot \text{m}^2/\text{C}$

The net flux is $\Phi_E = \Phi_{E_1} + \Phi_{E_2} = +37.5 \text{ N} \cdot \text{m}^2/\text{C} - 105.0 \text{ N} \cdot \text{m}^2/\text{C} = -67.5 \text{ N} \cdot \text{m}^2/\text{C}.$

The net flux is negative (inward), so the net charge enclosed is negative.

Apply Gauss's law: $\Phi_E = \dfrac{Q_{encl}}{\epsilon_0}$

$Q_{encl} = \Phi_E \epsilon_0 = (-67.5 \text{ N} \cdot \text{m}^2/\text{C})(8.854 \times 10^{-12} \text{ C}^2/\text{N} \cdot \text{m}^2) = -5.98 \times 10^{-10} \text{ C}.$

(b) EVALUATE: If there were no charge within the parallelpiped the net flux would be zero. This is not the case, so there is charge inside. The electric field lines that pass out through the surface of the parallelpiped must terminate on charges, so there also must be charges outside the parallelpiped.

22.43. **IDENTIFY:** First make a free-body diagram of the sphere. The electric force acts to the left on it since the electric field due to the sheet is horizontal. Since it hangs at rest, the sphere is in equilibrium so the forces on it add to zero, by Newton's first law. Balance horizontal and vertical force components separately.

SET UP: Call T the tension in the thread and E the electric field. Balancing horizontal forces gives $T \sin\theta = qE$. Balancing vertical forces we get $T \cos\theta = mg$. Combining these equations gives $\tan\theta = qE/mg$, which means that $\theta = \arctan(qE/mg)$. The electric field for a sheet of charge is $E = \sigma/2\varepsilon_0$.

EXECUTE: Substituting the numbers gives us

$E = \dfrac{\sigma}{2\epsilon_0} = \dfrac{2.50 \times 10^{-9} \text{ C/m}^2}{2(8.85 \times 10^{-12} \text{ C}^2/\text{N} \cdot \text{m}^2)} = 1.41 \times 10^2 \text{ N/C}.$ Then

$\theta = \arctan\left[\dfrac{(5.00 \times 10^{-8} \text{ C})(1.41 \times 10^2 \text{ N/C})}{(4.00 \times 10^{-6} \text{ kg})(9.80 \text{ m/s}^2)} \right] = 10.2°.$

EVALUATE: Increasing the field, or decreasing the mass of the sphere, would cause the sphere to hang at a larger angle.

22.45. **IDENTIFY:** Apply Gauss's law.

SET UP: Use a Gaussian surface that is a sphere of radius r and that is concentric with the charge distributions.

EXECUTE: **(a)** For $r < R$, $E = 0$, since these points are within the conducting material. For $R < r < 2R$,

$E = \dfrac{1}{4\pi\epsilon_0}\dfrac{Q}{r^2}$, since the charge enclosed is Q. The field is radially outward. For $r > 2R$, $E = \dfrac{1}{4\pi\epsilon_0}\dfrac{2Q}{r^2}$

since the charge enclosed is $2Q$. The field is radially outward.
(b) The graph of E versus r is sketched in Figure 22.45.
EVALUATE: For $r < 2R$ the electric field is unaffected by the presence of the charged shell.

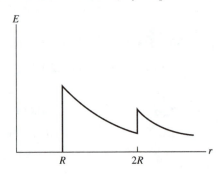

Figure 22.45

22.47.   **IDENTIFY:** Apply Gauss's law to a spherical Gaussian surface with radius r. Calculate the electric field at the surface of the Gaussian sphere.
(a) SET UP: (i) $r < a$: The Gaussian surface is sketched in Figure 22.47a.

EXECUTE: $\Phi_E = EA = E(4\pi r^2)$

$Q_{\text{encl}} = 0$; no charge is enclosed

$\Phi_E = \dfrac{Q_{\text{encl}}}{\epsilon_0}$ says

$E(4\pi r^2) = 0$ and $E = 0$.

Figure 22.47a

(ii) $a < r < b$: Points in this region are in the conductor of the small shell, so $E = 0$.
(iii) **SET UP:** $b < r < c$: The Gaussian surface is sketched in Figure 22.47b.
Apply Gauss's law to a spherical Gaussian surface with radius $b < r < c$.

EXECUTE: $\Phi_E = EA = E(4\pi r^2)$
The Gaussian surface encloses all of the small shell and none of the large shell, so $Q_{\text{encl}} = +2q$.

Figure 22.47b

$\Phi_E = \dfrac{Q_{\text{encl}}}{\epsilon_0}$ gives $E(4\pi r^2) = \dfrac{2q}{\epsilon_0}$ so $E = \dfrac{2q}{4\pi\epsilon_0 r^2}$. Since the enclosed charge is positive the electric field is

radially outward.
(iv) $c < r < d$: Points in this region are in the conductor of the large shell, so $E = 0$.
(v) **SET UP:** $r > d$: Apply Gauss's law to a spherical Gaussian surface with radius $r > d$, as shown in Figure 22.47c.

EXECUTE: $\Phi_E = EA = E(4\pi r^2)$

The Gaussian surface encloses all
of the small shell and all of the
large shell, so $Q_{encl} = +2q + 4q = 6q$.

Figure 22.47c

$\Phi_E = \dfrac{Q_{encl}}{\epsilon_0}$ gives $E(4\pi r^2) = \dfrac{6q}{\epsilon_0}$

$E = \dfrac{6q}{4\pi\epsilon_0 r^2}$. Since the enclosed charge is positive the electric field is radially outward.

The graph of E versus r is sketched in Figure 22.47d.

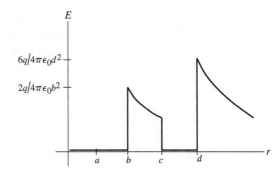

Figure 22.47d

(b) IDENTIFY and SET UP: Apply Gauss's law to a sphere that lies outside the surface of the shell for which we want to find the surface charge.

EXECUTE: (i) charge on inner surface of the small shell: Apply Gauss's law to a spherical Gaussian surface with radius $a < r < b$. This surface lies within the conductor of the small shell, where $E = 0$, so $\Phi_E = 0$. Thus by Gauss's law $Q_{encl} = 0$, so there is zero charge on the inner surface of the small shell.

(ii) charge on outer surface of the small shell: The total charge on the small shell is $+2q$. We found in part (i) that there is zero charge on the inner surface of the shell, so all $+2q$ must reside on the outer surface.

(iii) charge on inner surface of large shell: Apply Gauss's law to a spherical Gaussian surface with radius $c < r < d$. The surface lies within the conductor of the large shell, where $E = 0$, so $\Phi_E = 0$. Thus by Gauss's law $Q_{encl} = 0$. The surface encloses the $+2q$ on the small shell so there must be charge $-2q$ on the inner surface of the large shell to make the total enclosed charge zero.

(iv) charge on outer surface of large shell: The total charge on the large shell is $+4q$. We showed in part (iii) that the charge on the inner surface is $-2q$, so there must be $+6q$ on the outer surface.

EVALUATE: The electric field lines for $b < r < c$ originate from the surface charge on the outer surface of the inner shell and all terminate on the surface charge on the inner surface of the outer shell. These surface charges have equal magnitude and opposite sign. The electric field lines for $r > d$ originate from the surface charge on the outer surface of the outer sphere.

22.53.   **IDENTIFY:** We apply Gauss's law in (a) and take a spherical Gaussian surface because of the spherical symmetry of the charge distribution. In (b), the net field is the vector sum of the field due to q and the field due to the sphere.

(a) SET UP: $\rho(r) = \dfrac{\alpha}{r}$, $dV = 4\pi r^2 dr$, and $Q = \displaystyle\int_a^r \rho(r')dV$.

EXECUTE: For a Gaussian sphere of radius r, $Q_{encl} = \int_a^r \rho(r')dV = 4\pi\alpha \int_a^r r'dr' = 4\pi\alpha \frac{1}{2}(r^2 - a^2)$. Gauss's

law says that $E(4\pi r^2) = \frac{2\pi\alpha(r^2 - a^2)}{\epsilon_0}$, which gives $E = \frac{\alpha}{2\epsilon_0}\left(1 - \frac{a^2}{r^2}\right)$.

(b) The electric field of the point charge is $E_q = \frac{q}{4\pi\epsilon_0 r^2}$. The total electric field

is $E_{total} = \frac{\alpha}{2\epsilon_0} - \frac{\alpha}{2\epsilon_0}\frac{a^2}{r^2} + \frac{q}{4\pi\epsilon_0 r^2}$. For E_{total} to be constant, $-\frac{\alpha a^2}{2\epsilon_0} + \frac{q}{4\pi\epsilon_0} = 0$ and $q = 2\pi\alpha a^2$. The

constant electric field is $\frac{\alpha}{2\epsilon_0}$.

EVALUATE: The net field is constant, but not zero.

22.55.   **IDENTIFY:**   There is a force on each electron due to the other electron and a force due to the sphere of charge. Use Coulomb's law for the force between the electrons. Example 22.9 gives E inside a uniform sphere and Eq. (21.3) gives the force.

SET UP: The positions of the electrons are sketched in Figure 22.55a.

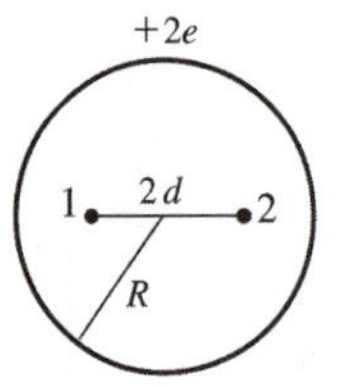

If the electrons are in equilibrium the net force on each one is zero.

Figure 22.55a

EXECUTE: Consider the forces on electron 2. There is a repulsive force F_1 due to the other electron, electron 1.

$$F_1 = \frac{1}{4\pi\epsilon_0}\frac{e^2}{(2d)^2}$$

The electric field inside the uniform distribution of positive charge is $E = \frac{Qr}{4\pi\epsilon_0 R^3}$ (Example 22.9), where

$Q = +2e$. At the position of electron 2, $r = d$. The force F_{cd} exerted by the positive charge distribution is

$F_{cd} = eE = \frac{e(2e)d}{4\pi\epsilon_0 R^3}$ and is attractive.

The force diagram for electron 2 is given in Figure 22.55b.

Figure 22.55b

Net force equals zero implies $F_1 = F_{cd}$ and $\frac{1}{4\pi\epsilon_0}\frac{e^2}{4d^2} = \frac{2e^2 d}{4\pi\epsilon_0 R^3}$.

Thus $(1/4d^2) = 2d/R^3$, so $d^3 = R^3/8$ and $d = R/2$.

EVALUATE: The electric field of the sphere is radially outward; it is zero at the center of the sphere and increases with distance from the center. The force this field exerts on one of the electrons is radially inward and increases as the electron is farther from the center. The force from the other electron is radially outward, is infinite when $d = 0$ and decreases as d increases. It is reasonable therefore for there to be a value of d for which these forces balance.

22.57. **(a) IDENTIFY** and **SET UP:** Consider the direction of the field for x slightly greater than and slightly less than zero. The slab is sketched in Figure 22.57a.

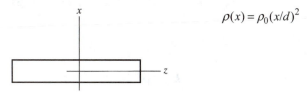

Figure 22.57a

EXECUTE: The charge distribution is symmetric about $x = 0$, so by symmetry $E(x) = E(-x)$. But for $x > 0$ the field is in the $+x$ direction and for $x < 0$ the field is in the $-x$ direction. At $x = 0$ the field can't be both in the $+x$ and $-x$ directions so must be zero. That is, $E_x(x) = -E_x(-x)$. At point $x = 0$ this gives $E_x(0) = -E_x(0)$ and this equation is satisfied only for $E_x(0) = 0$.

(b) IDENTIFY and **SET UP:** $|x| > d$ (outside the slab)

Apply Gauss's law to a cylindrical Gaussian surface whose axis is perpendicular to the slab and whose end caps have area A and are the same distance $|x| > d$ from $x = 0$, as shown in Figure 22.57b.

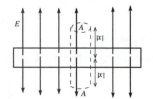

EXECUTE: $\Phi_E = 2EA$

Figure 22.57b

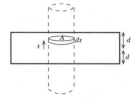

To find Q_{encl} consider a thin disk at coordinate x and with thickness dx, as shown in Figure 22.57c. The charge within this disk is

$$dq = \rho \, dV = \rho A \, dx = (\rho_0 A/d^2)\, x^2 dx.$$

Figure 22.57c

The total charge enclosed by the Gaussian cylinder is

$$Q_{encl} = 2\int_0^d dq = (2\rho_0 A/d^2)\int_0^d x^2 dx = (2\rho_0 A/d^2)(d^3/3) = \tfrac{2}{3}\rho_0 A d.$$

Then $\Phi_E = \dfrac{Q_{encl}}{\epsilon_0}$ gives $2EA = 2\rho_0 A d/3\epsilon_0$.

$$E = \rho_0 d/3\epsilon_0$$

$\vec{E}$ is directed away from $x = 0$, so $\vec{E} = (\rho_0 d/3\epsilon_0)(x/|x|)\hat{i}$.

(c) IDENTIFY and **SET UP:** $|x| < d$ (inside the slab)

Apply Gauss's law to a cylindrical Gaussian surface whose axis is perpendicular to the slab and whose end caps have area A and are the same distance $|x| < d$ from $x = 0$, as shown in Figure 22.57d

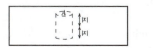

EXECUTE : $\Phi_E = 2EA$

Figure 22.57d

Q_{encl} is found as above, but now the integral on dx is only from 0 to x instead of 0 do d.

$$Q_{\text{encl}} = 2 \int_0^x dq = (2\rho_0 A/d^2) \int_0^x x^2 dx = (2\rho_0 A/d^2)(x^3/3).$$

Then $\Phi_E = \dfrac{Q_{\text{encl}}}{\epsilon_0}$ gives $2EA = 2\rho_0 Ax^3/3\epsilon_0 d^2$.

$$E = \rho_0 x^3/3\epsilon_0 d^2$$

$\vec{E}$ is directed away from $x = 0$, so $\vec{E} = (\rho_0 x^3/3\epsilon_0 d^2)\hat{\imath}$.

EVALUATE: Note that $E = 0$ at $x = 0$ as stated in part (a). Note also that the expressions for $|x| > d$ and $|x| < d$ agree for $x = d$.

22.61. **(a) IDENTIFY:** Use $\vec{E}(\vec{r})$ from Example (22.9) (inside the sphere) and relate the position vector of a point inside the sphere measured from the origin to that measured from the center of the sphere.
SET UP: For an insulating sphere of uniform charge density ρ and centered at the origin, the electric field inside the sphere is given by $E = Qr'/4\pi\epsilon_0 R^3$ (Example 22.9), where $\vec{r}'$ is the vector from the center of the sphere to the point where E is calculated.
But $\rho = 3Q/4\pi R^3$ so this may be written as $E = \rho r/3\epsilon_0$. And $\vec{E}$ is radially outward, in the direction of $\vec{r}'$, so $\vec{E} = \rho\vec{r}'/3\epsilon_0$.
For a sphere whose center is located by vector $\vec{b}$, a point inside the sphere and located by $\vec{r}$ is located by the vector $\vec{r}' = \vec{r} - \vec{b}$ relative to the center of the sphere, as shown in Figure 22.61.

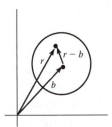

EXECUTE: Thus $\vec{E} = \dfrac{\rho(\vec{r} - \vec{b})}{3\epsilon_0}$

Figure 22.61

EVALUATE: When $b = 0$ this reduces to the result of Example 22.9. When $\vec{r} = \vec{b}$, this gives $E = 0$, which is correct since we know that $E = 0$ at the center of the sphere.
(b) IDENTIFY: The charge distribution can be represented as a uniform sphere with charge density ρ and centered at the origin added to a uniform sphere with charge density $-\rho$ and centered at $\vec{r} = \vec{b}$.
SET UP: $\vec{E} = \vec{E}_{\text{uniform}} + \vec{E}_{\text{hole}}$, where $\vec{E}_{\text{uniform}}$ is the field of a uniformly charged sphere with charge density ρ and $\vec{E}_{\text{hole}}$ is the field of a sphere located at the hole and with charge density $-\rho$. (Within the spherical hole the net charge density is $+\rho - \rho = 0$.)
EXECUTE: $\vec{E}_{\text{uniform}} = \dfrac{\rho\vec{r}}{3\epsilon_0}$, where $\vec{r}$ is a vector from the center of the sphere.

$\vec{E}_{\text{hole}} = \dfrac{-\rho(\vec{r} - \vec{b})}{3\epsilon_0}$, at points inside the hole.

Then $\vec{E} = \dfrac{\rho\vec{r}}{3\epsilon_0} + \left(\dfrac{-\rho(\vec{r} - \vec{b})}{3\epsilon_0} \right) = \dfrac{\rho\vec{b}}{3\epsilon_0}$.

EVALUATE: $\vec{E}$ is independent of $\vec{r}$ so is uniform inside the hole. The direction of $\vec{E}$ inside the hole is in the direction of the vector $\vec{b}$, the direction from the center of the insulating sphere to the center of the hole.

22.63.   **IDENTIFY:** The electric field at each point is the vector sum of the fields of the two charge distributions.

SET UP: Inside a sphere of uniform positive charge, $E = \dfrac{\rho r}{3\epsilon_0}$.

$\rho = \dfrac{Q}{\frac{4}{3}\pi R^3} = \dfrac{3Q}{4\pi R^3}$ so $E = \dfrac{Qr}{4\pi\epsilon_0 R^3}$, directed away from the center of the sphere. Outside a sphere of

uniform positive charge, $E = \dfrac{Q}{4\pi\epsilon_0 r^2}$, directed away from the center of the sphere.

EXECUTE: **(a)** $x = 0$. This point is inside sphere 1 and outside sphere 2. The fields are shown in Figure 22.63a.

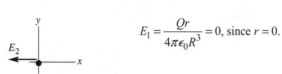

$E_1 = \dfrac{Qr}{4\pi\epsilon_0 R^3} = 0$, since $r = 0$.

Figure 22.63a

$E_2 = \dfrac{Q}{4\pi\epsilon_0 r^2}$ with $r = 2R$ so $E_2 = \dfrac{Q}{16\pi\epsilon_0 R^2}$, in the $-x$-direction.

Thus $\vec{E} = \vec{E}_1 + \vec{E}_2 = -\dfrac{Q}{16\pi\epsilon_0 R^2}\hat{i}$.

(b) $x = R/2$. This point is inside sphere 1 and outside sphere 2. Each field is directed away from the center of the sphere that produces it. The fields are shown in Figure 22.63b.

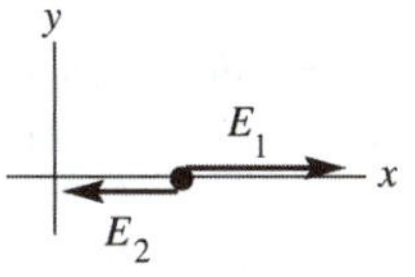

$E_1 = \dfrac{Qr}{4\pi\epsilon_0 R^3}$ with $r = R/2$ so

$E_1 = \dfrac{Q}{8\pi\epsilon_0 R^2}$

Figure 22.63b

$E_2 = \dfrac{Q}{4\pi\epsilon_0 r^2}$ with $r = 3R/2$ so $E_2 = \dfrac{Q}{9\pi\epsilon_0 R^2}$

$E = E_1 - E_2 = \dfrac{Q}{72\pi\epsilon_0 R^2}$, in the $+x$-direction and $\vec{E} = \dfrac{Q}{72\pi\epsilon_0 R^2}\hat{i}$

(c) $x = R$. This point is at the surface of each sphere. The fields have equal magnitudes and opposite directions, so $E = 0$.

(d) $x = 3R$. This point is outside both spheres. Each field is directed away from the center of the sphere that produces it. The fields are shown in Figure 22.63c.

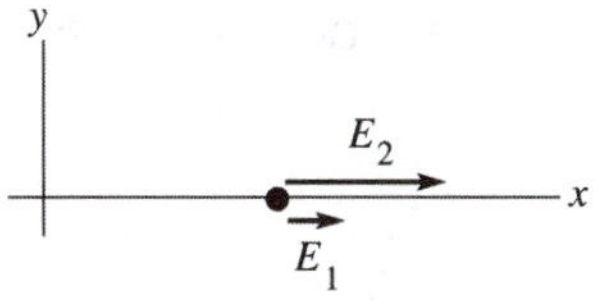

$E_1 = \dfrac{Q}{4\pi\epsilon_0 r^2}$ with $r = 3R$ so

$E_1 = \dfrac{Q}{36\pi\epsilon_0 R^2}$

Figure 22.63c

$$E_2 = \frac{Q}{4\pi\epsilon_0 r^2} \text{ with } r = R \text{ so } E_2 = \frac{Q}{4\pi\epsilon_0 R^2}$$

$$E = E_1 + E_2 = \frac{5Q}{18\pi\epsilon_0 R^2}, \text{ in the } +x\text{-direction and } \vec{E} = \frac{5Q}{18\pi\epsilon_0 R^2}\hat{i}$$

EVALUATE: The field of each sphere is radially outward from the center of the sphere. We must use the correct expression for $E(r)$ for each sphere, depending on whether the field point is inside or outside that sphere.

22.65. $\rho(r) = \rho_0(1 - r/R)$ for $r \leq R$ where $\rho_0 = 3Q/\pi R^3$. $\rho(r) = 0$ for $r \geq R$

(a) IDENTIFY: The charge density varies with r inside the spherical volume. Divide the volume up into thin concentric shells, of radius r and thickness dr. Find the charge dq in each shell and integrate to find the total charge.
SET UP: The thin shell is sketched in Figure 22.65a.

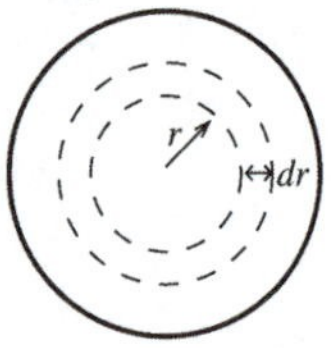

EXECUTE: The volume of such a shell is $dV = 4\pi r^2 dr$.
The charge contained within the shell is
$dq = \rho(r)dV = 4\pi r^2 \rho_0(1 - r/R)dr$.

Figure 22.65a

The total charge Q in the charge distribution is obtained by integrating dq over all such shells into which the sphere can be subdivided:

$$Q = \int dq = \int_0^R 4\pi r^2 \rho_0(1 - r/R)dr = 4\pi\rho_0 \int_0^R (r^2 - r^3/R)dr$$

$$Q = 4\pi\rho_0\left[\frac{r^3}{3} - \frac{r^4}{4R}\right]_0^R = 4\pi\rho_0\left(\frac{R^3}{3} - \frac{R^4}{4R}\right) = 4\pi\rho_0(R^3/12) = 4\pi(3Q/\pi R^3)(R^3/12) = Q, \text{ as was to be shown.}$$

(b) IDENTIFY: Apply Gauss's law to a spherical surface of radius r, where $r > R$.
SET UP: The Gaussian surface is shown in Figure 22.65b.

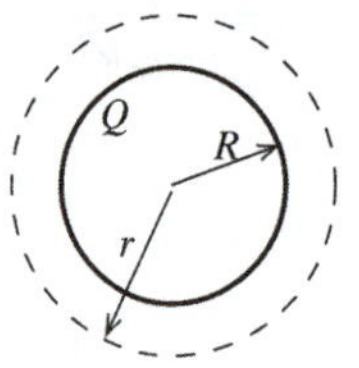

EXECUTE: $\Phi_E = \dfrac{Q_{encl}}{\epsilon_0}$

$$E(4\pi r^2) = \frac{Q}{\epsilon_0}$$

Figure 22.65b

$$E = \frac{Q}{4\pi\epsilon_0 r^2}; \text{ same as for point charge of charge } Q.$$

(c) IDENTIFY: Apply Gauss's law to a spherical surface of radius r, where $r < R$:
SET UP: The Gaussian surface is shown in Figure 22.65c.

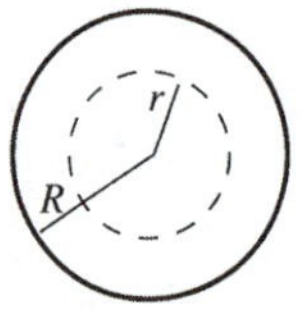

EXECUTE: $\Phi_E = \dfrac{Q_{encl}}{\epsilon_0}$

$$\Phi_E = E(4\pi r^2)$$

Figure 22.65c

To calculate the enclosed charge Q_{encl} use the same technique as in part (a), except integrate dq out to r rather than R. (We want the charge that is inside radius r.)

$$Q_{encl} = \int_0^r 4\pi r'^2 \rho_0 \left(1 - \frac{r'}{R}\right) dr' = 4\pi\rho_0 \int_0^r \left(r'^2 - \frac{r'^3}{R}\right) dr'$$

$$Q_{encl} = 4\pi\rho_0 \left[\frac{r'^3}{3} - \frac{r'^4}{4R}\right]_0^r = 4\pi\rho_0 \left(\frac{r^3}{3} - \frac{r^4}{4R}\right) = 4\pi\rho_0 r^3 \left(\frac{1}{3} - \frac{r}{4R}\right)$$

$$\rho_0 = \frac{3Q}{\pi R^3} \text{ so } Q_{encl} = 12Q\frac{r^3}{R^3}\left(\frac{1}{3} - \frac{r}{4R}\right) = Q\left(\frac{r^3}{R^3}\right)\left(4 - 3\frac{r}{R}\right).$$

Thus Gauss's law gives $E(4\pi r^2) = \frac{Q}{\epsilon_0}\left(\frac{r^3}{R^3}\right)\left(4 - 3\frac{r}{R}\right).$

$$E = \frac{Qr}{4\pi\epsilon_0 R^3}\left(4 - \frac{3r}{R}\right), r \le R$$

(d) The graph of E versus r is sketched in Figure 22.65d.

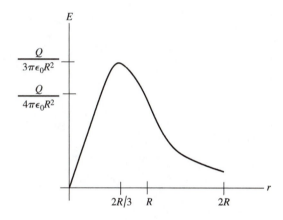

Figure 22.65d

(e) Where the electric field is a maximum, $\frac{dE}{dr} = 0$. Thus $\frac{d}{dr}\left(4r - \frac{3r^2}{R}\right) = 0$ so $4 - 6r/R = 0$ and $r = 2R/3$.

At this value of r, $E = \frac{Q}{4\pi\epsilon_0 R^3}\left(\frac{2R}{3}\right)\left(4 - \frac{3}{R}\frac{2R}{3}\right) = \frac{Q}{3\pi\epsilon_0 R^2}.$

EVALUATE: Our expressions for $E(r)$ for $r < R$ and for $r > R$ agree at $r = R$. The results of part (e) for the value of r where $E(r)$ is a maximum agrees with the graph in part (d).

ELECTRIC POTENTIAL

23.3. **IDENTIFY:** The work needed to assemble the nucleus is the sum of the electrical potential energies of the protons in the nucleus, relative to infinity.

SET UP: The total potential energy is the scalar sum of all the individual potential energies, where each potential energy is $U = (1/4\pi\epsilon_0)(qq_0/r)$. Each charge is e and the charges are equidistant from each other,

so the total potential energy is $U = \dfrac{1}{4\pi\epsilon_0}\left(\dfrac{e^2}{r} + \dfrac{e^2}{r} + \dfrac{e^2}{r}\right) = \dfrac{3e^2}{4\pi\epsilon_0 r}$.

EXECUTE: Adding the potential energies gives

$$U = \frac{3e^2}{4\pi\epsilon_0 r} = \frac{3(1.60\times10^{-19}\ \text{C})^2(9.00\times10^9\ \text{N}\cdot\text{m}^2/\text{C}^2)}{2.00\times10^{-15}\ \text{m}} = 3.46\times10^{-13}\ \text{J} = 2.16\ \text{MeV}$$

EVALUATE: This is a small amount of energy on a macroscopic scale, but on the scale of atoms, 2 MeV is quite a lot of energy.

23.5. **(a) IDENTIFY:** Use conservation of energy:

$$K_a + U_a + W_{\text{other}} = K_b + U_b$$

U for the pair of point charges is given by Eq. (23.9).

SET UP:

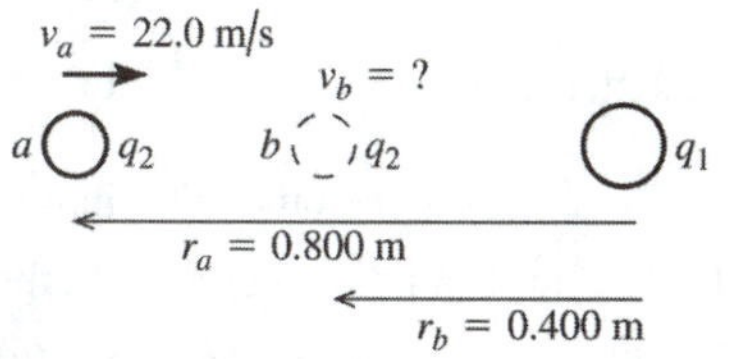

Let point a be where q_2 is 0.800 m from q_1 and point b be where q_2 is 0.400 m from q_1, as shown in Figure 23.5a.

Figure 23.5a

EXECUTE: Only the electric force does work, so $W_{\text{other}} = 0$ and $U = \dfrac{1}{4\pi\epsilon_0}\dfrac{q_1 q_2}{r}$.

$$K_a = \tfrac{1}{2}mv_a^2 = \tfrac{1}{2}(1.50\times10^{-3}\ \text{kg})(22.0\ \text{m/s})^2 = 0.3630\ \text{J}$$

$$U_a = \frac{1}{4\pi\epsilon_0}\frac{q_1 q_2}{r_a} = (8.988\times10^9\ \text{N}\cdot\text{m}^2/\text{C}^2)\frac{(-2.80\times10^{-6}\ \text{C})(-7.80\times10^{-6}\ \text{C})}{0.800\ \text{m}} = +0.2454\ \text{J}$$

$$K_b = \tfrac{1}{2}mv_b^2$$

$$U_b = \frac{1}{4\pi\epsilon_0}\frac{q_1 q_2}{r_b} = (8.988\times10^9\ \text{N}\cdot\text{m}^2/\text{C}^2)\frac{(-2.80\times10^{-6}\ \text{C})(-7.80\times10^{-6}\ \text{C})}{0.400\ \text{m}} = +0.4907\ \text{J}$$

The conservation of energy equation then gives $K_b = K_a + (U_a - U_b)$

$$\tfrac{1}{2}mv_b^2 = +0.3630 \text{ J} + (0.2454 \text{ J} - 0.4907 \text{ J}) = 0.1177 \text{ J}$$

$$v_b = \sqrt{\frac{2(0.1177 \text{ J})}{1.50 \times 10^{-3} \text{ kg}}} = 12.5 \text{ m/s}$$

EVALUATE: The potential energy increases when the two positively charged spheres get closer together, so the kinetic energy and speed decrease.

(b) IDENTIFY: Let point c be where q_2 has its speed momentarily reduced to zero. Apply conservation of energy to points a and c: $K_a + U_a + W_{\text{other}} = K_c + U_c$.

SET UP: Points a and c are shown in Figure 23.5b.

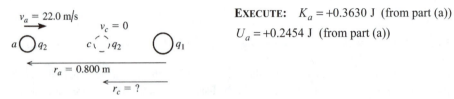

EXECUTE: $K_a = +0.3630 \text{ J}$ (from part (a))

$U_a = +0.2454 \text{ J}$ (from part (a))

Figure 23.5b

$K_c = 0$ (at distance of closest approach the speed is zero)

$$U_c = \frac{1}{4\pi\epsilon_0}\frac{q_1 q_2}{r_c}$$

Thus conservation of energy $K_a + U_a = U_c$ gives $\dfrac{1}{4\pi\epsilon_0}\dfrac{q_1 q_2}{r_c} = +0.3630 \text{ J} + 0.2454 \text{ J} = 0.6084 \text{ J}$

$$r_c = \frac{1}{4\pi\epsilon_0}\frac{q_1 q_2}{0.6084 \text{ J}} = (8.988 \times 10^9 \text{ N} \cdot \text{m}^2/\text{C}^2)\frac{(-2.80 \times 10^{-6} \text{ C})(-7.80 \times 10^{-6} \text{ C})}{+0.6084 \text{ J}} = 0.323 \text{ m}.$$

EVALUATE: $U \to \infty$ as $r \to 0$ so q_2 will stop no matter what its initial speed is.

23.7.   **IDENTIFY:** The total potential energy is the scalar sum of the individual potential energies of each pair of charges.

SET UP: For a pair of point charges the electrical potential energy is $U = k\dfrac{qq'}{r}$. In the O-H-O combination the O^- is 0.180 nm from the H^+ and 0.290 nm from the other O^-. In the N-H-N combination the N^- is 0.190 nm from the H^+ and 0.300 nm from the other N^-. In the O-H-N combination the O^- is 0.180 nm from the H^+ and 0.290 nm from the N^-. U is positive for like charges and negative for unlike charges.

EXECUTE: O-H-O $\text{O}^- - \text{H}^+$, $U = -1.28 \times 10^{-18}$ J; $\text{O}^- - \text{O}^-$, $U = +7.93 \times 10^{-19}$ J.

N-H-N $\text{N}^- - \text{H}^+$, $U = -1.21 \times 10^{-18}$ J; $\text{N}^- - \text{N}^-$, $U = +7.67 \times 10^{-19}$ J.

O-H-N $\text{O}^- - \text{H}^+$, $U = -1.28 \times 10^{-18}$ J; $\text{O}^- - \text{N}^-$, $U = +7.93 \times 10^{-19}$ J.

The total potential energy is -3.77×10^{-18} J $+ 2.35 \times 10^{-18}$ J $= -1.42 \times 10^{-18}$ J.

EVALUATE: For pairs of opposite sign the potential energy is negative and for pairs of the same sign the potential energy is positive. The net electrical potential energy is the algebraic sum of the potential energy of each pair.

23.9.   **IDENTIFY:** The protons repel each other and therefore accelerate away from one another. As they get farther and farther away, their kinetic energy gets greater and greater but their acceleration keeps decreasing. Conservation of energy and Newton's laws apply to these protons.

SET UP: Let a be the point when they are 0.750 nm apart and b be the point when they are very far apart. A proton has charge $+e$ and mass 1.67×10^{-27} kg. As they move apart the protons have equal kinetic energies and speeds. Their potential energy is $U = ke^2/r$ and $K = \tfrac{1}{2}mv^2$. $K_a + U_a = K_b + U_b$.

EXECUTE: **(a)** They have maximum speed when they are far apart and all their initial electrical potential energy has been converted to kinetic energy. $K_a + U_a = K_b + U_b$.

$K_a = 0$ and $U_b = 0$, so

$$K_b = U_a = k\frac{e^2}{r_a} = (8.99 \times 10^9 \text{ N} \cdot \text{m}^2/\text{C}^2)\frac{(1.60 \times 10^{-19} \text{ C})^2}{0.750 \times 10^{-9} \text{ m}} = 3.07 \times 10^{-19} \text{ J}.$$

$K_b = \frac{1}{2}mv_b^2 + \frac{1}{2}mv_b^2$, so $K_b = mv_b^2$ and $v_b = \sqrt{\frac{K_b}{m}} = \sqrt{\frac{3.07 \times 10^{-19} \text{ J}}{1.67 \times 10^{-27} \text{ kg}}} = 1.36 \times 10^4 \text{ m/s}$.

(b) Their acceleration is largest when the force between them is largest and this occurs at $r = 0.750$ nm, when they are closest.

$$F = k\frac{e^2}{r^2} = (8.99 \times 10^9 \text{ N} \cdot \text{m}^2/\text{C}^2)\left(\frac{1.60 \times 10^{-19} \text{ C}}{0.750 \times 10^{-9} \text{ m}}\right)^2 = 4.09 \times 10^{-10} \text{ N}.$$

$$a = \frac{F}{m} = \frac{4.09 \times 10^{-10} \text{ N}}{1.67 \times 10^{-27} \text{ kg}} = 2.45 \times 10^{17} \text{ m/s}^2.$$

EVALUATE: The acceleration of the protons decreases as they move farther apart, but the force between them is repulsive so they continue to increase their speeds and hence their kinetic energies.

23.11. **IDENTIFY:** Apply Eq. (23.2). The net work to bring the charges in from infinity is equal to the change in potential energy. The total potential energy is the sum of the potential energies of each pair of charges, calculated from Eq. (23.9).

SET UP: Let 1 be where all the charges are infinitely far apart. Let 2 be where the charges are at the corners of the triangle, as shown in Figure 23.11.

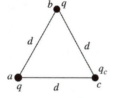

Let q_c be the third, unknown charge.

Figure 23.11

EXECUTE: $W = -\Delta U = -(U_2 - U_1)$, where W is the work done by the Coulomb force.

$$U_1 = 0$$

$$U_2 = U_{ab} + U_{ac} + U_{bc} = \frac{1}{4\pi\epsilon_0 d}(q^2 + 2qq_c)$$

Want $W = 0$, so $W = -(U_2 - U_1)$ gives $0 = -U_2$

$$0 = \frac{1}{4\pi\epsilon_0 d}(q^2 + 2qq_c)$$

$q^2 + 2qq_c = 0$ and $q_c = -q/2$.

EVALUATE: The potential energy for the two charges q is positive and for each q with q_c it is negative. There are two of the q, q_c terms so must have $q_c < q$.

23.13. **IDENTIFY** and **SET UP:** Apply conservation of energy to points A and B.

EXECUTE: $K_A + U_A = K_B + U_B$

$U = qV$, so $K_A + qV_A = K_B + qV_B$

$K_B = K_A + q(V_A - V_B) = 0.00250 \text{ J} + (-5.00 \times 10^{-6} \text{ C})(200 \text{ V} - 800 \text{ V}) = 0.00550 \text{ J}$

$v_B = \sqrt{2K_B/m} = 7.42 \text{ m/s}$

EVALUATE: It is faster at B; a negative charge gains speed when it moves to higher potential.

23.17.   **IDENTIFY:** The potential at any point is the scalar sum of the potentials due to individual charges.

SET UP: $V = kq/r$ and $W_{ab} = q(V_a - V_b)$.

EXECUTE: **(a)** $r_{a1} = r_{a2} = \frac{1}{2}\sqrt{(0.0300 \text{ m})^2 + (0.0300 \text{ m})^2} = 0.0212$ m. $V_a = k\left(\dfrac{q_1}{r_{a1}} + \dfrac{q_2}{r_{a2}}\right) = 0.$

(b) $r_{b1} = 0.0424$ m, $r_{b2} = 0.0300$ m.

$$V_b = k\left(\frac{q_1}{r_{b1}} + \frac{q_2}{r_{b2}}\right) = (8.99 \times 10^9 \text{ N} \cdot \text{m}^2/\text{C}^2)\left(\frac{+2.00 \times 10^{-6} \text{ C}}{0.0424 \text{ m}} + \frac{-2.00 \times 10^{-6} \text{ C}}{0.0300 \text{ m}}\right) = -1.75 \times 10^5 \text{ V.}$$

(c) $W_{ab} = q_3(V_a - V_b) = (-5.00 \times 10^{-6} \text{ C})[0 - (-1.75 \times 10^5 \text{ V})] = -0.875$ J.

EVALUATE: Since $V_b < V_a$, a positive charge would be pulled by the existing charges from a to b, so they would do positive work on this charge. But they would repel a negative charge and hence do negative work on it, as we found in part (c).

23.19.   **IDENTIFY:** $V = \dfrac{1}{4\pi\epsilon_0} \sum_i \dfrac{q_i}{r_i}$

SET UP: The locations of the changes and points A and B are sketched in Figure 23.19.

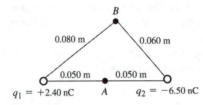

Figure 23.19

EXECUTE: **(a)** $V_A = \dfrac{1}{4\pi\epsilon_0}\left(\dfrac{q_1}{r_{A1}} + \dfrac{q_2}{r_{A2}}\right)$

$$V_A = (8.988 \times 10^9 \text{ N} \cdot \text{m}^2/\text{C}^2)\left(\frac{+2.40 \times 10^{-9} \text{ C}}{0.050 \text{ m}} + \frac{-6.50 \times 10^{-9} \text{ C}}{0.050 \text{ m}}\right) = -737 \text{ V}$$

(b) $V_B = \dfrac{1}{4\pi\epsilon_0}\left(\dfrac{q_1}{r_{B1}} + \dfrac{q_2}{r_{B2}}\right)$

$$V_B = (8.988 \times 10^9 \text{ N} \cdot \text{m}^2/\text{C}^2)\left(\frac{+2.40 \times 10^{-9} \text{ C}}{0.080 \text{ m}} + \frac{-6.50 \times 10^{-9} \text{ C}}{0.060 \text{ m}}\right) = -704 \text{ V}$$

(c) IDENTIFY and SET UP: Use Eq. (23.13) and the results of parts (a) and (b) to calculate W.

EXECUTE: $W_{B \to A} = q'(V_B - V_A) = (2.50 \times 10^{-9} \text{ C})(-704 \text{ V} - (-737 \text{ V})) = +8.2 \times 10^{-8}$ J

EVALUATE: The electric force does positive work on the positive charge when it moves from higher potential (point B) to lower potential (point A).

23.23.   **IDENTIFY and SET UP:** Apply conservation of energy, Eq. (23.3). Use Eq. (23.12) to express U in terms of V.

(a) EXECUTE: $K_1 + qV_1 = K_2 + qV_2$, $q(V_2 - V_1) = K_1 - K_2$; $q = -1.602 \times 10^{-19}$ C.

$K_1 = \frac{1}{2}m_e v_1^2 = 4.099 \times 10^{-18}$ J; $K_2 = \frac{1}{2}m_e v_2^2 = 2.915 \times 10^{-17}$ J. $\Delta V = V_2 - V_1 = \dfrac{K_1 - K_2}{q} = 156$ V.

EVALUATE: The electron gains kinetic energy when it moves to higher potential.

(b) EXECUTE: Now $K_1 = 2.915 \times 10^{-17}$ J, $K_2 = 0$. $V_2 - V_1 = \dfrac{K_1 - K_2}{q} = -182$ V.

EVALUATE: The electron loses kinetic energy when it moves to lower potential.

23.29.   **(a) IDENTIFY and SET UP:**   The electric field on the ring's axis is calculated in Example 21.9. The force on the electron exerted by this field is given by Eq. (21.3).

EXECUTE: When the electron is on either side of the center of the ring, the ring exerts an attractive force directed toward the center of the ring. This restoring force produces oscillatory motion of the electron along the axis of the ring, with amplitude 30.0 cm. The force on the electron is *not* of the form $F = -kx$ so the oscillatory motion is not simple harmonic motion.

(b) IDENTIFY: Apply conservation of energy to the motion of the electron.

SET UP: $K_a + U_a = K_b + U_b$ with a at the initial position of the electron and b at the center of the ring.

From Example 23.11, $V = \dfrac{1}{4\pi\epsilon_0} \dfrac{Q}{\sqrt{x^2 + R^2}}$, where R is the radius of the ring.

EXECUTE: $x_a = 30.0$ cm, $x_b = 0$.

$K_a = 0$ (released from rest), $K_b = \frac{1}{2}mv^2$

Thus $\frac{1}{2}mv^2 = U_a - U_b$

And $U = qV = -eV$ so $v = \sqrt{\dfrac{2e(V_b - V_a)}{m}}$.

$$V_a = \frac{1}{4\pi\epsilon_0} \frac{Q}{\sqrt{x_a^2 + R^2}} = (8.988 \times 10^9 \ \text{N} \cdot \text{m}^2/\text{C}^2) \frac{24.0 \times 10^{-9} \ \text{C}}{\sqrt{(0.300 \ \text{m})^2 + (0.150 \ \text{m})^2}}$$

$$V_a = 643 \ \text{V}$$

$$V_b = \frac{1}{4\pi\epsilon_0} \frac{Q}{\sqrt{x_b^2 + R^2}} = (8.988 \times 10^9 \ \text{N} \cdot \text{m}^2/\text{C}^2) \frac{24.0 \times 10^{-9} \ \text{C}}{0.150 \ \text{m}} = 1438 \ \text{V}$$

$$v = \sqrt{\frac{2e(V_b - V_a)}{m}} = \sqrt{\frac{2(1.602 \times 10^{-19} \ \text{C})(1438 \ \text{V} - 643 \ \text{V})}{9.109 \times 10^{-31} \ \text{kg}}} = 1.67 \times 10^7 \ \text{m/s}$$

EVALUATE: The positively charged ring attracts the negatively charged electron and accelerates it. The electron has its maximum speed at this point. When the electron moves past the center of the ring the force on it is opposite to its motion and it slows down.

23.33.   **IDENTIFY:**   For points outside the cylinder, its electric field behaves like that of a line of charge. Since a voltmeter reads potential difference, that is what we need to calculate.

SET UP: The potential difference is $\Delta V = \dfrac{\lambda}{2\pi\epsilon_0} \ln (r_b/r_a)$.

EXECUTE: **(a)** Substituting numbers gives

$$\Delta V = \frac{\lambda}{2\pi\epsilon_0} \ln (r_b/r_a) = (8.50 \times 10^{-6} \ \text{C/m})(2 \times 9.00 \times 10^9 \ \text{N} \cdot \text{m}^2/\text{C}^2) \ln\left(\frac{10.0 \ \text{cm}}{6.00 \ \text{cm}}\right)$$

$$\Delta V = 7.82 \times 10^4 \ \text{V} = 78{,}200 \ \text{V} = 78.2 \ \text{kV}$$

(b) $E = 0$ inside the cylinder, so the potential is constant there, meaning that the voltmeter reads zero.

EVALUATE: Caution! The fact that the voltmeter reads zero in part (b) does not mean that $V = 0$ inside the cylinder. The electric field is zero, but the potential is constant and equal to the potential at the surface.

23.35.   **IDENTIFY:**   The electric field of the line of charge does work on the sphere, increasing its kinetic energy.

SET UP: $K_1 + U_1 = K_2 + U_2$ and $K_1 = 0$. $U = qV$ so $qV_1 = K_2 + qV_2$. $V = \dfrac{\lambda}{2\pi\epsilon_0} \ln\left(\dfrac{r_0}{r}\right)$.

EXECUTE: $V_1 = \dfrac{\lambda}{2\pi\epsilon_0} \ln\left(\dfrac{r_0}{r_1}\right)$. $V_2 = \dfrac{\lambda}{2\pi\epsilon_0} \ln\left(\dfrac{r_0}{r_2}\right)$.

$$K_2 = q(V_1 - V_2) = \frac{\lambda}{2\pi\epsilon_0}\left(\ln\left(\frac{r_0}{r_1}\right) - \ln\left(\frac{r_0}{r_2}\right)\right) = \frac{\lambda q}{2\pi\epsilon_0}(\ln r_2 - \ln r_1) = \frac{\lambda q}{2\pi\epsilon_0} \ln\left(\frac{r_2}{r_1}\right).$$

$$K_2 = \frac{(3.00 \times 10^{-6} \text{ C/m})(8.00 \times 10^{-6} \text{ C})}{2\pi(8.854 \times 10^{-12} \text{ C}^2/(\text{N} \cdot \text{m}^2))} \ln\left(\frac{4.50}{1.50}\right) = 0.474 \text{ J}.$$

EVALUATE: The potential due to the line of charge does *not* go to zero at infinity but is defined to be zero at an arbitrary distance r_0 from the line.

23.37.   **IDENTIFY:** We can model the axon membrane as a large sheet having equal but opposite charges on its opposite faces.

SET UP: For two oppositely charged sheets of charge, $V_{ab} = Ed$. The positively charged sheet is the one at higher potential.

EXECUTE: (a) $E = \dfrac{V_{ab}}{d} = \dfrac{70 \times 10^{-3} \text{ V}}{7.5 \times 10^{-9} \text{ m}} = 9.3 \times 10^6$ V/m. The electric field is directed inward, toward the interior of the axon, since the outer surface of the membrane has positive charge and $\vec{E}$ points away from positive charge and toward negative charge.

(b) The outer surface has positive charge so it is at higher potential than the inner surface.

EVALUATE: The electric field is quite strong compared to ordinary laboratory fields in devices such as student oscilloscopes. The potential difference is only 70 mV, but it occurs over a distance of only 7.5 nm.

23.43.   **IDENTIFY:** Example 23.8 shows that the potential of a solid conducting sphere is the same at every point inside the sphere and is equal to its value $V = q/4\pi\epsilon_0 R$ at the surface. Use the given value of E to find q.

SET UP: For negative charge the electric field is directed toward the charge.

For points outside this spherical charge distribution the field is the same as if all the charge were concentrated at the center.

EXECUTE: $E = \dfrac{|q|}{4\pi\epsilon_0 r^2}$ and $|q| = 4\pi\epsilon_0 E r^2 = \dfrac{(3800 \text{ N/C})(0.200 \text{ m})^2}{8.99 \times 10^9 \text{ N} \cdot \text{m}^2/\text{C}^2} = 1.69 \times 10^{-8}$ C.

Since the field is directed inward, the charge must be negative. The potential of a point charge, taking ∞ as zero, is $V = \dfrac{q}{4\pi\epsilon_0 r} = \dfrac{(8.99 \times 10^9 \text{ N} \cdot \text{m}^2/\text{C}^2)(-1.69 \times 10^{-8} \text{ C})}{0.200 \text{ m}} = -760$ V at the surface of the sphere.

Since the charge all resides on the surface of a conductor, the field inside the sphere due to this symmetrical distribution is zero. No work is therefore done in moving a test charge from just inside the surface to the center, and the potential at the center must also be –760 V.

EVALUATE: Inside the sphere the electric field is zero and the potential is constant.

23.45.   **IDENTIFY** and **SET UP:** Use Eq. (23.19) to calculate the components of $\vec{E}$.

EXECUTE: $V = Axy - Bx^2 + Cy$

(a) $E_x = -\dfrac{\partial V}{\partial x} = -Ay + 2Bx$

$E_y = -\dfrac{\partial V}{\partial y} = -Ax - C$

$E_z = \dfrac{\partial V}{\partial z} = 0$

(b) $E = 0$ requires that $E_x = E_y = E_z = 0$.

$E_z = 0$ everywhere.

$E_y = 0$ at $x = -C/A$.

And E_x is also equal to zero for this x, any value of z and $y = 2Bx/A = (2B/A)(-C/A) = -2BC/A^2$.

EVALUATE: V doesn't depend on z so $E_z = 0$ everywhere.

23.51.   **IDENTIFY:** $U = k\left(\dfrac{q_1 q_2}{r_{12}} + \dfrac{q_1 q_3}{r_{13}} + \dfrac{q_2 q_3}{r_{23}}\right)$

SET UP: In part (a), $r_{12} = 0.200$ m, $r_{23} = 0.100$ m and $r_{13} = 0.100$ m. In part (b) let particle 3 have coordinate x, so $r_{12} = 0.200$ m, $r_{13} = x$ and $r_{23} = 0.200 - x$.

EXECUTE: **(a)** $U = k\left(\dfrac{(4.00\ \text{nC})(-3.00\ \text{nC})}{(0.200\ \text{m})} + \dfrac{(4.00\ \text{nC})(2.00\ \text{nC})}{(0.100\ \text{m})} + \dfrac{(-3.00\ \text{nC})(2.00\ \text{nC})}{(0.100\ \text{m})}\right) = -3.60 \times 10^{-7}\ \text{J}$

(b) If $U = 0$, then $0 = k\left(\dfrac{q_1 q_2}{r_{12}} + \dfrac{q_1 q_3}{x} + \dfrac{q_2 q_3}{r_{12} - x}\right)$. Solving for x we find:

$0 = -60 + \dfrac{8}{x} - \dfrac{6}{0.2 - x} \Rightarrow 60x^2 - 26x + 1.6 = 0 \Rightarrow x = 0.074\ \text{m}, 0.360\ \text{m}$. Therefore, $x = 0.074\ \text{m}$ since it is

the only value between the two charges.

EVALUATE: U_{13} is positive and both U_{23} and U_{12} are negative. If $U = 0$, then $|U_{13}| = |U_{23}| + |U_{12}|$. For

$x = 0.074\ \text{m}$, $U_{13} = +9.7 \times 10^{-7}\ \text{J}$, $U_{23} = -4.3 \times 10^{-7}\ \text{J}$ and $U_{12} = -5.4 \times 10^{-7}\ \text{J}$. It is true that $U = 0$ at

this x.

23.53. **IDENTIFY:** The remaining nucleus (radium minus the ejected alpha particle) repels the alpha particle, giving it 4.79 MeV of kinetic energy when it is far from the nucleus. The mechanical energy of the system is conserved.

SET UP: $U = k\dfrac{qq'}{r}$. $U_a + K_a = U_b + K_b$. The charge of the alpha particle is $+2e$ and the charge of the

radon nucleus is $+86e$.

EXECUTE: **(a)** The final energy of the alpha particle, 4.79 MeV, equals the electrical potential energy of the alpha-radon combination just before the decay. $U = 4.79\ \text{MeV} = 7.66 \times 10^{-13}\ \text{J}$.

(b) $r = \dfrac{kqq'}{U} = \dfrac{(8.99 \times 10^9\ \text{N} \cdot \text{m}^2/\text{C}^2)(2)(86)(1.60 \times 10^{-19}\ \text{C})^2}{7.66 \times 10^{-13}\ \text{J}} = 5.17 \times 10^{-14}\ \text{m}$.

EVALUATE: Although we have made some simplifying assumptions (such as treating the atomic nucleus as a spherically symmetric charge, even when very close to it), this result gives a fairly reasonable estimate for the size of a nucleus.

23.57. **IDENTIFY and SET UP:** Calculate the components of $\vec{E}$ from Eq. (23.19). Eq. (21.3) gives $\vec{F}$ from $\vec{E}$.

EXECUTE: **(a)** $V = Cx^{4/3}$

$$C = V/x^{4/3} = 240\ \text{V}/(13.0 \times 10^{-3}\ \text{m})^{4/3} = 7.85 \times 10^4\ \text{V/m}^{4/3}$$

(b) $E_x = -\dfrac{\partial V}{\partial x} = -\dfrac{4}{3}Cx^{1/3} = -(1.05 \times 10^5\ \text{V/m}^{4/3})x^{1/3}$

The minus sign means that E_x is in the $-x$-direction, which says that $\vec{E}$ points from the positive anode toward the negative cathode.

(c) $\vec{F} = q\vec{E}$ so $F_x = -eE_x = \frac{4}{3}eCx^{1/3}$

Halfway between the electrodes means $x = 6.50 \times 10^{-3}\ \text{m}$.

$$F_x = \frac{4}{3}(1.602 \times 10^{-19}\ \text{C})(7.85 \times 10^4\ \text{V/m}^{4/3})(6.50 \times 10^{-3}\ \text{m})^{1/3} = 3.13 \times 10^{-15}\ \text{N}$$

F_x is positive, so the force is directed toward the positive anode.

EVALUATE: V depends only on x, so $E_y = E_z = 0$. $\vec{E}$ is directed from high potential (anode) to low potential (cathode). The electron has negative charge, so the force on it is directed opposite to the electric field.

23.59. **IDENTIFY:** $U = \dfrac{kq_1 q_2}{r}$

SET UP: Eight charges means there are $8(8-1)/2 = 28$ pairs. There are 12 pairs of q and $-q$ separated by

d, 12 pairs of equal charges separated by $\sqrt{2}d$ and 4 pairs of q and $-q$ separated by $\sqrt{3}d$.

EXECUTE: **(a)** $U = kq^2\left(-\dfrac{12}{d} + \dfrac{12}{\sqrt{2}d} - \dfrac{4}{\sqrt{3}d}\right) = -\dfrac{12kq^2}{d}\left(1 - \dfrac{1}{\sqrt{2}} + \dfrac{1}{3\sqrt{3}}\right) = -1.46q^2/\pi\epsilon_0 d$

EVALUATE: **(b)** The fact that the electric potential energy is less than zero means that it is energetically favorable for the crystal ions to be together.

23.61.   **(a) IDENTIFY:** Use Eq. (23.10) for the electron and each proton.
SET UP: The positions of the particles are shown in Figure 23.61a.

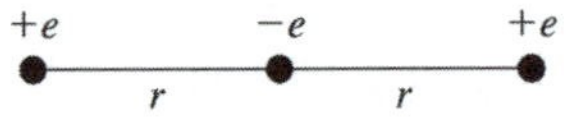

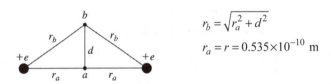

$$r = (1.07 \times 10^{-10} \text{ m})/2 = 0.535 \times 10^{-10} \text{ m}$$

Figure 23.61a

EXECUTE: The potential energy of interaction of the electron with each proton is $U = \dfrac{1}{4\pi\epsilon_0} \dfrac{(-e^2)}{r}$, so the

total potential energy is

$$U = -\frac{2e^2}{4\pi\epsilon_0 r} = -\frac{2(8.988 \times 10^9 \text{ N} \cdot \text{m}^2/\text{C}^2)(1.60 \times 10^{-19} \text{ C})^2}{0.535 \times 10^{-10} \text{ m}} = -8.60 \times 10^{-18} \text{ J}$$

$$U = -8.60 \times 10^{-18} \text{ J}(1 \text{ eV}/1.602 \times 10^{-19} \text{ J}) = -53.7 \text{ eV}$$

EVALUATE: The electron and proton have charges of opposite signs, so the potential energy of the system
is negative.
(b) IDENTIFY and SET UP: The positions of the protons and points a and b are shown in Figure 23.61b.

$$r_b = \sqrt{r_a^2 + d^2}$$

$$r_a = r = 0.535 \times 10^{-10} \text{ m}$$

Figure 23.61b

Apply $K_a + U_a + W_{\text{other}} = K_b + U_b$ with point a midway between the protons and point b where the
electron instantaneously has $v = 0$ (at its maximum displacement d from point a).
EXECUTE: Only the Coulomb force does work, so $W_{\text{other}} = 0$.

$U_a = -8.60 \times 10^{-18}$ J (from part (a))

$K_a = \frac{1}{2}mv^2 = \frac{1}{2}(9.109 \times 10^{-31} \text{ kg})(1.50 \times 10^6 \text{ m/s})^2 = 1.025 \times 10^{-18}$ J

$K_b = 0$

$U_b = -2ke^2/r_b$

Then $U_b = K_a + U_a - K_b = 1.025 \times 10^{-18}$ J $- 8.60 \times 10^{-18}$ J $= -7.575 \times 10^{-18}$ J.

$$r_b = -\frac{2ke^2}{U_b} = -\frac{2(8.988 \times 10^9 \text{ N} \cdot \text{m}^2/\text{C}^2)(1.60 \times 10^{-19} \text{ C})^2}{-7.575 \times 10^{-18} \text{ J}} = 6.075 \times 10^{-11} \text{ m}$$

Then $d = \sqrt{r_b^2 - r_a^2} = \sqrt{(6.075 \times 10^{-11} \text{ m})^2 - (5.35 \times 10^{-11} \text{ m})^2} = 2.88 \times 10^{-11}$ m.

EVALUATE: The force on the electron pulls it back toward the midpoint. The transverse distance the
electron moves is about 0.27 times the separation of the protons.

23.67.   **(a) IDENTIFY and SET UP:** Problem 23.63 derived that $E = \dfrac{V_{ab}}{\ln(b/a)} \dfrac{1}{r}$, where a is the radius of the inner

cylinder (wire) and b is the radius of the outer hollow cylinder. The potential difference between the two
cylinders is V_{ab}. Use this expression to calculate E at the specified r.
EXECUTE: Midway between the wire and the cylinder wall is at a radius of
$r = (a + b)/2 = (90.0 \times 10^{-6} \text{ m} + 0.140 \text{ m})/2 = 0.07004 \text{ m}.$

$$E = \frac{V_{ab}}{\ln(b/a)} \frac{1}{r} = \frac{50.0 \times 10^3 \text{ V}}{\ln(0.140 \text{ m}/90.0 \times 10^{-6} \text{ m})(0.07004 \text{ m})} = 9.71 \times 10^4 \text{ V/m}$$

(b) IDENTIFY and SET UP: The electric force is given by Eq. (21.3). Set this equal to ten times the weight of the particle and solve for $|q|$, the magnitude of the charge on the particle.

EXECUTE: $F_E = 10mg$

$$|q|E = 10mg \text{ and } |q| = \frac{10mg}{E} = \frac{10(30.0\times10^{-9} \text{ kg})(9.80 \text{ m/s}^2)}{9.71\times10^4 \text{ V/m}} = 3.03\times10^{-11} \text{ C}$$

EVALUATE: It requires only this modest net charge for the electric force to be much larger than the weight.

23.71. **IDENTIFY:** We must integrate to find the total energy because the energy to bring in more charge depends on the charge already present.

SET UP: If ρ is the uniform volume charge density, the charge of a spherical shell or radius r and thickness dr is $dq = \rho 4\pi r^2 \, dr$, and $\rho = Q/(4/3 \, \pi R^3)$. The charge already present in a sphere of radius r is $q = \rho(4/3 \, \pi r^3)$. The energy to bring the charge dq to the surface of the charge q is Vdq, where V is the potential due to q, which is $q/4\pi\epsilon_0 r$.

EXECUTE: The total energy to assemble the entire sphere of radius R and charge Q is sum (integral) of the tiny increments of energy.

$$U = \int Vdq = \int \frac{q}{4\pi\epsilon_0 r} \, dq = \int_0^R \frac{\rho \frac{4}{3}\pi r^3}{4\pi\epsilon_0 r}(\rho 4\pi r^2 dr) = \frac{3}{5}\left(\frac{1}{4\pi\epsilon_0}\frac{Q^2}{R}\right)$$

where we have substituted $\rho = Q/(4/3 \, \pi R^3)$ and simplified the result.

EVALUATE: For a point charge, $R \to 0$ so $U \to \infty$, which means that a point charge should have infinite self-energy. This suggests that either point charges are impossible, or that our present treatment of physics is not adequate at the extremely small scale, or both.

23.73. **IDENTIFY:** The sphere no longer behaves as a point charge because we are inside of it. We know how the electric field varies with distance from the center of the sphere and want to use this to find the potential difference between the center and surface, which requires integration.

SET UP: Use the result of Problem 23.72. For $r < R$, $V = \frac{kQ}{2R}\left(3 - \frac{r^2}{R^2}\right)$.

EXECUTE: At the center of the sphere, $r = 0$ and $V_1 = \frac{3kQ}{2R}$. At the surface of the sphere, $r = R$ and $V_2 = \frac{kQ}{R}$. The potential difference is $V_1 - V_2 = \frac{kQ}{2R} = \frac{(8.99\times10^9 \text{ N}\cdot\text{m}^2/\text{C}^2)(4.00\times10^{-6} \text{ C})}{2(0.0500 \text{ m})} = 3.60\times10^5 \text{ V}$.

EVALUATE: To check our answer, we could actually do the integration. We can use the fact that $E = \frac{kQr}{R^3}$ so $V_1 - V_2 = \int_0^R E dr = \frac{kQ}{R^3}\int_0^R rdr = \frac{kQ}{R^3}\left(\frac{R^2}{2}\right) = \frac{kQ}{2R}$.

23.79. **IDENTIFY:** Slice the rod into thin slices and use Eq. (23.14) to calculate the potential due to each slice. Integrate over the length of the rod to find the total potential at each point.

(a) SET UP: An infinitesimal slice of the rod and its distance from point P are shown in Figure 23.79a.

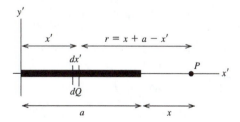

Figure 23.79a

Use coordinates with the origin at the left-hand end of the rod and one axis along the rod. Call the axes x' and y' so as not to confuse them with the distance x given in the problem.

EXECUTE: Slice the charged rod up into thin slices of width dx'. Each slice has charge $dQ = Q(dx'/a)$ and a distance $r = x + a - x'$ from point P. The potential at P due to the small slice dQ is

$$dV = \frac{1}{4\pi\epsilon_0}\left(\frac{dQ}{r}\right) = \frac{1}{4\pi\epsilon_0}\frac{Q}{a}\left(\frac{dx'}{x+a-x'}\right).$$

Compute the total V at P due to the entire rod by integrating dV over the length of the rod ($x' = 0$ to $x' = a$):

$$V = \int dV = \frac{Q}{4\pi\epsilon_0 a}\int_0^a \frac{dx'}{(x+a-x')} = \frac{Q}{4\pi\epsilon_0 a}[-\ln(x+a-x')]_0^a = \frac{Q}{4\pi\epsilon_0 a}\ln\left(\frac{x+a}{x}\right).$$

EVALUATE: As $x \rightarrow \infty$, $V \rightarrow \frac{Q}{4\pi\epsilon_0 a}\ln\left(\frac{x}{x}\right) = 0$.

(b) SET UP: An infinitesimal slice of the rod and its distance from point R are shown in Figure 23.79b.

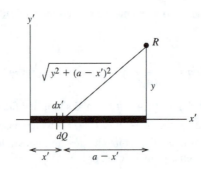

Figure 23.79b

$dQ = (Q/a)dx'$ as in part (a)

Each slice dQ is a distance $r = \sqrt{y^2 + (a-x')^2}$ from point R.

EXECUTE: The potential dV at R due to the small slice dQ is

$$dV = \frac{1}{4\pi\epsilon_0}\left(\frac{dQ}{r}\right) = \frac{1}{4\pi\epsilon_0}\frac{Q}{a}\frac{dx'}{\sqrt{y^2 + (a-x')^2}}.$$

$$V = \int dV = \frac{Q}{4\pi\epsilon_0 a}\int_0^a \frac{dx'}{\sqrt{y^2 + (a-x')^2}}.$$

In the integral make the change of variable $u = a - x'; du = -dx'$

$$V = -\frac{Q}{4\pi\epsilon_0 a}\int_a^0 \frac{du}{\sqrt{y^2 + u^2}} = -\frac{Q}{4\pi\epsilon_0 a}\left[\ln\left(u + \sqrt{y^2 + u^2}\right)\right]_a^0.$$

$$V = -\frac{Q}{4\pi\epsilon_0 a}\left[\ln y - \ln\left(a + \sqrt{y^2 + a^2}\right)\right] = \frac{Q}{4\pi\epsilon_0 a}\left[\ln\left(\frac{a + \sqrt{a^2 + y^2}}{y}\right)\right].$$

(The expression for the integral was found in Appendix B.)

EVALUATE: As $y \rightarrow \infty$, $V \rightarrow \frac{Q}{4\pi\epsilon_0 a}\ln\left(\frac{y}{y}\right) = 0$.

(c) SET UP: *part (a):* $V = \frac{Q}{4\pi\epsilon_0 a}\ln\left(\frac{x+a}{x}\right) = \frac{Q}{4\pi\epsilon_0 a}\ln\left(1 + \frac{a}{x}\right)$.

From Appendix B, $\ln(1+u) = u - u^2/2\ldots$, so $\ln(1 + a/x) = a/x - a^2/2x^2$ and this becomes a/x when x is large.

EXECUTE: Thus $V \to \dfrac{Q}{4\pi\epsilon_0 a}\left(\dfrac{a}{x}\right) = \dfrac{Q}{4\pi\epsilon_0 x}$. For large x, V becomes the potential of a point charge.

part (b): $V = \dfrac{Q}{4\pi\epsilon_0 a}\left[\ln\left(\dfrac{a+\sqrt{a^2+y^2}}{y}\right)\right] = \dfrac{Q}{4\pi\epsilon_0 a}\ln\left(\dfrac{a}{y}+\sqrt{1+\dfrac{a^2}{y^2}}\right)$.

From Appendix B, $\sqrt{1+a^2/y^2} = (1+a^2/y^2)^{1/2} = 1 + a^2/2y^2 + \ldots$

Thus $a/y + \sqrt{1+a^2/y^2} \to 1 + a/y + a^2/2y^2 + \ldots \to 1 + a/y$. And then using $\ln(1+u) \approx u$ gives

$$V \to \dfrac{Q}{4\pi\epsilon_0 a}\ln(1+a/y) \to \dfrac{Q}{4\pi\epsilon_0 a}\left(\dfrac{a}{y}\right) = \dfrac{Q}{4\pi\epsilon_0 y}.$$

EVALUATE: For large y, V becomes the potential of a point charge.

23.85.   **IDENTIFY:**   Apply conservation of energy: $E_1 = E_2$.

SET UP: In the collision the initial kinetic energy of the two particles is converted into potential energy at the distance of closest approach.

EXECUTE: **(a)** The two protons must approach to a distance of $2r_p$, where r_p is the radius of a proton.

$E_1 = E_2$ gives $2\left[\dfrac{1}{2}m_p v^2\right] = \dfrac{ke^2}{2r_p}$ and $v = \sqrt{\dfrac{k(1.60\times10^{-19}\ \text{C})^2}{2(1.2\times10^{-15}\ \text{m})(1.67\times10^{-27}\ \text{kg})}} = 7.58\times10^6\ \text{m/s}$.

(b) For a helium-helium collision, the charges and masses change from (a) and

$v = \sqrt{\dfrac{k(2(1.60\times10^{-19}\ \text{C}))^2}{(3.5\times10^{-15}\ \text{m})(2.99)(1.67\times10^{-27}\ \text{kg})}} = 7.26\times10^6\ \text{m/s}$.

(c) $K = \dfrac{3kT}{2} = \dfrac{mv^2}{2}$. $T_P = \dfrac{m_p v^2}{3k} = \dfrac{(1.67\times10^{-27}\ \text{kg})(7.58\times10^6\ \text{m/s})^2}{3(1.38\times10^{-23}\ \text{J/K})} = 2.3\times10^9\ \text{K}$.

$T_{He} = \dfrac{m_{He} v^2}{3k} = \dfrac{(2.99)(1.67\times10^{-27}\ \text{kg})(7.26\times10^6\ \text{m/s})^2}{3(1.38\times10^{-23}\ \text{J/K})} = 6.4\times10^9\ \text{K}$.

(d) These calculations were based on the particles' average speed. The distribution of speeds ensures that there is always a certain percentage with a speed greater than the average speed, and these particles can undergo the necessary reactions in the sun's core.

EVALUATE: The kinetic energies required for fusion correspond to very high temperatures.

23.87.   **IDENTIFY:**   Apply conservation of energy to the motion of the daughter nuclei.

SET UP: Problem 23.72 shows that the electrical potential energy of the two nuclei is the same as if all their charge was concentrated at their centers.

EXECUTE: **(a)** The two daughter nuclei have half the volume of the original uranium nucleus, so their radii are smaller by a factor of the cube root of 2: $r = \dfrac{7.4\times10^{-15}\ \text{m}}{\sqrt[3]{2}} = 5.9\times10^{-15}\ \text{m}$.

(b) $U = \dfrac{k(46e)^2}{2r} = \dfrac{k(46)^2(1.60\times10^{-19}\ \text{C})^2}{1.18\times10^{-14}\ \text{m}} = 4.14\times10^{-11}\ \text{J}$. $U = 2K$, where K is the final kinetic energy of each nucleus. $K = U/2 = (4.14\times10^{-11}\ \text{J})/2 = 2.07\times10^{-11}\ \text{J}$.

(c) If we have 10.0 kg of uranium, then the number of nuclei is

$n = \dfrac{10.0\ \text{kg}}{(236\ \text{u})(1.66\times10^{-27}\ \text{kg/u})} = 2.55\times10^{25}$ nuclei. And each releases energy U, so

$E = nU = (2.55\times10^{25})(4.14\times10^{-11}\ \text{J}) = 1.06\times10^{15}\ \text{J} = 253$ kilotons of TNT.

(d) We could call an atomic bomb an "electric" bomb since the electric potential energy provides the kinetic energy of the particles.

EVALUATE: This simple model considers only the electrical force between the daughter nuclei and neglects the nuclear force.

24

CAPACITANCE AND DIELECTRICS

24.3. **IDENTIFY** and **SET UP:** It is a parallel-plate air capacitor, so we can apply the equations of Section 24.1.

EXECUTE: (a) $C = \dfrac{Q}{V_{ab}}$ so $V_{ab} = \dfrac{Q}{C} = \dfrac{0.148 \times 10^{-6} \text{ C}}{245 \times 10^{-12} \text{ F}} = 604 \text{ V}$

(b) $C = \dfrac{\epsilon_0 A}{d}$ so $A = \dfrac{Cd}{\epsilon_0} = \dfrac{(245 \times 10^{-12} \text{ F})(0.328 \times 10^{-3} \text{ m})}{8.854 \times 10^{-12} \text{ C}^2/\text{N} \cdot \text{m}^2} = 9.08 \times 10^{-3} \text{ m}^2 = 90.8 \text{ cm}^2$

(c) $V_{ab} = Ed$ so $E = \dfrac{V_{ab}}{d} = \dfrac{604 \text{ V}}{0.328 \times 10^{-3} \text{ m}} = 1.84 \times 10^6 \text{ V/m}$

(d) $E = \dfrac{\sigma}{\epsilon_0}$ so $\sigma = E\epsilon_0 = (1.84 \times 10^6 \text{ V/m})(8.854 \times 10^{-12} \text{ C}^2/\text{N} \cdot \text{m}^2) = 1.63 \times 10^{-5} \text{ C/m}^2$

EVALUATE: We could also calculate σ directly as Q/A. $\sigma = \dfrac{Q}{A} = \dfrac{0.148 \times 10^{-6} \text{ C}}{9.08 \times 10^{-3} \text{ m}^2} = 1.63 \times 10^{-5} \text{ C/m}^2$,

which checks.

24.5. **IDENTIFY:** $C = \dfrac{Q}{V_{ab}}$. $C = \dfrac{\epsilon_0 A}{d}$.

SET UP: When the capacitor is connected to the battery, $V_{ab} = 12.0 \text{ V}$.

EXECUTE: (a) $Q = CV_{ab} = (10.0 \times 10^{-6} \text{ F})(12.0 \text{ V}) = 1.20 \times 10^{-4} \text{ C} = 120 \text{ } \mu\text{C}$

(b) When d is doubled C is halved, so Q is halved. $Q = 60 \text{ } \mu\text{C}$.

(c) If r is doubled, A increases by a factor of 4. C increases by a factor of 4 and Q increases by a factor of 4. $Q = 480 \text{ } \mu\text{C}$.

EVALUATE: When the plates are moved apart, less charge on the plates is required to produce the same potential difference. With the separation of the plates constant, the electric field must remain constant to produce the same potential difference. The electric field depends on the surface charge density, σ. To produce the same σ, more charge is required when the area increases.

24.11. **IDENTIFY:** Apply the results of Example 24.4. $C = Q/V$.

SET UP: $r_a = 0.50 \text{ mm}$, $r_b = 5.00 \text{ mm}$

EXECUTE: (a) $C = \dfrac{L2\pi\epsilon_0}{\ln(r_b/r_a)} = \dfrac{(0.180 \text{ m})2\pi\epsilon_0}{\ln(5.00/0.50)} = 4.35 \times 10^{-12} \text{ F}$.

(b) $V = Q/C = (10.0 \times 10^{-12} \text{ C})/(4.35 \times 10^{-12} \text{ F}) = 2.30 \text{ V}$

EVALUATE: $\dfrac{C}{L} = 24.2 \text{ pF}$. This value is similar to those in Example 24.4. The capacitance is determined entirely by the dimensions of the cylinders.

24.13. **IDENTIFY:** We can use the definition of capacitance to find the capacitance of the capacitor, and then relate the capacitance to geometry to find the inner radius.

(a) **SET UP:** By the definition of capacitance, $C = Q/V$.

EXECUTE: $C = \dfrac{Q}{V} = \dfrac{3.30 \times 10^{-9} \text{ C}}{2.20 \times 10^2 \text{ V}} = 1.50 \times 10^{-11} \text{ F} = 15.0 \text{ pF}$

(b) SET UP: The capacitance of a spherical capacitor is $C = 4\pi\epsilon_0 \dfrac{r_a r_b}{r_b - r_a}$.

EXECUTE: Solve for r_a and evaluate using $C = 15.0$ pF and $r_b = 4.00$ cm, giving $r_a = 3.09$ cm.

(c) SET UP: We can treat the inner sphere as a point charge located at its center and use Coulomb's law,

$E = \dfrac{1}{4\pi\epsilon_0}\dfrac{q}{r^2}$.

EXECUTE: $E = \dfrac{(9.00 \times 10^9 \text{ N} \cdot \text{m}^2/\text{C}^2)(3.30 \times 10^{-9} \text{ C})}{(0.0309 \text{ m})^2} = 3.12 \times 10^4 \text{ N/C}$

EVALUATE: Outside the capacitor, the electric field is zero because the charges on the spheres are equal in magnitude but opposite in sign.

24.17.   **IDENTIFY:**   Replace series and parallel combinations of capacitors by their equivalents. In each equivalent network apply the rules for Q and V for capacitors in series and parallel; start with the simplest network and work back to the original circuit.

SET UP: Do parts (a) and (b) together. The capacitor network is drawn in Figure 24.17a.

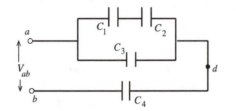

$C_1 = C_2 = C_3 = C_4 = 4.00 \ \mu\text{F}$

$V_{ab} = 28.0$ V

Figure 24.17a

EXECUTE: Simplify the circuit by replacing the capacitor combinations by their equivalents: C_1 and C_2 are in series and are equivalent to C_{12} (Figure 24.17b).

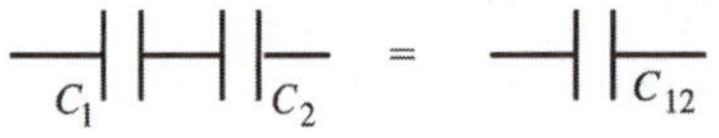

$$\dfrac{1}{C_{12}} = \dfrac{1}{C_1} + \dfrac{1}{C_2}$$

Figure 24.17b

$C_{12} = \dfrac{C_1 C_2}{C_1 + C_2} = \dfrac{(4.00 \times 10^{-6} \text{ F})(4.00 \times 10^{-6} \text{ F})}{4.00 \times 10^{-6} \text{ F} + 4.00 \times 10^{-6} \text{ F}} = 2.00 \times 10^{-6} \text{ F}$

C_{12} and C_3 are in parallel and are equivalent to C_{123} (Figure 24.17c).

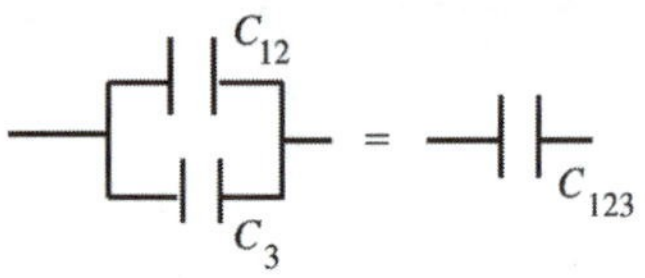

$C_{123} = C_{12} + C_3$

$C_{123} = 2.00 \times 10^{-6} \text{ F} + 4.00 \times 10^{-6} \text{ F}$

$C_{123} = 6.00 \times 10^{-6} \text{ F}$

Figure 24.17c

C_{123} and C_4 are in series and are equivalent to C_{1234} (Figure 24.17d).

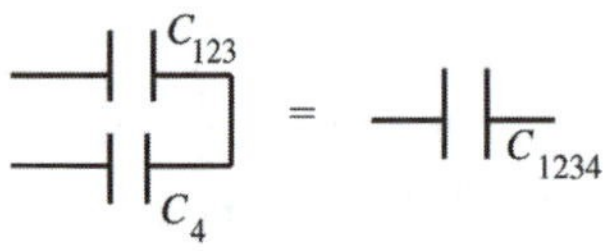

$\dfrac{1}{C_{1234}} = \dfrac{1}{C_{123}} + \dfrac{1}{C_4}$

Figure 24.17d

$$C_{1234} = \frac{C_{123}C_4}{C_{123} + C_4} = \frac{(6.00 \times 10^{-6}\ \text{F})(4.00 \times 10^{-6}\ \text{F})}{6.00 \times 10^{-6}\ \text{F} + 4.00 \times 10^{-6}\ \text{F}} = 2.40 \times 10^{-6}\ \text{F}$$

The circuit is equivalent to the circuit shown in Figure 24.17e.

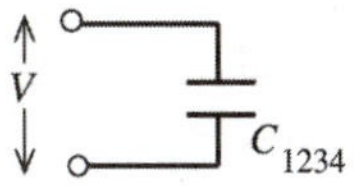

$V_{1234} = V = 28.0\ \text{V}$

$Q_{1234} = C_{1234}V = (2.40 \times 10^{-6}\ \text{F})(28.0\ \text{V}) = 67.2\ \mu\text{C}$

Figure 24.17e

Now build back up the original circuit, step by step. C_{1234} represents C_{123} and C_4 in series (Figure 24.17f).

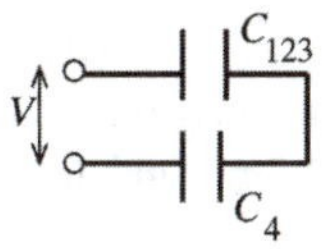

$Q_{123} = Q_4 = Q_{1234} = 67.2\ \mu\text{C}$

(charge same for capacitors in series)

Figure 24.17f

Then $V_{123} = \dfrac{Q_{123}}{C_{123}} = \dfrac{67.2\ \mu\text{C}}{6.00\ \mu\text{F}} = 11.2\ \text{V}$

$V_4 = \dfrac{Q_4}{C_4} = \dfrac{67.2\ \mu\text{C}}{4.00\ \mu\text{F}} = 16.8\ \text{V}$

Note that $V_4 + V_{123} = 16.8\ \text{V} + 11.2\ \text{V} = 28.0\ \text{V}$, as it should.

Next consider the circuit as written in Figure 24.17g.

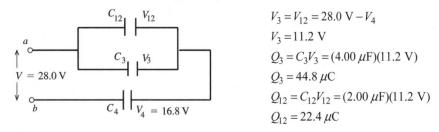

$V_3 = V_{12} = 28.0\ \text{V} - V_4$

$V_3 = 11.2\ \text{V}$

$Q_3 = C_3 V_3 = (4.00\ \mu\text{F})(11.2\ \text{V})$

$Q_3 = 44.8\ \mu\text{C}$

$Q_{12} = C_{12}V_{12} = (2.00\ \mu\text{F})(11.2\ \text{V})$

$Q_{12} = 22.4\ \mu\text{C}$

Figure 24.17g

Finally, consider the original circuit, as shown in Figure 24.17h.

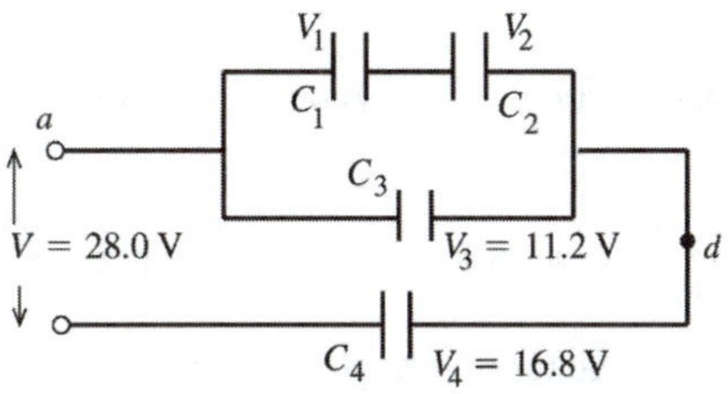

$Q_1 = Q_2 = Q_{12} = 22.4\ \mu\text{C}$

(charge same for capacitors in series)

$V_1 = \dfrac{Q_1}{C_1} = \dfrac{22.4\ \mu\text{C}}{4.00\ \mu\text{F}} = 5.6\ \text{V}$

$V_2 = \dfrac{Q_2}{C_2} = \dfrac{22.4\ \mu\text{C}}{4.00\ \mu\text{F}} = 5.6\ \text{V}$

Figure 24.17h

Note that $V_1 + V_2 = 11.2\ \text{V}$, which equals V_3 as it should.

Summary: $Q_1 = 22.4\ \mu\text{C}$, $V_1 = 5.6\ \text{V}$

$Q_2 = 22.4\ \mu\text{C},\ V_2 = 5.6\ \text{V}$

$Q_3 = 44.8\ \mu\text{C},\ V_3 = 11.2\ \text{V}$

$Q_4 = 67.2\ \mu\text{C},\ V_4 = 16.8\ \text{V}$

(c) $V_{ad} = V_3 = 11.2\ \text{V}$

EVALUATE: $V_1 + V_2 + V_4 = V$, or $V_3 + V_4 = V$. $Q_1 = Q_2, Q_1 + Q_3 = Q_4$ and $Q_4 = Q_{1234}$.

24.21.   **IDENTIFY:** Three of the capacitors are in series, and this combination is in parallel with the other two capacitors.

SET UP: For capacitors in series the voltages add and the charges are the same;

$\dfrac{1}{C_{eq}} = \dfrac{1}{C_1} + \dfrac{1}{C_2} + \cdots$. For capacitors in parallel the voltages are the same and the charges add;

$C_{eq} = C_1 + C_2 + \text{L}.\quad C = \dfrac{Q}{V}$.

EXECUTE: **(a)** The equivalent capacitance of the 18.0 nF, 30.0 nF and 10.0 nF capacitors in series is 5.29 nF. When these capacitors are replaced by their equivalent we get the network sketched in Figure 24.21. The equivalent capacitance of these three capacitors in parallel is 19.3 nF, and this is the equivalent capacitance of the original network.

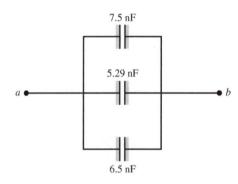

Figure 24.21

(b) $Q_{\text{tot}} = C_{eq}V = (19.3\ \text{nF})(25\ \text{V}) = 482\ \text{nC}$.

(c) The potential across each capacitor in the parallel network of Figure 24.21 is 25 V.

$Q_{6.5} = C_{6.5}V_{6.5} = (6.5\ \text{nF})(25\ \text{V}) = 162\ \text{nC}$.

d) 25 V.

EVALUATE: As with most circuits, we must go through a series of steps to simplify it as we solve for the unknowns.

24.23.   **IDENTIFY:** Refer to Figure 24.10b in the textbook. For capacitors in parallel, $C_{eq} = C_1 + C_2 + \cdots$. For

capacitors in series, $\dfrac{1}{C_{eq}} = \dfrac{1}{C_1} + \dfrac{1}{C_2} + \cdots$.

SET UP: The $11\ \mu\text{F}$, $4\ \mu\text{F}$ and replacement capacitor are in parallel and this combination is in series with the $9.0\ \mu\text{F}$ capacitor.

EXECUTE: $\dfrac{1}{C_{eq}} = \dfrac{1}{8.0\ \mu\text{F}} = \left(\dfrac{1}{(11 + 4.0 + x)\mu\text{F}} + \dfrac{1}{9.0\ \mu\text{F}} \right)$. $(15 + x)\mu\text{F} = 72\ \mu\text{F}$ and $x = 57\ \mu\text{F}$.

EVALUATE: Increasing the capacitance of the one capacitor by a large amount makes a small increase in the equivalent capacitance of the network.

24.25.   **IDENTIFY and SET UP:** The energy density is given by Eq. (24.11): $u = \frac{1}{2}\epsilon_0 E^2$. Use $V = Ed$ to solve for E.

EXECUTE: Calculate E: $E = \dfrac{V}{d} = \dfrac{400\ \text{V}}{5.00 \times 10^{-3}\ \text{m}} = 8.00 \times 10^4\ \text{V/m}$.

Then $u = \frac{1}{2}\epsilon_0 E^2 = \frac{1}{2}(8.854 \times 10^{-12} \text{ C}^2/\text{N} \cdot \text{m}^2)(8.00 \times 10^4 \text{ V/m})^2 = 0.0283 \text{ J/m}^3$.

EVALUATE: E is smaller than the value in Example 24.8 by about a factor of 6 so u is smaller by about a factor of $6^2 = 36$.

24.33. **IDENTIFY:** $U = \frac{1}{2}QV$. Solve for Q. $C = Q/V$.

SET UP: Example 24.4 shows that for a cylindrical capacitor, $\frac{C}{L} = \frac{2\pi\epsilon_0}{\ln(r_b/r_a)}$.

EXECUTE: **(a)** $U = \frac{1}{2}QV$ gives $Q = \frac{2U}{V} = \frac{2(3.20 \times 10^{-9} \text{ J})}{4.00 \text{ V}} = 1.60 \times 10^{-9} \text{ C}$.

(b) $\frac{C}{L} = \frac{2\pi\epsilon_0}{\ln(r_b/r_a)}$.

$\frac{r_b}{r_a} = \exp(2\pi\epsilon_0 L/C) = \exp(2\pi\epsilon_0 LV/Q) = \exp(2\pi\epsilon_0 (15.0 \text{ m})(4.00 \text{ V})/(1.60 \times 10^{-9} \text{ C})) = 8.05$.

The radius of the outer conductor is 8.05 times the radius of the inner conductor.

EVALUATE: When the ratio r_b/r_a increases, C/L decreases and less charge is stored for a given potential difference.

24.35. **IDENTIFY:** $C = KC_0$. $U = \frac{1}{2}CV^2$.

SET UP: $C_0 = 12.5 \ \mu\text{F}$ is the value of the capacitance without the dielectric present.

EXECUTE: **(a)** With the dielectric, $C = (3.75)(12.5 \ \mu\text{F}) = 46.9 \ \mu\text{F}$.

before: $U = \frac{1}{2}C_0V^2 = \frac{1}{2}(12.5 \times 10^{-6} \text{ F})(24.0 \text{ V})^2 = 3.60 \text{ mJ}$

after: $U = \frac{1}{2}CV^2 = \frac{1}{2}(46.9 \times 10^{-6} \text{ F})(24.0 \text{ V})^2 = 13.5 \text{ mJ}$

(b) $\Delta U = 13.5 \text{ mJ} - 3.6 \text{ mJ} = 9.9 \text{ mJ}$. The energy increased.

EVALUATE: The power supply must put additional charge on the plates to maintain the same potential difference when the dielectric is inserted. $U = \frac{1}{2}QV$, so the stored energy increases.

24.37. **IDENTIFY and SET UP:** Q is constant so we can apply Eq. (24.14). The charge density on each surface of the dielectric is given by Eq. (24.16).

EXECUTE: $E = \frac{E_0}{K}$ so $K = \frac{E_0}{E} = \frac{3.20 \times 10^5 \text{ V/m}}{2.50 \times 10^5 \text{ V/m}} = 1.28$

(a) $\sigma_i = \sigma(1 - 1/K)$

$\sigma = \epsilon_0 E_0 = (8.854 \times 10^{-12} \text{ C}^2/\text{N} \cdot \text{m}^2)(3.20 \times 10^5 \text{ N/C}) = 2.833 \times 10^{-6} \text{ C/m}^2$

$\sigma_i = (2.833 \times 10^{-6} \text{ C/m}^2)(1 - 1/1.28) = 6.20 \times 10^{-7} \text{ C/m}^2$

(b) As calculated above, $K = 1.28$.

EVALUATE: The surface charges on the dielectric produce an electric field that partially cancels the electric field produced by the charges on the capacitor plates.

24.39. **IDENTIFY and SET UP:** For a parallel-plate capacitor with a dielectric we can use the equation $C = K\epsilon_0 A/d$. Minimum A means smallest possible d. d is limited by the requirement that E be less than $1.60 \times 10^7 \text{ V/m}$ when V is as large as 5500 V.

EXECUTE: $V = Ed$ so $d = \frac{V}{E} = \frac{5500 \text{ V}}{1.60 \times 10^7 \text{ V/m}} = 3.44 \times 10^{-4} \text{ m}$

Then $A = \frac{Cd}{K\epsilon_0} = \frac{(1.25 \times 10^{-9} \text{ F})(3.44 \times 10^{-4} \text{ m})}{(3.60)(8.854 \times 10^{-12} \text{ C}^2/\text{N} \cdot \text{m}^2)} = 0.0135 \text{ m}^2$.

EVALUATE: The relation $V = Ed$ applies with or without a dielectric present. A would have to be larger if there were no dielectric.

24.41.   **IDENTIFY:** The permittivity ϵ of a material is related to its dielectric constant by $\epsilon = K\epsilon_0$. The maximum voltage is related to the maximum possible electric field before dielectric breakdown by $V_{max} = E_{max}d$.

$E = \dfrac{E_0}{K} = \dfrac{\sigma}{K\epsilon_0}$, where σ is the surface charge density on each plate. The induced surface charge density on the surface of the dielectric is given by $\sigma_i = \sigma(1 - 1/K)$.

SET UP: From Table 24.2, for polystyrene $K = 2.6$ and the dielectric strength (maximum allowed electric field) is 2×10^7 V/m.

EXECUTE: (a) $\epsilon = K\epsilon_0 = (2.6)\epsilon_0 = 2.3 \times 10^{-11}$ C^2/N·m^2

(b) $V_{max} = E_{max}d = (2.0 \times 10^7 \text{ V/m})(2.0 \times 10^{-3} \text{ m}) = 4.0 \times 10^4$ V

(c) $E = \dfrac{\sigma}{K\epsilon_0}$ and $\sigma = \epsilon E = (2.3 \times 10^{-11}$ C^2/N·m$^2)(2.0 \times 10^7$ V/m$) = 0.46 \times 10^{-3}$ C/m^2.

$\sigma_i = \sigma\left(1 - \dfrac{1}{K}\right) = (0.46 \times 10^{-3} \text{ C/m}^2)(1 - 1/2.6) = 2.8 \times 10^{-4}$ C/m^2.

EVALUATE: The net surface charge density is $\sigma_{net} = \sigma - \sigma_i = 1.8 \times 10^{-4}$ C/m^2 and the electric field between the plates is $E = \sigma_{net}/\epsilon_0$.

24.43.   **(a) IDENTIFY and SET UP:** Since the capacitor remains connected to the power supply the potential difference doesn't change when the dielectric is inserted. Use Eq. (24.9) to calculate V and combine it with Eq. (24.12) to obtain a relation between the stored energies and the dielectric constant and use this to calculate K.

EXECUTE: Before the dielectric is inserted $U_0 = \frac{1}{2}C_0V^2$ so $V = \sqrt{\dfrac{2U_0}{C_0}} = \sqrt{\dfrac{2(1.85 \times 10^{-5} \text{ J})}{360 \times 10^{-9} \text{ F}}} = 10.1$ V.

(b) $K = C/C_0$

$U_0 = \frac{1}{2}C_0V^2$, $U = \frac{1}{2}CV^2$ so $C/C_0 = U/U_0$

$K = \dfrac{U}{U_0} = \dfrac{1.85 \times 10^{-5} \text{ J} + 2.32 \times 10^{-5} \text{ J}}{1.85 \times 10^{-5} \text{ J}} = 2.25$

EVALUATE: K increases the capacitance and then from $U = \frac{1}{2}CV^2$, with V constant an increase in C gives an increase in U.

24.51.   **IDENTIFY:** Both the electric field and the potential difference depend on the linear charge density on the cylinders. We can use this fact to relate the field to the potential difference between the cylinders.

SET UP: $E = \dfrac{\lambda}{2\pi\epsilon_0 r}$ and $V = \dfrac{\lambda}{2\pi\epsilon_0}\ln(r_b/r_a)$, so $E = \dfrac{V}{r\ln(r_b/r_a)}$.

EXECUTE: $E = \dfrac{V}{r\ln(r_b/r_a)} = \dfrac{80.0 \text{ V}}{(2.80 \times 10^{-3} \text{ m})(\ln(3.10/2.50))} = 1.33 \times 10^5$ V/m.

EVALUATE: At any point between the cylinders, E is directly proportional to V because V is proportional to the charge on the inner cylinder. This is the charge that causes the electric field.

24.53.   **IDENTIFY:** Some of the charge from the original capacitor flows onto the uncharged capacitor until the potential differences across the two capacitors are the same.

SET UP: $C = \dfrac{Q}{V_{ab}}$. Let $C_1 = 20.0 \ \mu$F and $C_2 = 10.0 \ \mu$F. The energy stored in a capacitor is

$\frac{1}{2}QV_{ab} = \frac{1}{2}CV_{ab}^2 = \frac{1}{2}\dfrac{Q^2}{C}$.

EXECUTE: (a) The initial charge on the 20.0 μF capacitor is

$Q = C_1(800 \text{ V}) = (20.0 \times 10^{-6} \text{ F})(800 \text{ V}) = 0.0160$ C.

(b) In the final circuit, charge Q is distributed between the two capacitors and $Q_1 + Q_2 = Q$. The final circuit contains only the two capacitors, so the voltage across each is the same, $V_1 = V_2$. $V = \frac{Q}{C}$ so $V_1 = V_2$

gives $\frac{Q_1}{C_1} = \frac{Q_2}{C_2}$. $Q_1 = \frac{C_1}{C_2} Q_2 = 2Q_2$. Using this in $Q_1 + Q_2 = 0.0160$ C gives $3Q_2 = 0.0160$ C and

$Q_2 = 5.33 \times 10^{-3}$ C. $Q = 2Q_2 = 1.066 \times 10^{-2}$ C. $V_1 = \frac{Q_1}{C_1} = \frac{1.066 \times 10^{-2} \text{ C}}{20.0 \times 10^{-6} \text{ F}} = 533$ V.

$V_2 = \frac{Q_2}{C_2} = \frac{5.33 \times 10^{23} \text{ C}}{10.0 \times 10^{26} \text{ F}} = 533$ V. The potential differences across the capacitors are the same, as they should be.

(c) Energy $= \frac{1}{2} C_1 V^2 + \frac{1}{2} C_2 V^2 = \frac{1}{2}(C_1 + C_2) V^2$ gives

Energy $= \frac{1}{2}(20.0 \times 10^{-6} \text{ F} + 10.0 \times 10^{-6} \text{ F})(533 \text{ V})^2 = 4.26$ J.

(d) The $20.0 \ \mu F$ capacitor initially has energy $= \frac{1}{2} C_1 V^2 = \frac{1}{2}(20.0 \times 10^{-6} \text{ F})(800 \text{ V})^2 = 6.40$ J. The decrease in stored energy that occurs when the capacitors are connected is $6.40 \text{ J} - 4.26 \text{ J} = 2.14$ J.

EVALUATE: The decrease in stored energy is because of conversion of electrical energy to other forms during the motion of the charge when it becomes distributed between the two capacitors. Thermal energy is generated by the current in the wires and energy is emitted in electromagnetic waves.

24.55. **IDENTIFY:** Simplify the network by replacing series and parallel combinations by their equivalent. The stored energy in a capacitor is $U = \frac{1}{2} CV^2$.

SET UP: For capacitors in series the voltages add and the charges are the same; $\frac{1}{C_{eq}} = \frac{1}{C_1} + \frac{1}{C_2} + \cdots$. For capacitors in parallel the voltages are the same and the charges add; $C_{eq} = C_1 + C_2 + \cdots$ $C = \frac{Q}{V}$.

$U = \frac{1}{2} CV^2$.

EXECUTE: **(a)** Find C_{eq} for the network by replacing each series or parallel combination by its equivalent. The successive simplified circuits are shown in Figure 24.55a–c.

$U_{tot} = \frac{1}{2} C_{eq} V^2 = \frac{1}{2}(2.19 \times 10^{-6} \text{ F})(12.0 \text{ V})^2 = 1.58 \times 10^{-4} \text{ J} = 158 \ \mu J$

(b) From Figure 24.55c, $Q_{tot} = C_{eq} V = (2.19 \times 10^{-6} \text{ F})(12.0 \text{ V}) = 2.63 \times 10^{-5}$ C. From Figure 24.55b,

$Q_{4.8} = 2.63 \times 10^{-5}$ C. $V_{4.8} = \frac{Q_{4.8}}{C_{4.8}} = \frac{2.63 \times 10^{-5} \text{ C}}{4.80 \times 10^{-6} \text{ F}} = 5.48$ V.

$U_{4.8} = \frac{1}{2} CV^2 = \frac{1}{2}(4.80 \times 10^{-6} \text{ F})(5.48 \text{ V})^2 = 7.21 \times 10^{-5} \text{ J} = 72.1 \ \mu J$

This one capacitor stores nearly half the total stored energy.

EVALUATE: $U = \frac{Q^2}{2C}$. For capacitors in series the capacitor with the smallest C stores the greatest amount of energy.

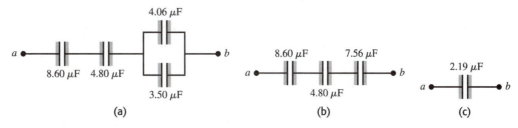

4.06 μF

a 8.60 μF 4.80 μF 3.50 μF b

a 8.60 μF 7.56 μF 4.80 μF b

a 2.19 μF b

(a)   (b)   (c)

Figure 24.55

24.57. (a) IDENTIFY: Replace series and parallel combinations of capacitors by their equivalents.
SET UP: The network is sketched in Figure 24.57a.

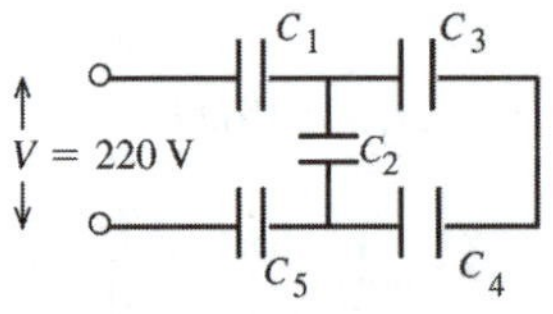

$$C_1 = C_5 = 8.4 \; \mu F$$
$$C_2 = C_3 = C_4 = 4.2 \; \mu F$$

Figure 24.57a

EXECUTE: Simplify the circuit by replacing the capacitor combinations by their equivalents: C_3 and C_4 are in series and can be replaced by C_{34} (Figure 24.57b):

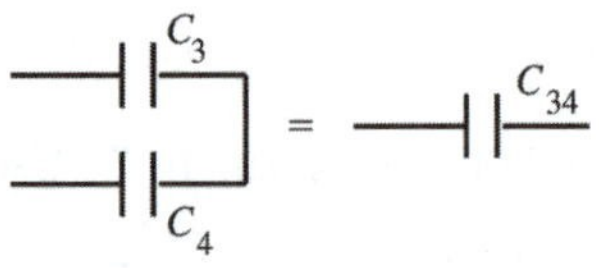

$$\frac{1}{C_{34}} = \frac{1}{C_3} + \frac{1}{C_4}$$
$$\frac{1}{C_{34}} = \frac{C_3 + C_4}{C_3 C_4}$$

Figure 24.57b

$$C_{34} = \frac{C_3 C_4}{C_3 + C_4} = \frac{(4.2 \; \mu F)(4.2 \; \mu F)}{4.2 \; \mu F + 4.2 \; \mu F} = 2.1 \; \mu F$$

C_2 and C_{34} are in parallel and can be replaced by their equivalent (Figure 24.57c):

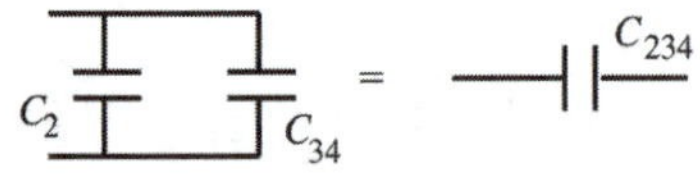

$$C_{234} = C_2 + C_{34}$$
$$C_{234} = 4.2 \; \mu F + 2.1 \; \mu F$$
$$C_{234} = 6.3 \; \mu F$$

Figure 24.57c

C_1, C_5 and C_{234} are in series and can be replaced by C_{eq} (Figure 24.57d):

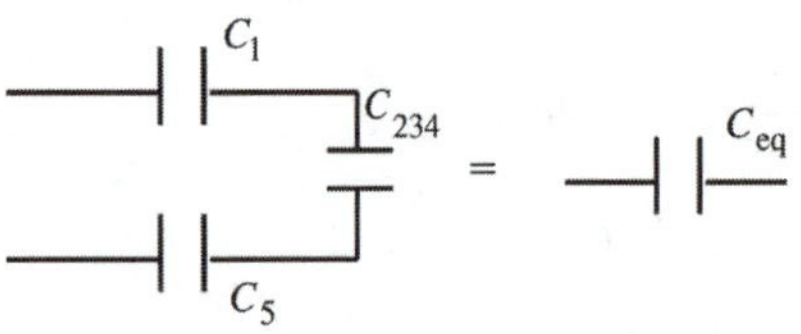

$$\frac{1}{C_{eq}} = \frac{1}{C_1} + \frac{1}{C_5} + \frac{1}{C_{234}}$$
$$\frac{1}{C_{eq}} = \frac{2}{8.4 \; \mu F} + \frac{1}{6.3 \; \mu F}$$
$$C_{eq} = 2.5 \; \mu F$$

Figure 24.57d

EVALUATE: For capacitors in series the equivalent capacitor is smaller than any of those in series. For capacitors in parallel the equivalent capacitance is larger than any of those in parallel.
(b) IDENTIFY and SET UP: In each equivalent network apply the rules for Q and V for capacitors in series and parallel; start with the simplest network and work back to the original circuit.
EXECUTE: The equivalent circuit is drawn in Figure 24.57e.

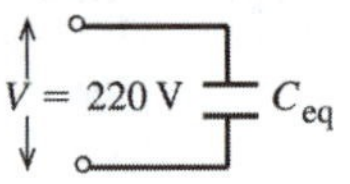

$$Q_{eq} = C_{eq} V$$
$$Q_{eq} = (2.5 \; \mu F)(220 \; V) = 550 \; \mu C$$

Figure 24.57e

$Q_1 = Q_5 = Q_{234} = 550\ \mu C$ (capacitors in series have same charge)

$V_1 = \dfrac{Q_1}{C_1} = \dfrac{550\ \mu C}{8.4\ \mu F} = 65$ V

$V_5 = \dfrac{Q_5}{C_5} = \dfrac{550\ \mu C}{8.4\ \mu F} = 65$ V

$V_{234} = \dfrac{Q_{234}}{C_{234}} = \dfrac{550\ \mu C}{6.3\ \mu F} = 87$ V

Now draw the network as in Figure 24.57f.

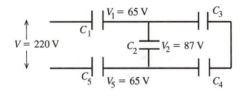

$V_2 = V_{34} = V_{234} = 87$ V

capacitors in parallel have the same potential

Figure 24.57f

$Q_2 = C_2 V_2 = (4.2\ \mu F)(87\ V) = 370\ \mu C$

$Q_{34} = C_{34} V_{34} = (2.1\ \mu F)(87\ V) = 180\ \mu C$

Finally, consider the original circuit (Figure 24.57g).

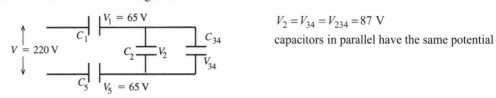

$Q_3 = Q_4 = Q_{34} = 180\ \mu C$

capacitors in series have the same charge

Figure 24.57g

$V_3 = \dfrac{Q_3}{C_3} = \dfrac{180\ \mu C}{4.2\ \mu F} = 43$ V

$V_4 = \dfrac{Q_4}{C_4} = \dfrac{180\ \mu C}{4.2\ \mu F} = 43$ V

Summary: $Q_1 = 550\ \mu C$, $V_1 = 65$ V

$Q_2 = 370\ \mu C$, $V_2 = 87$ V

$Q_3 = 180\ \mu C$, $V_3 = 43$ V

$Q_4 = 180\ \mu C$, $V_4 = 43$ V

$Q_5 = 550\ \mu C$, $V_5 = 65$ V

EVALUATE: $V_3 + V_4 = V_2$ and $V_1 + V_2 + V_5 = 220$ V (apart from some small rounding error)

$Q_1 = Q_2 + Q_3$ and $Q_5 = Q_2 + Q_4$

24.59.   **IDENTIFY:** Capacitors in series carry the same charge, while capacitors in parallel have the same potential difference across them.

SET UP: $V_{ab} = 150$ V, $Q_1 = 150\ \mu C$, $Q_3 = 450\ \mu C$, and $V = Q/C$.

EXECUTE: $C_1 = 3.00\ \mu F$ so $V_1 = \dfrac{Q_1}{C_1} = \dfrac{150\ \mu C}{3.00\ \mu F} = 50.0$ V and $V_1 = V_2 = 50.0$ V. $V_1 + V_3 = V_{ab}$ so

$V_3 = 100$ V. $C_3 = \dfrac{Q_3}{V_3} = \dfrac{450\ \mu C}{100\ V} = 4.50\ \mu F$. $Q_1 + Q_2 = Q_3$ so $Q_2 = Q_3 - Q_1 = 450\ \mu C - 150\ \mu C = 300\ \mu C$

and $C_2 = \dfrac{Q_2}{V_2} = \dfrac{300\ \mu C}{50.0\ V} = 6.00\ \mu F$.

EVALUATE: Capacitors in parallel only carry the same charge if they have the same capacitance.

24.61.   **(a) IDENTIFY:**   Replace the three capacitors in series by their equivalent. The charge on the equivalent capacitor equals the charge on each of the original capacitors.

SET UP: The three capacitors can be replaced by their equivalent as shown in Figure 24.61a.

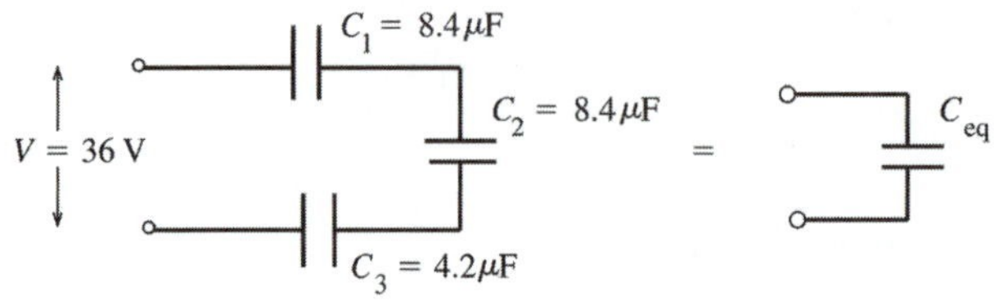

Figure 24.61a

EXECUTE: $C_3 = C_1/2$ so $\dfrac{1}{C_{eq}} = \dfrac{1}{C_1} + \dfrac{1}{C_2} + \dfrac{1}{C_3} = \dfrac{4}{8.4\,\mu F}$ and $C_{eq} = 8.4\,\mu F/4 = 2.1\,\mu F$

$Q = C_{eq}V = (2.1\,\mu F)(36\ V) = 76\,\mu C$

The three capacitors are in series so they each have the same charge: $Q_1 = Q_2 = Q_3 = 76\,\mu C$

EVALUATE: The equivalent capacitance for capacitors in series is smaller than each of the original capacitors.

(b) IDENTIFY and SET UP: Use $U = \frac{1}{2}QV$. We know each Q and we know that $V_1 + V_2 + V_3 = 36$ V.

EXECUTE: $U = \frac{1}{2}Q_1V_1 + \frac{1}{2}Q_2V_2 + \frac{1}{2}Q_3V_3$

But $Q_1 = Q_2 = Q_3 = Q$ so $U = \frac{1}{2}Q(V_1 + V_2 + V_3)$

But also $V_1 + V_2 + V_3 = V = 36$ V, so $U = \frac{1}{2}QV = \frac{1}{2}(76\,\mu C)(36\ V) = 1.4 \times 10^{-3}$ J.

EVALUATE: We could also use $U = Q^2/2C$ and calculate U for each capacitor.

(c) IDENTIFY: The charges on the plates redistribute to make the potentials across each capacitor the same.
SET UP: The capacitors before and after they are connected are sketched in Figure 24.61b.

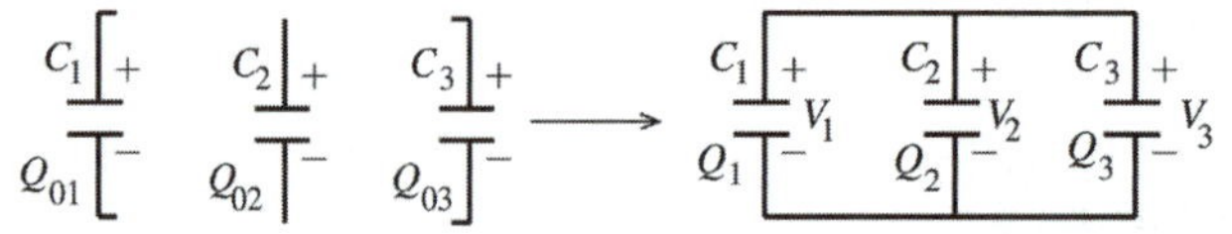

Figure 24.61b

EXECUTE: The total positive charge that is available to be distributed on the upper plates of the three capacitors is $Q_0 = Q_{01} + Q_{02} + Q_{03} = 3(76\,\mu C) = 228\,\mu C$. Thus $Q_1 + Q_2 + Q_3 = 228\,\mu C$. After the circuit is completed the charge distributes to make $V_1 = V_2 = V_3$. $V = Q/C$ and $V_1 = V_2$ so $Q_1/C_1 = Q_2/C_2$ and then $C_1 = C_2$ says $Q_1 = Q_2$. $V_1 = V_3$ says $Q_1/C_1 = Q_3/C_3$ and $Q_1 = Q_3(C_1/C_3) = Q_3(8.4\,\mu F/4.2\,\mu F) = 2Q_3$. Using $Q_2 = Q_1$ and $Q_1 = 2Q_3$ in the above equation gives $2Q_3 + 2Q_3 + Q_3 = 228\,\mu C$.

$5Q_3 = 228\,\mu C$ and $Q_3 = 45.6\,\mu C$, $Q_1 = Q_2 = 91.2\,\mu C$

Then $V_1 = \dfrac{Q_1}{C_1} = \dfrac{91.2\,\mu C}{8.4\,\mu F} = 11$ V, $V_2 = \dfrac{Q_2}{C_2} = \dfrac{91.2\,\mu C}{8.4\,\mu F} = 11$ V, and $V_3 = \dfrac{Q_3}{C_3} = \dfrac{45.6\,\mu C}{4.2\,\mu F} = 11$ V.

The voltage across each capacitor in the parallel combination is 11 V.

(d) $U = \frac{1}{2}Q_1V_1 + \frac{1}{2}Q_2V_2 + \frac{1}{2}Q_3V_3$.

But $V_1 = V_2 = V_3$ so $U = \frac{1}{2}V_1(Q_1 + Q_2 + Q_3) = \frac{1}{2}(11\ V)(228\,\mu C) = 1.3 \times 10^{-3}$ J.

EVALUATE: This is less than the original energy of 1.4×10^{-3} J. The stored energy has decreased, as in Example 24.7.

24.63. **IDENTIFY:** Replace series and parallel combinations of capacitors by their equivalents. In each equivalent network apply the rules for Q and V for capacitors in series and parallel; start with the simplest network and work back to the original circuit.

(a) SET UP: The network is sketched in Figure 24.63a.

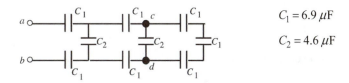

$C_1 = 6.9 \ \mu F$

$C_2 = 4.6 \ \mu F$

Figure 24.63a

EXECUTE: Simplify the network by replacing the capacitor combinations by their equivalents. Make the replacement shown in Figure 24.63b.

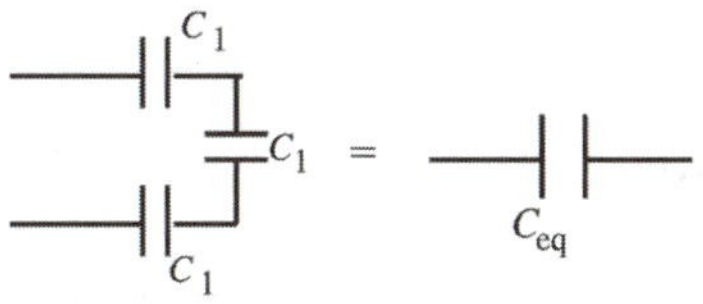

$$\frac{1}{C_{eq}} = \frac{3}{C_1}$$

$$C_{eq} = \frac{C_1}{3} = \frac{6.9 \ \mu F}{3} = 2.3 \ \mu F$$

Figure 24.63b

Next make the replacement shown in Figure 24.63c.

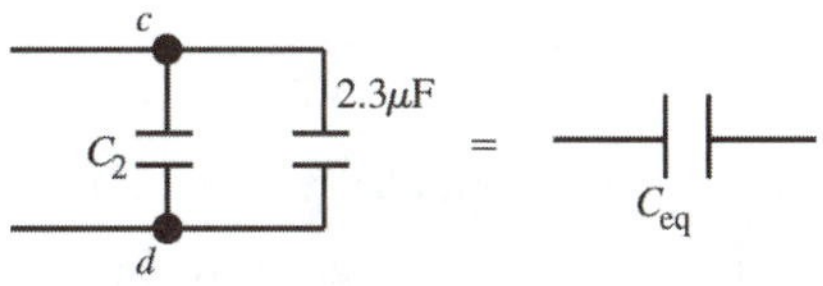

$$C_{eq} = 2.3 \ \mu F + C_2$$

$$C_{eq} = 2.3 \ \mu F + 4.6 \ \mu F = 6.9 \ \mu F$$

Figure 24.63c

Make the replacement shown in Figure 24.63d.

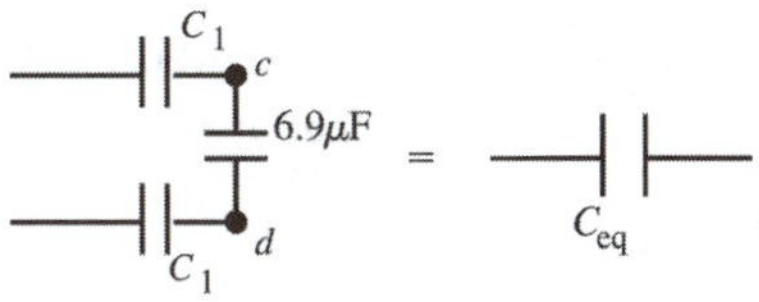

$$\frac{1}{C_{eq}} = \frac{2}{C_1} + \frac{1}{6.9 \ \mu F} = \frac{3}{6.9 \ \mu F}$$

$$C_{eq} = 2.3 \ \mu F$$

Figure 24.63d

Make the replacement shown in Figure 24.63e.

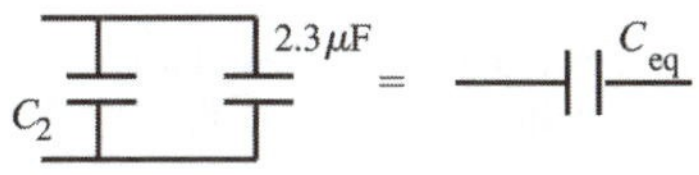

$$C_{eq} = C_2 + 2.3 \ \mu F = 4.6 \ \mu F + 2.3 \ \mu F$$

$$C_{eq} = 6.9 \ \mu F$$

Figure 24.63e

Make the replacement shown in Figure 24.63f.

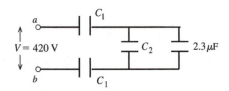

$$\frac{1}{C_{eq}} = \frac{2}{C_1} + \frac{1}{6.9 \; \mu F} = \frac{3}{6.9 \; \mu F}$$

$$C_{eq} = 2.3 \; \mu F$$

Figure 24.63f

(b) Consider the network as drawn in Figure 24.63g.

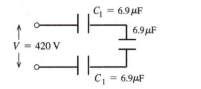

From part (a) $2.3 \; \mu F$ is the equivalent capacitance of the rest of the network.

Figure 24.63g

The equivalent network is shown in Figure 24.63h.

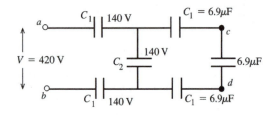

The capacitors are in series, so all three capacitors have the same Q.

Figure 24.63h

But here all three have the same C, so by $V = Q/C$ all three must have the same V. The three voltages must add to 420 V, so each capacitor has $V = 140$ V. The $6.9 \; \mu F$ to the right is the equivalent of C_2 and the $2.3 \; \mu F$ capacitor in parallel, so $V_2 = 140$ V. (Capacitors in parallel have the same potential difference.) Hence $Q_1 = C_1 V_1 = (6.9 \; \mu F)(140 \; V) = 9.7 \times 10^{-4}$ C and $Q_2 = C_2 V_2 = (4.6 \; \mu F)(140 \; V) = 6.4 \times 10^{-4}$ C.

(c) From the potentials deduced in part (b) we have the situation shown in Figure 24.63i.

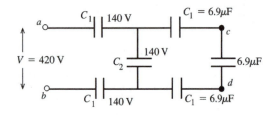

From part (a) $6.9 \; \mu F$ is the equivalent capacitance of the rest of the network.

Figure 24.63i

The three right-most capacitors are in series and therefore have the same charge. But their capacitances are also equal, so by $V = Q/C$ they each have the same potential difference. Their potentials must sum to 140 V, so the potential across each is 47 V and $V_{cd} = 47$ V.

EVALUATE: In each capacitor network the rules for combining V for capacitors in series and parallel are obeyed. Note that $V_{cd} < V$, in fact $V - 2(140 \; V) - 2(47 \; V) = V_{cd}$.

24.71.   **IDENTIFY:** The increase in temperature of the wire allows us to find out how much heat it gained, which is the energy initially stored in the capacitor. We can use this energy to find the capacitance of the capacitor.

SET UP: The heat in the wire is $Q = mc\Delta T$ and the energy stored in the capacitor is $U = \frac{1}{2}CV^2$, which is equal to the heat Q.

EXECUTE: The heat that goes into the wire is

$Q = mc\Delta T = (12.0 \times 10^{-3} \text{ kg})(910 \text{ J/(kg} \cdot \text{K)})(11.2 \text{ K}) = 122.3 \text{ J}$. For the capacitor, $U = \frac{1}{2}CV^2$.

$C = \dfrac{2U}{V^2} = \dfrac{2(122.3 \text{ J})}{(2.25 \times 10^3 \text{ V})^2} = 4.83 \times 10^{-5} \text{ F} = 48.3 \ \mu\text{F}$.

EVALUATE: A capacitance of 48.3 μF is quite reasonable for ordinary laboratory capacitors.

24.73. **IDENTIFY:** The two slabs of dielectric are in series with each other.

SET UP: The capacitor is equivalent to C_1 and C_2 in series, so $\dfrac{1}{C_1} + \dfrac{1}{C_2} = \dfrac{1}{C}$, which gives $C = \dfrac{C_1 C_2}{C_1 + C_2}$.

EXECUTE: With $d = 1.90$ mm, $C_1 = \dfrac{K_1 \epsilon_0 A}{d}$ and $C_2 = \dfrac{K_2 \epsilon_0 A}{d}$.

$C = \left(\dfrac{K_1 K_2}{K_1 + K_2}\right)\dfrac{\epsilon_0 A}{d} = \dfrac{(8.854 \times 10^{-12} \text{ C}^2/(\text{N} \cdot \text{m}^2)(0.0800 \text{ m})^2}{1.90 \times 10^{-3} \text{ m}}\left(\dfrac{(4.7)(2.6)}{4.7 + 2.6}\right) = 4.992 \times 10^{-11} \text{ F}$.

$U = \dfrac{1}{2}CV^2 = \dfrac{1}{2}(4.992 \times 10^{-11} \text{ F})(86.0 \text{ V})^2 = 1.85 \times 10^{-7} \text{ J}$.

EVALUATE: The dielectrics increase the capacitance, allowing the capacitor to store more energy than if it were air-filled.

24.75. **IDENTIFY:** The object is equivalent to two identical capacitors in parallel, where each has the same area A, plate separation d and dielectric with dielectric constant K.

SET UP: For each capacitor in the parallel combination, $C = \dfrac{\epsilon_0 A}{d}$.

EXECUTE: **(a)** The charge distribution on the plates is shown in Figure 24.75.

(b) $C = 2\left(\dfrac{\epsilon_0 A}{d}\right) = \dfrac{2(4.2)\epsilon_0 (0.120 \text{ m})^2}{4.5 \times 10^{-4} \text{ m}} = 2.38 \times 10^{-9} \text{ F}$.

EVALUATE: If two of the plates are separated by both sheets of paper to form a capacitor,

$C = \dfrac{\epsilon_0 A}{2d} = \dfrac{2.38 \times 10^{-9} \text{ F}}{4}$, smaller by a factor of 4 compared to the capacitor in the problem.

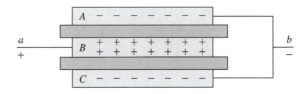

Figure 24.75

25

CURRENT, RESISTANCE, AND ELECTROMOTIVE FORCE

25.5. **IDENTIFY** and **SET UP:** Use Eq. (25.3) to calculate the drift speed and then use that to find the time to travel the length of the wire.

EXECUTE: **(a)** Calculate the drift speed v_d:

$$J = \frac{I}{A} = \frac{I}{\pi r^2} = \frac{4.85 \text{ A}}{\pi (1.025 \times 10^{-3} \text{ m})^2} = 1.469 \times 10^6 \text{ A/m}^2$$

$$v_d = \frac{J}{n|q|} = \frac{1.469 \times 10^6 \text{ A/m}^2}{(8.5 \times 10^{28}/\text{m}^3)(1.602 \times 10^{-19} \text{ C})} = 1.079 \times 10^{-4} \text{ m/s}$$

$$t = \frac{L}{v_d} = \frac{0.710 \text{ m}}{1.079 \times 10^{-4} \text{ m/s}} = 6.58 \times 10^3 \text{ s} = 110 \text{ min.}$$

(b) $v_d = \dfrac{I}{\pi r^2 n|q|}$

$$t = \frac{L}{v_d} = \frac{\pi r^2 n |q| L}{I}$$

t is proportional to r^2 and hence to d^2 where $d = 2r$ is the wire diameter.

$$t = (6.58 \times 10^3 \text{ s}) \left(\frac{4.12 \text{ mm}}{2.05 \text{ mm}} \right)^2 = 2.66 \times 10^4 \text{ s} = 440 \text{ min.}$$

(c) **EVALUATE:** The drift speed is proportional to the current density and therefore it is inversely proportional to the square of the diameter of the wire. Increasing the diameter by some factor decreases the drift speed by the square of that factor.

25.7. **IDENTIFY** and **SET UP:** Apply Eq. (25.1) to find the charge dQ in time dt. Integrate to find the total charge in the whole time interval.

EXECUTE: **(a)** $dQ = I \, dt$

$$Q = \int_0^{8.0\text{s}} (55 \text{ A} - (0.65 \text{ A/s}^2) t^2) dt = \left[(55 \text{ A}) t - (0.217 \text{ A/s}^2) t^3 \right]_0^{8.0 \text{ s}}$$

$$Q = (55 \text{ A})(8.0 \text{ s}) - (0.217 \text{ A/s}^2)(8.0 \text{ s})^3 = 330 \text{ C}$$

(b) $I = \dfrac{Q}{t} = \dfrac{330 \text{ C}}{8.0 \text{ s}} = 41 \text{ A}$

EVALUATE: The current decreases from 55 A to 13.4 A during the interval. The decrease is not linear and the average current is not equal to $(55\text{A} + 13.4 \text{ A})/2$.

25.11. **IDENTIFY:** First use Ohm's law to find the resistance at 20.0°C; then calculate the resistivity from the resistance. Finally use the dependence of resistance on temperature to calculate the temperature coefficient of resistance.

SET UP: Ohm's law is $R = V/I$, $R = \rho L/A$, $R = R_0[1 + \alpha(T - T_0)]$, and the radius is one-half the diameter.

EXECUTE: **(a)** At 20.0°C, $R = V/I = (15.0 \text{ V})/(18.5 \text{ A}) = 0.811 \, \Omega$. Using $R = \rho L/A$ and solving for ρ gives $\rho = RA/L = R\pi(D/2)^2/L = (0.811 \, \Omega)\pi[(0.00500 \text{ m})/2]^2/(1.50 \text{ m}) = 1.06 \times 10^{-5} \, \Omega \cdot \text{m}$.

(b) At 92.0°C, $R = V/I = (15.0 \text{ V})/(17.2 \text{ A}) = 0.872 \, \Omega$. Using $R = R_0[1 + \alpha(T - T_0)]$ with T_0 taken as 20.0°C, we have $0.872 \, \Omega = (0.811 \, \Omega)[1 + \alpha(92.0°C - 20.0°C)]$. This gives $\alpha = 0.00105 \, (\text{C}°)^{-1}$.

EVALUATE: The results are typical of ordinary metals.

25.15. **(a) IDENTIFY:** Start with the definition of resistivity and use its dependence on temperature to find the electric field.

SET UP: $E = \rho J = \rho_{20}[1 + \alpha(T - T_0)]\dfrac{I}{\pi r^2}$

EXECUTE: $E = (5.25 \times 10^{-8} \, \Omega \cdot \text{m})[1 + (0.0045/\text{C}°)(120°\text{C} - 20°\text{C})](12.5 \text{ A})/[\pi(0.000500 \text{ m})^2] = 1.21 \text{ V/m}$.

(Note that the resistivity at 120°C turns out to be $7.61 \times 10^{-8} \, \Omega \cdot \text{m}$.)

EVALUATE: This result is fairly large because tungsten has a larger resistivity than copper.

(b) IDENTIFY: Relate resistance and resistivity.

SET UP: $R = \rho L/A = \rho L/\pi r^2$

EXECUTE: $R = (7.61 \times 10^{-8} \, \Omega \cdot \text{m})(0.150 \text{ m})/[\pi(0.000500 \text{ m})^2] = 0.0145 \, \Omega$

EVALUATE: Most metals have very low resistance.

(c) IDENTIFY: The potential difference is proportional to the length of wire.

SET UP: $V = EL$

EXECUTE: $V = (1.21 \text{ V/m})(0.150 \text{ m}) = 0.182 \text{ V}$

EVALUATE: We could also calculate $V = IR = (12.5 \text{ A})(0.0145 \, \Omega) = 0.181 \text{ V}$, in agreement with part (c).

25.19. **IDENTIFY** and **SET UP:** Use Eq. (25.10) to calculate A. Find the volume of the wire and use the density to calculate the mass.

EXECUTE: Find the volume of one of the wires:

$R = \dfrac{\rho L}{A}$ so $A = \dfrac{\rho L}{R}$ and

$\text{Volume} = AL = \dfrac{\rho L^2}{R} = \dfrac{(1.72 \times 10^{-8} \, \Omega \cdot \text{m})(3.50 \text{ m})^2}{0.125 \, \Omega} = 1.686 \times 10^{-6} \text{ m}^3$

$m = (\text{density})V = (8.9 \times 10^3 \text{ kg/m}^3)(1.686 \times 10^{-6} \text{ m}^3) = 15 \text{ g}$

EVALUATE: The mass we calculated is reasonable for a wire.

25.23. **IDENTIFY** and **SET UP:** Eq. (25.5) relates the electric field that is given to the current density. $V = EL$ gives the potential difference across a length L of wire and Eq. (25.11) allows us to calculate R.

EXECUTE: **(a)** Eq. (25.5): $\rho = E/J$ so $J = E/\rho$

From Table 25.1 the resistivity for gold is $2.44 \times 10^{-8} \, \Omega \cdot \text{m}$.

$J = \dfrac{E}{\rho} = \dfrac{0.49 \text{ V/m}}{2.44 \times 10^{-8} \, \Omega \cdot \text{m}} = 2.008 \times 10^7 \text{ A/m}^2$

$I = JA = J\pi r^2 = (2.008 \times 10^7 \text{ A/m}^2)\pi(0.42 \times 10^{-3} \text{ m})^2 = 11 \text{ A}$

(b) $V = EL = (0.49 \text{ V/m})(6.4 \text{ m}) = 3.1 \text{ V}$

(c) We can use Ohm's law (Eq. (25.11)): $V = IR$.

$R = \dfrac{V}{I} = \dfrac{3.1 \text{ V}}{11 \text{ A}} = 0.28 \, \Omega$

EVALUATE: We can also calculate R from the resistivity and the dimensions of the wire (Eq. 25.10):

$R = \dfrac{\rho L}{A} = \dfrac{\rho L}{\pi r^2} = \dfrac{(2.44 \times 10^{-8} \, \Omega \cdot \text{m})(6.4 \text{ m})}{\pi(0.42 \times 10^{-3} \text{ m})^2} = 0.28 \, \Omega$, which checks.

25.31. **IDENTIFY:** The terminal voltage of the battery is $V_{ab} = \mathcal{E} - Ir$. The voltmeter reads the potential difference between its terminals.

SET UP: An ideal voltmeter has infinite resistance.

EXECUTE: **(a)** Since an ideal voltmeter has infinite resistance, so there would be NO current through the $2.0\,\Omega$ resistor.

(b) $V_{ab} = \mathcal{E} = 5.0$ V; Since there is no current there is no voltage lost over the internal resistance.

(c) The voltmeter reading is therefore 5.0 V since with no current flowing there is no voltage drop across either resistor.

EVALUATE: This not the proper way to connect a voltmeter. If we wish to measure the terminal voltage of the battery in a circuit that does not include the voltmeter, then connect the voltmeter across the terminals of the battery.

25.33.   **IDENTIFY:** The voltmeter reads the potential difference V_{ab} between the terminals of the battery.

SET UP: <u>open circuit</u> $I = 0$. The circuit is sketched in Figure 25.33a.

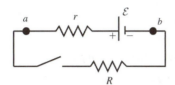

EXECUTE: $V_{ab} = \mathcal{E} = 3.08$ V

Figure 25.33a

SET UP: <u>switch closed</u> The circuit is sketched in Figure 25.33b.

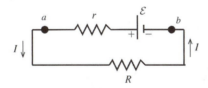

EXECUTE: $V_{ab} = \mathcal{E} - Ir = 2.97$ V

$$r = \frac{\mathcal{E} - 2.97 \text{ V}}{I}$$

$$r = \frac{3.08 \text{ V} - 2.97 \text{ V}}{1.65 \text{ A}} = 0.067\,\Omega$$

Figure 25.33b

And $V_{ab} = IR$ so $R = \dfrac{V_{ab}}{I} = \dfrac{2.97 \text{ V}}{1.65 \text{ A}} = 1.80\,\Omega.$

EVALUATE: When current flows through the battery there is a voltage drop across its internal resistance and its terminal voltage V is less than its emf.

25.35.   **(a) IDENTIFY and SET UP:** Assume that the current is clockwise. The circuit is sketched in Figure 25.35a.

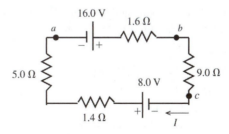

Figure 25.35a

Add up the potential rises and drops as travel clockwise around the circuit.

EXECUTE: $16.0 \text{ V} - I(1.6\,\Omega) - I(9.0\,\Omega) + 8.0 \text{ V} - I(1.4\,\Omega) - I(5.0\,\Omega) = 0$

$$I = \frac{16.0 \text{ V} + 8.0 \text{ V}}{9.0\,\Omega + 1.4\,\Omega + 5.0\,\Omega + 1.6\,\Omega} = \frac{24.0 \text{ V}}{17.0\,\Omega} = 1.41 \text{ A, clockwise}$$

EVALUATE: The 16.0-V battery and the 8.0-V battery both drive the current in the same direction.

(b) IDENTIFY and SET UP: Start at point a and travel through the battery to point b, keeping track of the potential changes. At point b the potential is V_b.

EXECUTE: $V_a + 16.0 \text{ V} - I(1.6\,\Omega) = V_b$

$V_a - V_b = -16.0 \text{ V} + (1.41 \text{ A})(1.6\,\Omega)$

$V_{ab} = -16.0 \text{ V} + 2.3 \text{ V} = -13.7 \text{ V}$ (point a is at lower potential; it is the negative terminal). Therefore,

$V_{ba} = 13.7 \text{ V}.$

EVALUATE: Could also go counterclockwise from a to b:

$V_a + (1.41 \text{ A})(5.0\,\Omega) + (1.41 \text{ A})(1.4\,\Omega) - 8.0 \text{ V} + (1.41 \text{ A})(9.0\,\Omega) = V_b$

$V_{ab} = -13.7 \text{ V},$ which checks.

(c) IDENTIFY and SET UP: Start at point a and travel through the battery to point c, keeping track of the potential changes.

EXECUTE: $V_a + 16.0 \text{ V} - I(1.6\,\Omega) - I(9.0\,\Omega) = V_c$

$V_a - V_c = -16.0 \text{ V} + (1.41 \text{ A})(1.6\,\Omega + 9.0\,\Omega)$

$V_{ac} = -16.0 \text{ V} + 15.0 \text{ V} = -1.0 \text{ V}$ (point a is at lower potential than point c)

EVALUATE: Could also go counterclockwise from a to c:

$V_a + (1.41 \text{ A})(5.0\,\Omega) + (1.41 \text{ A})(1.4\,\Omega) - 8.0 \text{ V} = V_c$

$V_{ac} = -1.0 \text{ V},$ which checks.

(d) Call the potential zero at point a. Travel clockwise around the circuit. The graph is sketched in Figure 25.35b.

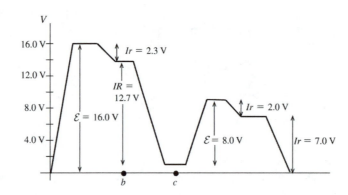

Figure 25.35b

25.41.   IDENTIFY: A "100-W" European bulb dissipates 100 W when used across 220 V.
(a) SET UP: Take the ratio of the power in the U.S. to the power in Europe, as in the alternative method for Problem 25.40, using $P = V^2/R$.

EXECUTE: $\dfrac{P_{US}}{P_E} = \dfrac{V_{US}^2/R}{V_E^2/R} = \left(\dfrac{V_{US}}{V_E}\right)^2 = \left(\dfrac{120 \text{ V}}{220 \text{ V}}\right)^2$. This gives $P_{US} = (100 \text{ W})\left(\dfrac{120 \text{ V}}{220 \text{ V}}\right)^2 = 29.8 \text{ W}.$

(b) SET UP: Use $P = IV$ to find the current.
EXECUTE: $I = P/V = (29.8 \text{ W})/(120 \text{ V}) = 0.248 \text{ A}$

EVALUATE: The bulb draws considerably less power in the U.S., so it would be much dimmer than in Europe.

25.47.   **(a)** IDENTIFY and SET UP: $P = VI$ and energy = (power)$\times$(time).

EXECUTE: $P = VI = (12 \text{ V})(60 \text{ A}) = 720 \text{ W}$

The battery can provide this for 1.0 h, so the energy the battery has stored is

$U = Pt = (720 \text{ W})(3600 \text{ s}) = 2.6 \times 10^6 \text{ J}.$

(b) IDENTIFY and SET UP: For gasoline the heat of combustion is $L_c = 46 \times 10^6$ J/kg. Solve for the mass m required to supply the energy calculated in part (a) and use density $\rho = m/V$ to calculate V.

EXECUTE: The mass of gasoline that supplies 2.6×10^6 J is $m = \dfrac{2.6 \times 10^6 \text{ J}}{46 \times 10^6 \text{ J/kg}} = 0.0565$ kg.

The volume of this mass of gasoline is

$V = \dfrac{m}{\rho} = \dfrac{0.0565 \text{ kg}}{900 \text{ kg/m}^3} = 6.3 \times 10^{-5} \text{ m}^3 \left(\dfrac{1000 \text{ L}}{1 \text{ m}^3} \right) = 0.063$ L.

(c) IDENTIFY and SET UP: Energy $=$ (power)$\times$(time); the energy is that calculated in part (a).

EXECUTE: $U = Pt, t = \dfrac{U}{P} = \dfrac{2.6 \times 10^6 \text{ J}}{450 \text{ W}} = 5800$ s $= 97$ min $= 1.6$ h.

EVALUATE: The battery discharges at a rate of 720 W (for 60 A) and is charged at a rate of 450 W, so it takes longer to charge than to discharge.

25.49. **IDENTIFY:** Some of the power generated by the internal emf of the battery is dissipated across the battery's internal resistance, so it is not available to the bulb.

SET UP: Use $P = I^2 R$ and take the ratio of the power dissipated in the internal resistance r to the total power.

EXECUTE: $\dfrac{P_r}{P_{\text{Total}}} = \dfrac{I^2 r}{I^2 (r + R)} = \dfrac{r}{r + R} = \dfrac{3.5 \, \Omega}{28.5 \, \Omega} = 0.123 = 12.3\%$

EVALUATE: About 88% of the power of the battery goes to the bulb. The rest appears as heat in the internal resistance.

25.57. **IDENTIFY and SET UP:** With the voltmeter connected across the terminals of the battery there is no current through the battery and the voltmeter reading is the battery emf; $\mathcal{E} = 12.6$ V.

With a wire of resistance R connected to the battery current I flows and $\mathcal{E} - Ir - IR = 0$, where r is the internal resistance of the battery. Apply this equation to each piece of wire to get two equations in the two unknowns.

EXECUTE: Call the resistance of the 20.0-m piece R_1; then the resistance of the 40.0-m piece is $R_2 = 2R_1$.

$\mathcal{E} - I_1 r - I_1 R_1 = 0; \quad 12.6 \text{ V} - (7.00 \text{ A})r - (7.00 \text{ A})R_1 = 0$

$\mathcal{E} - I_2 r - I_2 (2R_1) = 0; \quad 12.6 \text{ V} - (4.20 \text{ A})r - (4.20 \text{ A})(2R_1) = 0$

Solving these two equations in two unknowns gives $R_1 = 1.20 \, \Omega$. This is the resistance of 20.0 m, so the resistance of one meter is $[1.20 \, \Omega/(20.0 \text{ m})](1.00 \text{ m}) = 0.060 \, \Omega$.

EVALUATE: We can also solve for r and we get $r = 0.600 \, \Omega$. When measuring small resistances, the internal resistance of the battery has a large effect.

25.59. **IDENTIFY:** Conservation of charge requires that the current be the same in both sections of the wire.

$E = \rho J = \dfrac{\rho I}{A}$. For each section, $V = IR = JAR = \left(\dfrac{EA}{\rho} \right) \left(\dfrac{\rho L}{A} \right) = EL$. The voltages across each section add.

SET UP: $A = (\pi/4)D^2$, where D is the diameter.

EXECUTE: (a) The current must be the same in both sections of the wire, so the current in the thin end is 2.5 mA.

(b) $E_{1.6\text{mm}} = \rho J = \dfrac{\rho I}{A} = \dfrac{(1.72 \times 10^{-8} \, \Omega \cdot \text{m})(2.5 \times 10^{-3} \text{ A})}{(\pi/4)(1.6 \times 10^{-3} \text{ m})^2} = 2.14 \times 10^{-5}$ V/m.

(c) $E_{0.8\text{mm}} = \rho J = \dfrac{\rho I}{A} = \dfrac{(1.72 \times 10^{-8} \, \Omega \cdot \text{m})(2.5 \times 10^{-3} \text{ A})}{(\pi/4)(0.80 \times 10^{-3} \text{ m})^2} = 8.55 \times 10^{-5}$ V/m. This is $4E_{1.6\text{mm}}$.

(d) $V = E_{1.6\text{ mm}} L_{1.6 \text{ mm}} + E_{0.8 \text{ mm}} L_{0.8 \text{ mm}}. V = (2.14 \times 10^{-5} \text{ V/m})(1.20 \text{ m}) + (8.55 \times 10^{-5} \text{ V/m})(1.80 \text{ m}) = 1.80 \times 10^{-4}$ V.

EVALUATE: The currents are the same but the current density is larger in the thinner section and the electric field is larger there.

25.61.   **IDENTIFY:**   The current generates heat in the nichrome heating element. This heat increases the temperature of the water and its aluminum container.

SET UP: The rate of heating in the nichrome is $P = I^2 R$, the power is Q/t, and the current in the circuit is $I = \dfrac{\varepsilon}{R+r}$, where ε is the internal emf of the battery.

EXECUTE: $I = \dfrac{\varepsilon}{R+r} = \dfrac{96.0 \text{ V}}{28.0 \ \Omega + 1.2 \ \Omega} = 3.288 \text{ A}.$

$P = I^2 R = (3.288 \text{ A})^2 (28.0 \ \Omega) = 302.6 \text{ W}.$ The total heat needed is:

Cup: $Q = mc\Delta T = (0.130 \text{ kg})(910 \text{ J/(kg} \cdot \text{K}))(34.5°\text{C} - 21.2°\text{C}) = 1573 \text{ J}.$

Water: $Q = mc\Delta T = (0.200 \text{ kg})(4190 \text{ J/(kg} \cdot \text{K}))(34.5°\text{C} - 21.2°\text{C}) = 11{,}145 \text{ J}.$

Total: $Q = 12{,}718 \text{ J}.$ $t = \dfrac{Q}{P} = \dfrac{12{,}718 \text{ J}}{302.6 \text{ W}} = 42.0 \text{ s}.$

EVALUATE: A current of about 3 A is rather large and would generate heat at a considerable rate. It could reasonably change the temperature of the water and aluminum by about 13 C° in 42 s.

25.63.   **IDENTIFY:**   Knowing the current and the time for which it lasts, plus the resistance of the body, we can calculate the energy delivered.

SET UP: Electrical energy is deposited in his body at the rate $P = I^2 R$. Heat energy Q produces a temperature change ΔT according to $Q = mc\Delta T$, where $c = 4190 \text{ J/kg} \cdot \text{C}°$.

EXECUTE: **(a)** $P = I^2 R = (25{,}000 \text{ A})^2 (1.0 \text{ k}\Omega) = 6.25 \times 10^{11} \text{ W}.$ The energy deposited is

$Pt = (6.15 \times 10^{11} \text{ W})(40 \times 10^{-6} \text{ s}) = 2.5 \times 10^7 \text{ J}.$ Find ΔT when $Q = 2.5 \times 10^7 \text{ J}.$

$\Delta T = \dfrac{Q}{mc} = \dfrac{2.5 \times 10^7 \text{ J}}{(75 \text{ kg})(4190 \text{ J/kg} \cdot \text{C}°)} = 80 \text{ C}°.$

(b) An increase of only 63 C° brings the water in the body to the boiling point; part of the person's body will be vaporized.

EVALUATE: Even this approximate calculation shows that being hit by lightning is very dangerous.

25.65.   **(a) IDENTIFY:**   Apply Eq. (25.10) to calculate the resistance of each thin disk and then integrate over the truncated cone to find the total resistance.

SET UP:

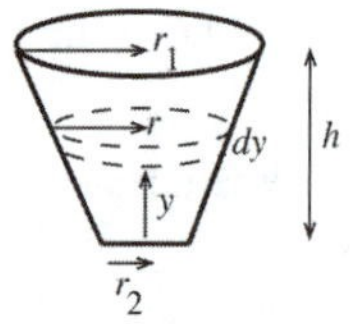

EXECUTE: The radius of a truncated cone a distance y above the bottom is given by $r = r_2 + (y/h)(r_1 - r_2) = r_2 + y\beta$ with $\beta = (r_1 - r_2)/h.$

Figure 25.65

Consider a thin slice a distance y above the bottom. The slice has thickness dy and radius r (see Figure 25.65.) The resistance of the slice is $dR = \dfrac{\rho \, dy}{A} = \dfrac{\rho \, dy}{\pi r^2} = \dfrac{\rho \, dy}{\pi (r_2 + \beta y)^2}.$

The total resistance of the cone if obtained by integrating over these thin slices:

$$R = \int dR = \frac{\rho}{\pi} \int_0^h \frac{dy}{(r_2 + \beta y)^2} = \frac{\rho}{\pi} \left[-\frac{1}{\beta}(r_2 + y\beta)^{-1} \right]_0^h = -\frac{\rho}{\pi \beta}\left[\frac{1}{r_2 + h\beta} - \frac{1}{r_2} \right]$$

But $r_2 + h\beta = r_1$

$$R = \frac{\rho}{\pi \beta}\left[\frac{1}{r_2} - \frac{1}{r_1} \right] = \frac{\rho}{\pi}\left(\frac{h}{r_1 - r_2} \right)\left(\frac{r_1 - r_2}{r_1 r_2} \right) = \frac{\rho h}{\pi r_1 r_2}$$

(b) EVALUATE: Let $r_1 = r_2 = r$. Then $R = \rho h / \pi r^2 = \rho L / A$ where $A = \pi r^2$ and $L = h$. This agrees with Eq. (25.10).

25.69.   **IDENTIFY:**   In each case write the terminal voltage in terms of $\mathcal{E}, I$ and r. Since I is known, this gives two equations in the two unknowns $\mathcal{E}$ and r.

SET UP: The battery with the 1.50-A current is sketched in Figure 25.69a.

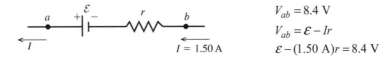

$V_{ab} = 8.4$ V

$V_{ab} = \mathcal{E} - Ir$

$\mathcal{E} - (1.50 \text{ A})r = 8.4$ V

Figure 25.69a

The battery with the 3.50-A current is sketched in Figure 25.69b.

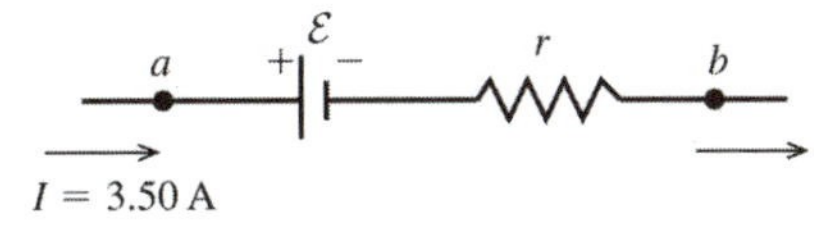

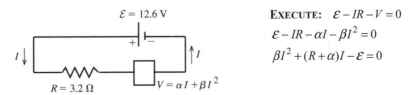

$V_{ab} = 9.4$ V

$V_{ab} = \mathcal{E} + Ir$

$\mathcal{E} + (3.5 \text{ A})r = 9.4$ V

Figure 25.69b

EXECUTE: **(a)** Solve the first equation for $\mathcal{E}$ and use that result in the second equation:

$\mathcal{E} = 8.4$ V $+ (1.50 \text{ A})r$

8.4 V $+ (1.50 \text{ A})r + (3.50 \text{ A})r = 9.4$ V

$(5.00 \text{ A})r = 1.0$ V so $r = \dfrac{1.0 \text{ V}}{5.00 \text{ A}} = 0.20 \ \Omega$

(b) Then $\mathcal{E} = 8.4$ V $+ (1.50 \text{ A})r = 8.4$ V $+ (1.50 \text{ A})(0.20 \ \Omega) = 8.7$ V

EVALUATE: When the current passes through the emf in the direction from $-$ to $+$, the terminal voltage is less than the emf and when it passes through from $+$ to $-$, the terminal voltage is greater than the emf.

25.73.   **IDENTIFY:**   Set the sum of the potential rises and drops around the circuit equal to zero and solve for I.

SET UP: The circuit is sketched in Figure 25.73.

$\mathcal{E} = 12.6$ V

$I \downarrow$ $\uparrow I$

$R = 3.2 \ \Omega$ $V = \alpha I + \beta I^2$

EXECUTE: $\mathcal{E} - IR - V = 0$

$\mathcal{E} - IR - \alpha I - \beta I^2 = 0$

$\beta I^2 + (R + \alpha)I - \mathcal{E} = 0$

Figure 25.73

The quadratic formula gives $I = (1/2\beta)\left[-(R+\alpha) \pm \sqrt{(R+\alpha)^2 + 4\beta\mathcal{E}}\right]$

I must be positive, so take the $+$ sign

$I = (1/2\beta)\left[-(R+\alpha) + \sqrt{(R+\alpha)^2 + 4\beta\mathcal{E}}\right]$

$I = -2.692$ A $+ 4.116$ A $= 1.42$ A

EVALUATE: For this I the voltage across the thermistor is 8.0 V. The voltage across the resistor must then be 12.6 V $- 8.0$ V $= 4.6$ V, and this agrees with Ohm's law for the resistor.

25.75.   **IDENTIFY:**   The ammeter acts as a resistance in the circuit loop. Set the sum of the potential rises and drops around the circuit equal to zero.

(a) SET UP: The circuit with the ammeter is sketched in Figure 25.75a.

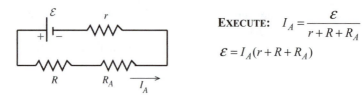

EXECUTE: $I_A = \dfrac{\mathcal{E}}{r + R + R_A}$

$\mathcal{E} = I_A(r + R + R_A)$

Figure 25.75a

SET UP: The circuit with the ammeter removed is sketched in Figure 25.75b.

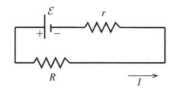

EXECUTE: $I = \dfrac{\mathcal{E}}{R + r}$

Figure 25.75b

Combining the two equations gives

$$I = \left(\frac{1}{R + r}\right)I_A(r + R + R_A) = I_A\left(1 + \frac{R_A}{r + R}\right)$$

(b) Want $I_A = 0.990I$. Use this in the result for part (a).

$$I = 0.990I\left(1 + \frac{R_A}{r + R}\right)$$

$$0.010 = 0.990\left(\frac{R_A}{r + R}\right)$$

$$R_A = (r + R)(0.010/0.990) = (0.45\ \Omega + 3.80\ \Omega)(0.010/0.990) = 0.0429\ \Omega$$

(c) $I - I_A = \dfrac{\mathcal{E}}{r + R} - \dfrac{\mathcal{E}}{r + R + R_A}$

$$I - I_A = \mathcal{E}\left(\frac{r + R + R_A - r - R}{(r + R)(r + R + R_A)}\right) = \frac{\mathcal{E}R_A}{(r + R)(r + R + R_A)}.$$

EVALUATE: The difference between I and I_A increases as R_A increases. If R_A is larger than the value calculated in part (b) then I_A differs from I by more than 1.0%.

25.77. **IDENTIFY:** The power supplied to the house is $P = VI$. The rate at which electrical energy is dissipated in the wires is I^2R, where $R = \dfrac{\rho L}{A}$.

SET UP: For copper, $\rho = 1.72 \times 10^{-8}\ \Omega \cdot \text{m}$.

EXECUTE: **(a)** The line voltage, current to be drawn, and wire diameter are what must be considered in household wiring.

(b) $P = VI$ gives $I = \dfrac{P}{V} = \dfrac{4200\ \text{W}}{120\ \text{V}} = 35\ \text{A}$, so the 8-gauge wire is necessary, since it can carry up to 40 A.

(c) $P = I^2R = \dfrac{I^2\rho L}{A} = \dfrac{(35\ \text{A})^2\,(1.72 \times 10^{-8}\ \Omega \cdot \text{m})(42.0\ \text{m})}{(\pi/4)(0.00326\ \text{m})^2} = 106\ \text{W}.$

(d) If 6-gauge wire is used, $P = \dfrac{I^2\rho L}{A} = \dfrac{(35\ \text{A})^2\,(1.72 \times 10^{-8}\ \Omega \cdot \text{m})\,(42\ \text{m})}{(\pi/4)\,(0.00412\ \text{m})^2} = 66\ \text{W}.$ The decrease in energy consumption is $\Delta E = \Delta Pt = (40\ \text{W})(365\ \text{days/yr})\,(12\ \text{h/day}) = 175\ \text{kWh/yr}$ and the savings is $(175\ \text{kWh/yr})(\$0.11/\text{kWh}) = \$19.25$ per year.

EVALUATE: The cost of the 4200 W used by the appliances is \$2020. The savings is about 1%.

25.79. **(a) IDENTIFY:** Set the sum of the potential rises and drops around the circuit equal to zero and solve the resulting equation for the current I. Apply Eq. (25.17) to each circuit element to find the power associated with it.
SET UP: The circuit is sketched in Figure 25.79.

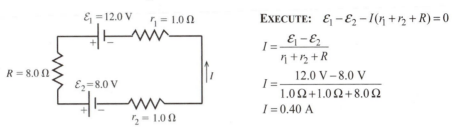

EXECUTE: $\mathcal{E}_1 - \mathcal{E}_2 - I(r_1 + r_2 + R) = 0$

$$I = \frac{\mathcal{E}_1 - \mathcal{E}_2}{r_1 + r_2 + R}$$

$$I = \frac{12.0\text{ V} - 8.0\text{ V}}{1.0\ \Omega + 1.0\ \Omega + 8.0\ \Omega}$$

$$I = 0.40\text{ A}$$

Figure 25.79

(b) $P = I^2 R + I^2 r_1 + I^2 r_2 = I^2(R + r_1 + r_2) = (0.40\text{ A})^2(8.0\ \Omega + 1.0\ \Omega + 1.0\ \Omega)$
$P = 1.6\text{ W}$

(c) Chemical energy is converted to electrical energy in a battery when the current goes through the battery from the negative to the positive terminal, so the electrical energy of the charges increases as the current passes through. This happens in the 12.0-V battery, and the rate of production of electrical energy is
$P = \mathcal{E}_1 I = (12.0\text{ V})(0.40\text{ A}) = 4.8\text{ W}.$

(d) Electrical energy is converted to chemical energy in a battery when the current goes through the battery from the positive to the negative terminal, so the electrical energy of the charges decreases as the current passes through. This happens in the 8.0-V battery, and the rate of consumption of electrical energy is
$P = \mathcal{E}_2 I = (8.0\text{ V})(0.40\text{ V}) = 3.2\text{ W}.$

(e) EVALUATE: Total rate of production of electrical energy $= 4.8\text{ W}.$ Total rate of consumption of electrical energy $= 1.6\text{ W} + 3.2\text{ W} = 4.8\text{ W},$ which equals the rate of production, as it must.

25.81. **IDENTIFY and SET UP:** The terminal voltage is $V_{ab} = \mathcal{E} - Ir = IR,$ where R is the resistance connected to the battery. During the charging the terminal voltage is $V_{ab} = \mathcal{E} + Ir.$ $P = VI$ and energy is $E = Pt.$ $I^2 r$ is the rate at which energy is dissipated in the internal resistance of the battery.
EXECUTE: **(a)** $V_{ab} = \mathcal{E} + Ir = 12.0\text{ V} + (10.0\text{ A})(0.24\ \Omega) = 14.4\text{ V}.$

(b) $E = Pt = IVt = (10\text{ A})(14.4\text{ V})(5)(3600\text{ s}) = 2.59 \times 10^6\text{ J}.$

(c) $E_{\text{diss}} = P_{\text{diss}}t = I^2 rt = (10\text{ A})^2(0.24\ \Omega)(5)(3600\text{ s}) = 4.32 \times 10^5\text{ J}.$

(d) Discharged at 10 A: $I = \dfrac{\mathcal{E}}{r + R} \Rightarrow R = \dfrac{\mathcal{E} - Ir}{I} = \dfrac{12.0\text{ V} - (10\text{ A})(0.24\ \Omega)}{10\text{ A}} = 0.96\ \Omega.$

(e) $E = Pt = IVt = (10\text{ A})(9.6\text{ V})(5)(3600\text{ s}) = 1.73 \times 10^6\text{ J}.$

(f) Since the current through the internal resistance is the same as before, there is the same energy dissipated as in (c): $E_{\text{diss}} = 4.32 \times 10^5\text{ J}.$

EVALUATE: **(g)** Part of the energy originally supplied was stored in the battery and part was lost in the internal resistance. So the stored energy was less than what was supplied during charging. Then when discharging, even more energy is lost in the internal resistance, and only what is left is dissipated by the external resistor.

25.83. **IDENTIFY:** No current flows through the capacitor when it is fully charged.

SET UP: With the capacitor fully charged, $I = \dfrac{\mathcal{E}}{R_1 + R_2}.$ $V_R = IR$ and $V_C = Q/C.$

EXECUTE: $V_C = \dfrac{Q}{C} = \dfrac{36.0\ \mu\text{C}}{9.00\ \mu\text{F}} = 4.00\text{ V}.$ $V_{R_1} = V_C = 4.00\text{ V}$ and $I = \dfrac{V_{R_1}}{R_1} = \dfrac{4.00\text{ V}}{6.00\ \Omega} = 0.667\text{ A}.$

$V_{R_2} = IR_2 = (0.667\text{ A})(4.00\ \Omega) = 2.668\text{ V}.$ $\mathcal{E} = V_{R_1} + V_{R_2} = 4.00\text{ V} + 2.668\text{ V} = 6.67\text{ V}.$

EVALUATE: When a capacitor is fully charged, it acts like an open circuit and prevents any current from flowing though it.

DIRECT-CURRENT CIRCUITS

26.7. **IDENTIFY:** First do as much series-parallel reduction as possible.

SET UP: The 45.0-Ω and 15.0-Ω resistors are in parallel, so first reduce them to a single equivalent resistance. Then find the equivalent series resistance of the circuit.

EXECUTE: $1/R_p = 1/(45.0\ \Omega) + 1/(15.0\ \Omega)$ and $R_p = 11.25\ \Omega$. The total equivalent resistance is $18.0\ \Omega + 11.25\ \Omega + 3.26\ \Omega = 32.5\ \Omega$. Ohm's law gives $I = (25.0\ \text{V})/(32.5\ \Omega) = 0.769\ \text{A}$.

EVALUATE: The circuit appears complicated until we realize that the 45.0-Ω and 15.0-Ω resistors are in parallel.

26.11. **IDENTIFY and SET UP:** Ohm's law applies to the resistors, the potential drop across resistors in parallel is the same for each of them, and at a junction the currents in must equal the currents out.

EXECUTE: **(a)** $V_2 = I_2 R_2 = (4.00\ \text{A})(6.00\ \Omega) = 24.0\ \text{V}$. $V_1 = V_2 = 24.0\ \text{V}$.

$I_1 = \dfrac{V_1}{R_1} = \dfrac{24.0\ \text{V}}{3.00\ \Omega} = 8.00\ \text{A}$. $I_3 = I_1 + I_2 = 4.00\ \text{A} + 8.00\ \text{A} = 12.0\ \text{A}$.

(b) $V_3 = I_3 R_3 = (12.0\ \text{A})(5.00\ \Omega) = 60.0\ \text{V}$. $\varepsilon = V_1 + V_3 = 24.0\ \text{V} + 60.0\ \text{V} = 84.0\ \text{V}$.

EVALUATE: Series/parallel reduction was not necessary in this case.

26.13. **IDENTIFY:** For resistors in parallel, the voltages are the same and the currents add. $\dfrac{1}{R_{eq}} = \dfrac{1}{R_1} + \dfrac{1}{R_2}$ so

$R_{eq} = \dfrac{R_1 R_2}{R_1 + R_2}$, For resistors in series, the currents are the same and the voltages add. $R_{eq} = R_1 + R_2$.

SET UP: The rules for combining resistors in series and parallel lead to the sequences of equivalent circuits shown in Figure 26.13.

EXECUTE: $R_{eq} = 5.00\ \Omega$. In Figure 26.13c, $I = \dfrac{60.0\ \text{V}}{5.00\ \Omega} = 12.0\ \text{A}$. This is the current through each of the resistors in Figure 26.13b. $V_{12} = IR_{12} = (12.0\ \text{A})(2.00\ \Omega) = 24.0\ \text{V}$.

$V_{34} = IR_{34} = (12.0\ \text{A})(3.00\ \Omega) = 36.0\ \text{V}$. Note that $V_{12} + V_{34} = 60.0\ \text{V}$. V_{12} is the voltage across R_1 and

across R_2, so $I_1 = \dfrac{V_{12}}{R_1} = \dfrac{24.0\ \text{V}}{3.00\ \Omega} = 8.00\ \text{A}$ and $I_2 = \dfrac{V_{12}}{R_2} = \dfrac{24.0\ \text{V}}{6.00\ \Omega} = 4.00\ \text{A}$. V_{34} is the voltage across R_3

and across R_4, so $I_3 = \dfrac{V_{34}}{R_3} = \dfrac{36.0\ \text{V}}{12.0\ \Omega} = 3.00\ \text{A}$ and $I_4 = \dfrac{V_{34}}{R_4} = \dfrac{36.0\ \text{V}}{4.00\ \Omega} = 9.00\ \text{A}$.

EVALUATE: Note that $I_1 + I_2 = I_3 + I_4$.

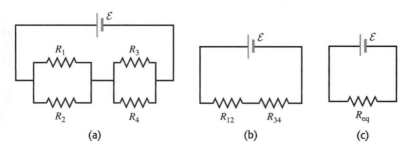

Figure 26.13

26.15.   **IDENTIFY:** In both circuits, with and without R_4, replace series and parallel combinations of resistors by their equivalents. Calculate the currents and voltages in the equivalent circuit and infer from this the currents and voltages in the original circuit. Use $P = I^2 R$ to calculate the power dissipated in each bulb.
(a) SET UP: The circuit is sketched in Figure 26.15a.

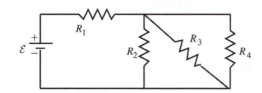

EXECUTE: $R_2, R_3,$ and R_4 are in parallel, so their equivalent resistance R_{eq} is given by

$$\frac{1}{R_{eq}} = \frac{1}{R_2} + \frac{1}{R_3} + \frac{1}{R_4}.$$

Figure 26.15a

$$\frac{1}{R_{eq}} = \frac{3}{4.50\,\Omega} \text{ and } R_{eq} = 1.50\,\Omega.$$

The equivalent circuit is drawn in Figure 26.15b.

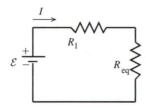

$$\varepsilon - I(R_1 + R_{eq}) = 0$$

$$I = \frac{\varepsilon}{R_1 + R_{eq}}$$

Figure 26.15b

$$I = \frac{9.00\text{ V}}{4.50\,\Omega + 1.50\,\Omega} = 1.50\text{ A and } I_1 = 1.50\text{ A}$$

Then $V_1 = I_1 R_1 = (1.50\text{ A})(4.50\,\Omega) = 6.75\text{ V}$

$I_{eq} = 1.50\text{ A}, V_{eq} = I_{eq} R_{eq} = (1.50\text{ A})(1.50\,\Omega) = 2.25\text{ V}$

For resistors in parallel the voltages are equal and are the same as the voltage across the equivalent resistor, so $V_2 = V_3 = V_4 = 2.25\text{ V}$.

$$I_2 = \frac{V_2}{R_2} = \frac{2.25\text{ V}}{4.50\,\Omega} = 0.500\text{ A}, I_3 = \frac{V_3}{R_3} = 0.500\text{ A}, I_4 = \frac{V_4}{R_4} = 0.500\text{ A}$$

EVALUATE: Note that $I_2 + I_3 + I_4 = 1.50\text{ A}$, which is I_{eq}. For resistors in parallel the currents add and their sum is the current through the equivalent resistor.
(b) SET UP: $P = I^2 R$
EXECUTE: $P_1 = (1.50\text{ A})^2 (4.50\,\Omega) = 10.1\text{ W}$

$P_2 = P_3 = P_4 = (0.500\text{ A})^2 (4.50\,\Omega) = 1.125\text{ W},$ which rounds to 1.12 W. R_1 glows brightest.

EVALUATE: Note that $P_2 + P_3 + P_4 = 3.37$ W. This equals $P_{eq} = I_{eq}^2 R_{eq} = (1.50 \text{ A})^2 (1.50 \ \Omega) = 3.37$ W, the power dissipated in the equivalent resistor.

(c) SET UP: With R_4 removed the circuit becomes the circuit in Figure 26.15c.

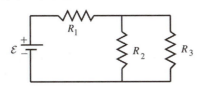

EXECUTE: R_2 and R_3 are in parallel and their equivalent resistance R_{eq} is given by

$$\frac{1}{R_{eq}} = \frac{1}{R_2} + \frac{1}{R_3} = \frac{2}{4.50 \ \Omega} \text{ and } R_{eq} = 2.25 \ \Omega.$$

Figure 26.15c

The equivalent circuit is shown in Figure 26.15d.

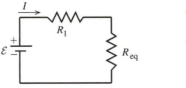

$$\varepsilon - I(R_1 + R_{eq}) = 0$$

$$I = \frac{\varepsilon}{R_1 + R_{eq}}$$

$$I = \frac{9.00 \text{ V}}{4.50 \ \Omega + 2.25 \ \Omega} = 1.333 \text{ A}$$

Figure 26.15d

$I_1 = 1.33$ A, $V_1 = I_1 R_1 = (1.333 \text{ A})(4.50 \ \Omega) = 6.00$ V

$I_{eq} = 1.33$ A, $V_{eq} = I_{eq} R_{eq} = (1.333 \text{ A})(2.25 \ \Omega) = 3.00$ V and $V_2 = V_3 = 3.00$ V.

$I_2 = \dfrac{V_2}{R_2} = \dfrac{3.00 \text{ V}}{4.50 \ \Omega} = 0.667$ A, $I_3 = \dfrac{V_3}{R_3} = 0.667$ A

(d) SET UP: $P = I^2 R$

EXECUTE: $P_1 = (1.333 \text{ A})^2 (4.50 \ \Omega) = 8.00$ W

$P_2 = P_3 = (0.667 \text{ A})^2 (4.50 \ \Omega) = 2.00$ W.

(e) EVALUATE: When R_4 is removed, P_1 decreases and P_2 and P_3 increase. Bulb R_1 glows less brightly and bulbs R_2 and R_3 glow more brightly. When R_4 is removed the equivalent resistance of the circuit increases and the current through R_1 decreases. But in the parallel combination this current divides into two equal currents rather than three, so the currents through R_2 and R_3 increase. Can also see this by noting that with R_4 removed and less current through R_1 the voltage drop across R_1 is less so the voltage drop across R_2 and across R_3 must become larger.

26.23. **IDENTIFY** and **SET UP:** Replace series and parallel combinations of resistors by their equivalents until the circuit is reduced to a single loop. Use the loop equation to find the current through the 20.0-Ω resistor. Set $P = I^2 R$ for the 20.0-Ω resistor equal to the rate Q/t at which heat goes into the water and set $Q = mc\Delta T$.

EXECUTE: Replace the network by the equivalent resistor, as shown in Figure 26.23.

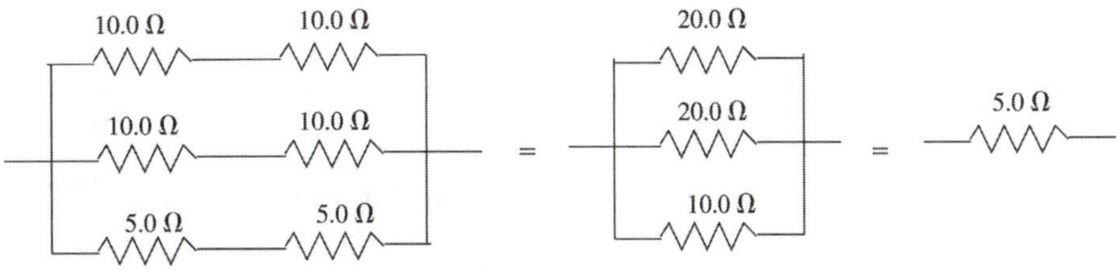

Figure 26.23

$30.0 \text{ V} - I(20.0\ \Omega + 5.0\ \Omega + 5.0\ \Omega) = 0;\ I = 1.00 \text{ A}$

For the 20.0-Ω resistor thermal energy is generated at the rate $P = I^2 R = 20.0$ W. $Q = Pt$ and $Q = mc\Delta T$

gives $t = \dfrac{mc\Delta T}{P} = \dfrac{(0.100 \text{ kg})(4190 \text{ J/kg} \cdot \text{K})(48.0 \text{ C}^\circ)}{20.0 \text{ W}} = 1.01 \times 10^3 \text{ s}$

EVALUATE: The battery is supplying heat at the rate $P = \varepsilon I = 30.0$ W. In the series circuit, more energy is dissipated in the larger resistor $(20.0\ \Omega)$ than in the smaller ones $(5.00\ \Omega)$.

26.25.   **IDENTIFY:** Apply Kirchhoff's point rule at point a to find the current through R. Apply Kirchhoff's loop rule to loops (1) and (2) shown in Figure 26.25a to calculate R and ε. Travel around each loop in the direction shown.
(a) SET UP:

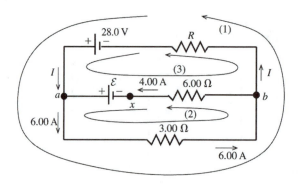

Figure 26.25a

EXECUTE: Apply Kirchhoff's point rule to point a: $\sum I = 0$ so $I + 4.00 \text{ A} - 6.00 \text{ A} = 0$
$I = 2.00$ A (in the direction shown in the diagram).
(b) Apply Kirchhoff's loop rule to loop (1): $-(6.00 \text{ A})(3.00\ \Omega) - (2.00 \text{ A})R + 28.0 \text{ V} = 0$
$-18.0 \text{ V} - (2.00\ \Omega)R + 28.0 \text{ V} = 0$

$R = \dfrac{28.0 \text{ V} - 18.0 \text{ V}}{2.00 \text{ A}} = 5.00\ \Omega$

(c) Apply Kirchhoff's loop rule to loop (2): $-(6.00 \text{ A})(3.00\ \Omega) - (4.00 \text{ A})(6.00\ \Omega) + \varepsilon = 0$
$\varepsilon = 18.0 \text{ V} + 24.0 \text{ V} = 42.0 \text{ V}$
EVALUATE: Can check that the loop rule is satisfied for loop (3), as a check of our work:
$28.0 \text{ V} - \varepsilon + (4.00 \text{ A})(6.00\ \Omega) - (2.00 \text{ A})R = 0$
$28.0 \text{ V} - 42.0 \text{ V} + 24.0 \text{ V} - (2.00 \text{ A})(5.00\ \Omega) = 0$
$52.0 \text{ V} = 42.0 \text{ V} + 10.0 \text{ V}$
$52.0 \text{ V} = 52.0 \text{ V},$ so the loop rule is satisfied for this loop.

(d) IDENTIFY: If the circuit is broken at point x there can be no current in the 6.00-Ω resistor. There is now only a single current path and we can apply the loop rule to this path.
SET UP: The circuit is sketched in Figure 26.25b.

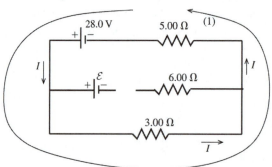

Figure 26.25b

EXECUTE: $+28.0 \text{ V} - (3.00 \text{ }\Omega)I - (5.00 \text{ }\Omega)I = 0$

$$I = \frac{28.0 \text{ V}}{8.00 \text{ }\Omega} = 3.50 \text{ A}$$

EVALUATE: Breaking the circuit at x removes the 42.0-V emf from the circuit and the current through the 3.00-Ω resistor is reduced.

26.27.   **IDENTIFY:**   Apply the junction rule at points a, b, c and d to calculate the unknown currents. Then apply the loop rule to three loops to calculate $\varepsilon_1, \varepsilon_2$ and R.

(a) **SET UP:**   The circuit is sketched in Figure 26.27.

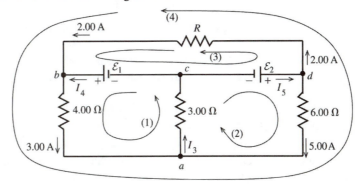

Figure 26.27

EXECUTE: Apply the junction rule to point a: $3.00 \text{ A} + 5.00 \text{ A} - I_3 = 0$

$I_3 = 8.00 \text{ A}$

Apply the junction rule to point b: $2.00 \text{ A} + I_4 - 3.00 \text{ A} = 0$

$I_4 = 1.00 \text{ A}$

Apply the junction rule to point c: $I_3 - I_4 - I_5 = 0$

$I_5 = I_3 - I_4 = 8.00 \text{ A} - 1.00 \text{ A} = 7.00 \text{ A}$

EVALUATE: As a check, apply the junction rule to point d: $I_5 - 2.00 \text{ A} - 5.00 \text{ A} = 0$

$I_5 = 7.00 \text{ A}$

(b) **EXECUTE:**   Apply the loop rule to loop (1): $\varepsilon_1 - (3.00 \text{ A})(4.00 \text{ }\Omega) - I_3(3.00 \text{ }\Omega) = 0$

$\varepsilon_1 = 12.0 \text{ V} + (8.00 \text{ A})(3.00 \text{ }\Omega) = 36.0 \text{ V}$

Apply the loop rule to loop (2): $\varepsilon_2 - (5.00 \text{ A})(6.00 \text{ }\Omega) - I_3(3.00 \text{ }\Omega) = 0$

$\varepsilon_2 = 30.0 \text{ V} + (8.00 \text{ A})(3.00 \text{ }\Omega) = 54.0 \text{ V}$

(c) Apply the loop rule to loop (3): $-(2.00 \text{ A})R - \varepsilon_1 + \varepsilon_2 = 0$

$$R = \frac{\varepsilon_2 - \varepsilon_1}{2.00 \text{ A}} = \frac{54.0 \text{ V} - 36.0 \text{ V}}{2.00 \text{ A}} = 9.00 \text{ }\Omega$$

EVALUATE: Apply the loop rule to loop (4) as a check of our calculations:

$-(2.00 \text{ A})R - (3.00 \text{ A})(4.00 \text{ }\Omega) + (5.00 \text{ A})(6.00 \text{ }\Omega) = 0$

$-(2.00 \text{ A})(9.00 \text{ }\Omega) - 12.0 \text{ V} + 30.0 \text{ V} = 0$

$-18.0 \text{ V} + 18.0 \text{ V} = 0$

26.31.   (a) **IDENTIFY:**   With the switch open, the circuit can be solved using series-parallel reduction.
SET UP: Find the current through the unknown battery using Ohm's law. Then use the equivalent resistance of the circuit to find the emf of the battery.
EXECUTE: The 30.0-Ω and 50.0-Ω resistors are in series, and hence have the same current. Using Ohm's law $I_{50} = (15.0 \text{ V})/(50.0 \text{ }\Omega) = 0.300 \text{ A} = I_{30}$. The potential drop across the 75.0-Ω resistor is the same as the potential drop across the 80.0-Ω series combination. We can use this fact to find the current through the 75.0-Ω resistor using Ohm's law: $V_{75} = V_{80} = (0.300 \text{ A})(80.0 \text{ }\Omega) = 24.0 \text{ V}$ and

$I_{75} = (24.0 \text{ V})/(75.0 \text{ }\Omega) = 0.320 \text{ A}.$

The current through the unknown battery is the sum of the two currents we just found:

$$I_{\text{Total}} = 0.300 \text{ A} + 0.320 \text{ A} = 0.620 \text{ A}$$

The equivalent resistance of the resistors in parallel is $1/R_{\text{p}} = 1/(75.0 \ \Omega) + 1/(80.0 \ \Omega)$. This gives

$R_{\text{p}} = 38.7 \ \Omega$. The equivalent resistance "seen" by the battery is $R_{\text{equiv}} = 20.0 \ \Omega + 38.7 \ \Omega = 58.7 \ \Omega$.

Applying Ohm's law to the battery gives $\varepsilon = R_{\text{equiv}} I_{\text{Total}} = (58.7 \ \Omega)(0.620 \text{ A}) = 36.4 \text{ V}$.

(b) IDENTIFY: With the switch closed, the 25.0-V battery is connected across the 50.0-Ω resistor.
SET UP: Take a loop around the right part of the circuit.
EXECUTE: Ohm's law gives $I = (25.0 \text{ V})/(50.0 \ \Omega) = 0.500 \text{ A}$.

EVALUATE: The current through the 50.0-Ω resistor, and the rest of the circuit, depends on whether or not the switch is open.

26.33.   **(a) IDENTIFY:** With the switch open, we have a series circuit with two batteries.
SET UP: Take a loop to find the current, then use Ohm's law to find the potential difference between a and b.
EXECUTE: Taking the loop: $I = (40.0 \text{ V})/(175 \ \Omega) = 0.229 \text{ A}$. The potential difference between a and b is

$V_b - V_a = +15.0 \text{ V} - (75.0 \ \Omega)(0.229 \text{ A}) = -2.14 \text{ V}$.

EVALUATE: The minus sign means that a is at a higher potential than b.
(b) IDENTIFY: With the switch closed, the ammeter part of the circuit divides the original circuit into two circuits. We can apply Kirchhoff's rules to both parts.
SET UP: Take loops around the left and right parts of the circuit, and then look at the current at the junction.
EXECUTE: The left-hand loop gives $I_{100} = (25.0 \text{ V})/(100.0 \ \Omega) = 0.250 \text{ A}$. The right-hand loop gives

$I_{75} = (15.0 \text{ V})/(75.0 \ \Omega) = 0.200 \text{ A}$. At the junction just above the switch we have $I_{100} = 0.250 \text{ A}$ (in) and

$I_{75} = 0.200 \text{ A}$ (out), so $I_A = 0.250 \text{ A} - 0.200 \text{ A} = 0.050 \text{ A}$, downward. The voltmeter reads zero because the potential difference across it is zero with the switch closed.
EVALUATE: The ideal ammeter acts like a short circuit, making a and b at the same potential. Hence the voltmeter reads zero.

26.37.   **IDENTIFY:** The meter introduces resistance into the circuit, which affects the current through the 5.00-kΩ resistor and hence the potential drop across it.
SET UP: Use Ohm's law to find the current through the 5.00-kΩ resistor and then the potential drop across it.
EXECUTE: **(a)** The parallel resistance with the voltmeter is 3.33 kΩ, so the total equivalent resistance across the battery is 9.33 kΩ, giving $I = (50.0 \text{ V})/(9.33 \text{ k}\Omega) = 5.36 \text{ mA}$. Ohm's law gives the potential

drop across the 5.00-kΩ resistor: $V_{5 \text{ k}\Omega} = (3.33 \text{ k}\Omega)(5.36 \text{ mA}) = 17.9 \text{ V}$

(b) The current in the circuit is now $I = (50.0 \text{ V})/(11.0 \text{ k}\Omega) = 4.55 \text{ mA}$.

$V_{5 \text{ k}\Omega} = (5.00 \text{ k}\Omega)(4.55 \text{ mA}) = 22.7 \text{ V}$.

(c) % error $= (22.7 \text{ V} - 17.9 \text{ V})/(22.7 \text{ V}) = 0.214 = 21.4\%$. (We carried extra decimal places for accuracy since we had to subtract our answers.)
EVALUATE: The presence of the meter made a very large percent error in the reading of the "true" potential across the resistor.

26.39.   **IDENTIFY:** Apply $\varepsilon = IR_{\text{total}}$ to relate the resistance R_x to the current in the circuit.

SET UP: R, R_x and the meter are in series, so $R_{\text{total}} = R + R_x + R_M$, where $R_M = 65.0 \ \Omega$ is the resistance

of the meter. $I_{\text{fsd}} = 2.50 \text{ mA}$ is the current required for full-scale deflection.

EXECUTE: **(a)** When the wires are shorted, the full-scale deflection current is obtained: $\varepsilon = IR_{\text{total}}$.

$1.52 \text{ V} = (2.50 \times 10^{-3} \text{ A})(65.0 \ \Omega + R)$ and $R = 543 \ \Omega$.

(b) If the resistance $R_x = 200 \ \Omega$: $I = \dfrac{V}{R_{\text{total}}} = \dfrac{1.52 \text{ V}}{65.0 \ \Omega + 543 \ \Omega + R_x} = 1.88 \text{ mA}$.

(c) $I_x = \dfrac{\varepsilon}{R_{\text{total}}} = \dfrac{1.52 \text{ V}}{65.0 \ \Omega + 543 \ \Omega + R_x}$ and $R_x = \dfrac{1.52 \text{ V}}{I_x} - 608 \ \Omega.$ For each value of I_x we have:

For $I_x = \dfrac{1}{4} I_{\text{fsd}} = 6.25 \times 10^{-4}$ A, $R_x = \dfrac{1.52 \text{ V}}{6.25 \times 10^{-4} \text{ A}} - 608 \ \Omega = 1824 \ \Omega.$

For $I_x = \dfrac{1}{2} I_{\text{fsd}} = 1.25 \times 10^{-3}$ A, $R_x = \dfrac{1.52 \text{ V}}{1.25 \times 10^{-3} \text{ A}} - 608 \ \Omega = 608 \ \Omega.$

For $I_x = \dfrac{3}{4} I_{\text{fsd}} = 1.875 \times 10^{-3}$ A, $R_x = \dfrac{1.52 \text{ V}}{1.875 \times 10^{-3} \text{ A}} - 608 \ \Omega = 203 \ \Omega.$

EVALUATE: The deflection of the meter increases when the resistance R_x decreases.

26.43. **IDENTIFY:** The capacitors, which are in parallel, will discharge exponentially through the resistors.
SET UP: Since V is proportional to Q, V must obey the same exponential equation as Q, $V = V_0 e^{-t/RC}$. The current is $I = (V_0/R) e^{-t/RC}$.
EXECUTE: **(a)** Solve for time when the potential across each capacitor is 10.0 V:

$$t = -RC \ln(V/V_0) = -(80.0 \ \Omega)(35.0 \ \mu\text{F}) \ln(10/45) = 4210 \ \mu\text{s} = 4.21 \text{ ms}$$

(b) $I = (V_0/R) e^{-t/RC}$. Using the above values, with $V_0 = 45.0$ V, gives $I = 0.125$ A.

EVALUATE: Since the current and the potential both obey the same exponential equation, they are both reduced by the same factor (0.222) in 4.21 ms.

26.45. **IDENTIFY and SET UP:** Apply the loop rule. The voltage across the resistor depends on the current through it and the voltage across the capacitor depends on the charge on its plates.
EXECUTE: $\varepsilon - V_R - V_C = 0$

$\varepsilon = 120$ V, $V_R = IR = (0.900 \text{ A})(80.0 \ \Omega) = 72$ V, so $V_C = 48$ V

$Q = CV = (4.00 \times 10^{-6} \text{ F})(48 \text{ V}) = 192 \ \mu\text{C}$

EVALUATE: The initial charge is zero and the final charge is $C\varepsilon = 480 \ \mu\text{C}$. Since current is flowing at the instant considered in the problem the capacitor is still being charged and its charge has not reached its final value.

26.47. **IDENTIFY:** The stored energy is proportional to the square of the charge on the capacitor, so it will obey an exponential equation, but not the same equation as the charge.
SET UP: The energy stored in the capacitor is $U = Q^2/2C$ and the charge on the plates is $Q_0 e^{-t/RC}$. The current is $I = I_0 e^{-t/RC}$.
EXECUTE: $U = Q^2/2C = (Q_0 e^{-t/RC})^2/2C = U_0 e^{-2t/RC}$. When the capacitor has lost 80% of its stored energy, the energy is 20% of the initial energy, which is $U_0/5$. $U_0/5 = U_0 \ e^{-2t/RC}$ gives

$t = (RC/2) \ln 5 = (25.0 \ \Omega)(4.62 \text{ pF})(\ln 5)/2 = 92.9 \text{ ps}.$

At this time, the current is $I = I_0 \ e^{-t/RC} = (Q_0/RC) e^{-t/RC}$, so

$$I = (3.5 \text{ nC})/[(25.0 \ \Omega)(4.62 \text{ pF})] \ e^{-(92.9 \text{ ps})/[(25.0 \ \Omega)(4.62 \text{ pF})]} = 13.6 \text{ A}.$$

EVALUATE: When the energy is reduced by 80%, neither the current nor the charge are reduced by that percent.

26.49. **IDENTIFY:** In both cases, simplify the complicated circuit by eliminating the appropriate circuit elements. The potential across an uncharged capacitor is initially zero, so it behaves like a short circuit. A fully charged capacitor allows no current to flow through it.
(a) SET UP: Just after closing the switch, the uncharged capacitors all behave like short circuits, so any resistors in parallel with them are eliminated from the circuit.
EXECUTE: The equivalent circuit consists of 50 Ω and 25 Ω in parallel, with this combination in series with 75 Ω, 15 Ω, and the 100-V battery. The equivalent resistance is $90 \ \Omega + 16.7 \ \Omega = 106.7 \ \Omega$, which gives $I = (100\text{V})/(106.\Omega) = 0.937$ A.

(b) SET UP: Long after closing the switch, the capacitors are essentially charged up and behave like open circuits since no charge can flow through them. They effectively eliminate any resistors in series with them since no current can flow through these resistors.

EXECUTE: The equivalent circuit consists of resistances of 75 Ω, 15 Ω and three 25-Ω resistors, all in series with the 100-V battery, for a total resistance of 165 Ω. Therefore $I = (100\text{V})/(165\ \Omega) = 0.606$ A

EVALUATE: The initial and final behavior of the circuit can be calculated quite easily using simple series-parallel circuit analysis. Intermediate times would require much more difficult calculations!

26.51. **IDENTIFY:** For each circuit apply the loop rule to relate the voltages across the circuit elements.

(a) SET UP: With the switch in position 2 the circuit is the charging circuit shown in Figure 26.51a.

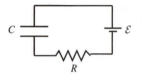

At $t = 0, q = 0$.

Figure 26.51a

EXECUTE: The charge q on the capacitor is given as a function of time by Eq. (26.12):

$q = C\varepsilon\left(1 - e^{-t/RC}\right)$

$Q_\text{f} = C\varepsilon = (1.50\times10^{-5}\ \text{F})(18.0\ \text{V}) = 2.70\times10^{-4}\ \text{C}$.

$RC = (980\ \Omega)(1.50\times10^{-5}\ \text{F}) = 0.0147$ s

Thus, at $t = 0.0100$ s, $q = (2.70\times10^{-4}\ \text{C})(1 - e^{-(0.0100\ \text{s})/(0.0147\ \text{s})}) = 133\ \mu\text{C}$.

(b) $v_C = \dfrac{q}{C} = \dfrac{133\ \mu\text{C}}{1.50\times10^{-5}\ \text{F}} = 8.87$ V

The loop rule says $\varepsilon - v_C - v_R = 0$.

$v_R = \varepsilon - v_C = 18.0\ \text{V} - 8.87\ \text{V} = 9.13$ V

(c) SET UP: Throwing the switch back to position 1 produces the discharging circuit shown in Figure 26.51b.

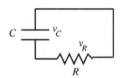

The initial charge Q_0 is the charge calculated in part (b), $Q_0 = 133\ \mu\text{C}$.

Figure 26.51b

EXECUTE: $v_C = \dfrac{q}{C} = \dfrac{133\ \mu\text{C}}{1.50\times10^{-5}\ \text{F}} = 8.87$ V, the same as just before the switch is thrown. But now

$v_C - v_R = 0$, so $v_R = v_C = 8.87$ V.

(d) SET UP: In the discharging circuit the charge on the capacitor as a function of time is given by Eq. (26.16): $q = Q_0 e^{-t/RC}$.

EXECUTE: $RC = 0.0147$ s, the same as in part (a). Thus at

$t = 0.0100$ s, $q = (133\ \mu\text{C})e^{-((0.0100\ \text{s})/(0.0147\ \text{s}))} = 67.4\ \mu\text{C}$.

EVALUATE: $t = 10.0$ ms is less than one time constant, so at the instant described in part (a) the capacitor is not fully charged; its voltage (8.87 V) is less than the emf. There is a charging current and a voltage drop across the resistor. In the discharging circuit the voltage across the capacitor starts at 8.87 V and decreases. After $t = 10.0$ ms it has decreased to $v_C = q/C = 4.49$ V.

26.53. **IDENTIFY and SET UP:** The heater and hair dryer are in parallel so the voltage across each is 120 V and the current through the fuse is the sum of the currents through each appliance. As the power consumed by the dryer increases, the current through it increases. The maximum power setting is the highest one for which the current through the fuse is less than 20 A.

EXECUTE: Find the current through the heater. $P = VI$ so $I = P/V = 1500\ \text{W}/120\ \text{V} = 12.5$ A. The maximum total current allowed is 20 A, so the current through the dryer must be less than $20\ \text{A} - 12.5\ \text{A} = 7.5$ A. The power dissipated by the dryer if the current has this value is $P = VI = (120\ \text{V})(7.5\ \text{A}) = 900$ W. For P at this value or larger the circuit breaker trips.

EVALUATE: $P = V^2/R$ and for the dryer V is a constant 120 V. The higher power settings correspond to a smaller resistance R and larger current through the device.

26.55. **IDENTIFY:** We need to do series/parallel reduction to solve this circuit.

SET UP: $P = \dfrac{\varepsilon^2}{R}$, where R is the equivalent resistance of the network. For resistors in series, $R_{eq} = R_1 + R_2$, and for resistors in parallel $1/R_p = 1/R_1 + 1/R_2$.

EXECUTE: $R = \dfrac{\varepsilon^2}{P} = \dfrac{(48.0 \text{ V})^2}{295 \text{ W}} = 7.810\,\Omega.$ $R_{12} = R_1 + R_2 = 8.00\,\Omega.$ $R = R_{123} + R_4.$

$R_{123} = R - R_4 = 7.810\,\Omega - 3.00\,\Omega = 4.810\,\Omega.$ $\dfrac{1}{R_{12}} + \dfrac{1}{R_3} = \dfrac{1}{R_{123}}.$ $\dfrac{1}{R_3} = \dfrac{1}{R_{123}} - \dfrac{1}{R_{12}} = \dfrac{R_{12} - R_{123}}{R_{123}R_{12}}.$

$R_3 = \dfrac{R_{123}R_{12}}{R_{12} - R_{123}} = \dfrac{(4.810\,\Omega)(8.00\,\Omega)}{8.00\,\Omega - 4.810\,\Omega} = 12.1\,\Omega.$

EVALUATE: The resistance R_3 is greater than R, since the equivalent parallel resistance is less than any of the resistors in parallel.

26.57. **IDENTIFY:** The Cu and Ni cables are in parallel. For each, $R = \dfrac{\rho L}{A}.$

SET UP: The composite cable is sketched in Figure 26.57. The cross-sectional area of the nickel segment is πa^2 and the area of the copper portion is $\pi(b^2 - a^2).$ For nickel $\rho = 7.8 \times 10^{-8}\,\Omega \cdot \text{m}.$ and for copper $\rho = 1.72 \times 10^{-8}\,\Omega \cdot \text{m}.$

EXECUTE: **(a)** $\dfrac{1}{R_{\text{Cable}}} = \dfrac{1}{R_{\text{Ni}}} + \dfrac{1}{R_{\text{Cu}}}.$ $R_{\text{Ni}} = \rho_{\text{Ni}} L/A = \rho_{\text{Ni}} \dfrac{L}{\pi a^2}$ and $R_{\text{Cu}} = \rho_{\text{Cu}}\, L/A = \rho_{\text{Cu}} \dfrac{L}{\pi(b^2 - a^2)}.$

Therefore, $\dfrac{1}{R_{\text{cable}}} = \dfrac{\pi a^2}{\rho_{\text{Ni}} L} + \dfrac{\pi(b^2 - a^2)}{\rho_{\text{Cu}} L}.$

$\dfrac{1}{R_{\text{cable}}} = \dfrac{\pi}{L}\left(\dfrac{a^2}{\rho_{\text{Ni}}} + \dfrac{b^2 - a^2}{\rho_{\text{Cu}}} \right) = \dfrac{\pi}{20 \text{ m}}\left[\dfrac{(0.050 \text{ m})^2}{7.8 \times 10^{-8}\,\Omega \cdot \text{m}} + \dfrac{(0.100 \text{ m})^2 - (0.050 \text{ m})^2}{1.72 \times 10^{-8}\,\Omega \cdot \text{m}} \right]$ and

$R_{\text{Cable}} = 13.6 \times 10^{-6}\,\Omega = 13.6\,\mu\Omega.$

(b) $R = \rho_{\text{eff}} \dfrac{L}{A} = \rho_{\text{eff}} \dfrac{L}{\pi b^2}.$ This gives $\rho_{\text{eff}} = \dfrac{\pi b^2 R}{L} = \dfrac{\pi(0.10 \text{ m})^2 (13.6 \times 10^{-6}\,\Omega)}{20 \text{ m}} = 2.14 \times 10^{-8}\,\Omega \cdot \text{m}$

EVALUATE: The effective resistivity of the cable is about 25% larger than the resistivity of copper. If nickel had infinite resitivity and only the copper portion conducted, the resistance of the cable would be 14.6 $\mu\Omega$, which is not much larger than the resistance calculated in part (a).

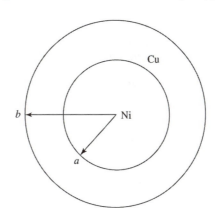

Figure 26.57

26.59.   **IDENTIFY:**   The terminal voltage of the battery depends on the current through it and therefore on the equivalent resistance connected to it. The power delivered to each bulb is $P = I^2R$, where I is the current through it.

SET UP: The terminal voltage of the source is $\varepsilon - Ir$.

EXECUTE: **(a)** The equivalent resistance of the two bulbs is $1.0\ \Omega$. This equivalent resistance is in series with the internal resistance of the source, so the current through the battery is

$$I = \frac{V}{R_{\text{total}}} = \frac{8.0\ \text{V}}{1.0\ \Omega + 0.80\ \Omega} = 4.4\ \text{A}.$$ and the current through each bulb is 2.2 A. The voltage applied to

each bulb is $\varepsilon - Ir = 8.0\ \text{V} - (4.4\ \text{A})(0.80\ \Omega) = 4.4\ \text{V}$. Therefore, $P_{\text{bulb}} = I^2R = (2.2\ \text{A})^2(2.0\ \Omega) = 9.7\ \text{W}$.

(b) If one bulb burns out, then $I = \dfrac{V}{R_{\text{total}}} = \dfrac{8.0\ \text{V}}{2.0\ \Omega + 0.80\ \Omega} = 2.9\ \text{A}$. The current through the remaining bulb

is 2.9 A, and $P = I^2R = (2.9\ \text{A})^2(2.0\ \Omega) = 16.3\ \text{W}$. The remaining bulb is brighter than before, because it is consuming more power.

EVALUATE: In Example 26.2 the internal resistance of the source is negligible and the brightness of the remaining bulb doesn't change when one burns out.

26.61.   **IDENTIFY:**   The ohmmeter reads the equivalent resistance between points a and b. Replace series and parallel combinations by their equivalent.

SET UP: For resistors in parallel, $\dfrac{1}{R_{\text{eq}}} = \dfrac{1}{R_1} + \dfrac{1}{R_2}$. For resistors in series, $R_{\text{eq}} = R_1 + R_2$.

EXECUTE: Circuit (a): The $75.0\text{-}\Omega$ and $40.0\text{-}\Omega$ resistors are in parallel and have equivalent resistance $26.09\ \Omega$. The $25.0\text{-}\Omega$ and $50.0\text{-}\Omega$ resistors are in parallel and have an equivalent resistance of $16.67\ \Omega$.

The equivalent network is given in Figure 26.61a. $\dfrac{1}{R_{\text{eq}}} = \dfrac{1}{100.0\ \Omega} + \dfrac{1}{23.05\ \Omega}$, so $R_{\text{eq}} = 18.7\ \Omega$.

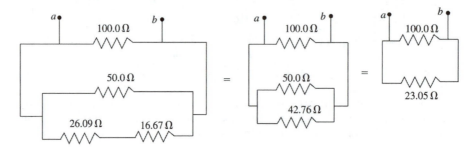

Figure 26.61a

Circuit (b): The $30.0\text{-}\Omega$ and $45.0\text{-}\Omega$ resistors are in parallel and have equivalent resistance $18.0\ \Omega$. The

equivalent network is given in Figure 26.61b. $\dfrac{1}{R_{\text{eq}}} = \dfrac{1}{10.0\ \Omega} + \dfrac{1}{30.3\ \Omega}$, so $R_{\text{eq}} = 7.5\ \Omega$.

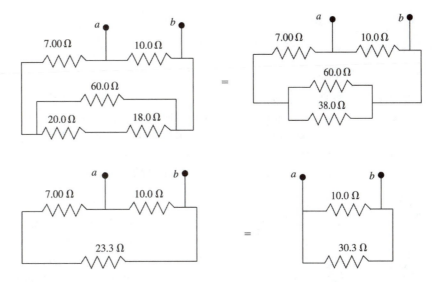

Figure 26.61b

EVALUATE: In circuit (a) the resistance along one path between a and b is $100.0\ \Omega$, but that is not the equivalent resistance between these points. A similar comment can be made about circuit (b).

26.63.   **IDENTIFY:**   Apply the junction rule to express the currents through the 5.00-Ω and 8.00-Ω resistors in terms of I_1, I_2 and I_3. Apply the loop rule to three loops to get three equations in the three unknown currents.

SET UP: The circuit is sketched in Figure 26.63.

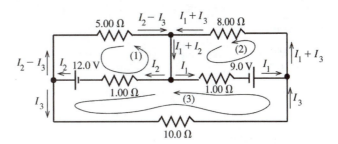

Figure 26.63

The current in each branch has been written in terms of I_1, I_2 and I_3 such that the junction rule is satisfied at each junction point.

EXECUTE: Apply the loop rule to loop (1).

$-12.0\text{ V} + I_2(1.00\ \Omega) + (I_2 - I_3)(5.00\ \Omega) = 0$

$I_2(6.00\ \Omega) - I_3(5.00\ \Omega) = 12.0\text{ V}$ eq. (1)

Apply the loop rule to loop (2).

$-I_1(1.00\ \Omega) + 9.00\text{ V} - (I_1 + I_3)(8.00\ \Omega) = 0$

$I_1(9.00\ \Omega) + I_3(8.00\ \Omega) = 9.00\text{ V}$ eq. (2)

Apply the loop rule to loop (3).

$-I_3(10.0\ \Omega) - 9.00\text{ V} + I_1(1.00\ \Omega) - I_2(1.00\ \Omega) + 12.0\text{ V} = 0$

$-I_1(1.00\ \Omega) + I_2(1.00\ \Omega) + I_3(10.0\ \Omega) = 3.00\text{ V}$ eq. (3)

Eq. (1) gives $I_2 = 2.00\text{ A} + \frac{5}{6}I_3$; eq. (2) gives $I_1 = 1.00\text{ A} - \frac{8}{9}I_3$

Using these results in eq. (3) gives $-(1.00 \text{ A} - \frac{8}{9}I_3)(1.00 \, \Omega) + (2.00 \text{ A} + \frac{5}{6}I_3)(1.00 \, \Omega) + I_3(10.0 \, \Omega) = 3.00$ V

$(\frac{16+15+180}{18})I_3 = 2.00$ A; $I_3 = \frac{18}{211}(2.00 \text{ A}) = 0.171$ A

Then $I_2 = 2.00 \text{ A} + \frac{5}{6}I_3 = 2.00 \text{ A} + \frac{5}{6}(0.171 \text{ A}) = 2.14$ A and

$I_1 = 1.00 \text{ A} - \frac{8}{9}I_3 = 1.00 \text{ A} - \frac{8}{9}(0.171 \text{ A}) = 0.848$ A.

EVALUATE: We could check that the loop rule is satisfied for a loop that goes through the 5.00-Ω, 8.00-Ω and 10.0-Ω resistors. Going around the loop clockwise: $-(I_2 - I_3)(5.00 \, \Omega) +$ $(I_1 + I_3)(8.00 \, \Omega) + I_3(10.0 \, \Omega) = -9.85 \text{ V} + 8.15 \text{ V} + 1.71 \text{ V},$ which does equal zero, apart from rounding.

26.65.   **IDENTIFY** and **SET UP:**   The circuit is sketched in Figure 26.65.

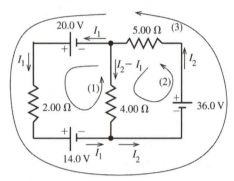

Two unknown currents I_1 (through the 2.00-Ω resistor) and I_2 (through the 5.00-Ω resistor) are labeled on the circuit diagram. The current through the 4.00-Ω resistor has been written as $I_2 - I_1$ using the junction rule.

Figure 26.65

Apply the loop rule to loops (1) and (2) to get two equations for the unknown currents, I_1 and I_2. Loop (3) can then be used to check the results.

EXECUTE: loop (1): $+20.0 \text{ V} - I_1(2.00 \, \Omega) - 14.0 \text{ V} + (I_2 - I_1)(4.00 \, \Omega) = 0$

$6.00I_1 - 4.00I_2 = 6.00$ A

$3.00I_1 - 2.00I_2 = 3.00$ A   eq. (1)

loop (2): $+36.0 \text{ V} - I_2(5.00 \, \Omega) - (I_2 - I_1)(4.00 \, \Omega) = 0$

$-4.00I_1 + 9.00I_2 = 36.0$ A   eq. (2)

Solving eq. (1) for I_1 gives $I_1 = 1.00 \text{ A} + \frac{2}{3}I_2$

Using this in eq. (2) gives $-4.00(1.00 \text{ A} + \frac{2}{3}I_2) + 9.00I_2 = 36.0$ A

$(-\frac{8}{3} + 9.00)I_2 = 40.0$ A and $I_2 = 6.32$ A.

Then $I_1 = 1.00 \text{ A} + \frac{2}{3}I_2 = 1.00 \text{ A} + \frac{2}{3}(6.32 \text{ A}) = 5.21$ A.

In summary then

Current through the 2.00-Ω resistor: $I_1 = 5.21$ A.

Current through the 5.00-Ω resistor: $I_2 = 6.32$ A.

Current through the 4.00-Ω resistor: $I_2 - I_1 = 6.32 \text{ A} - 5.21 \text{ A} = 1.11$ A.

EVALUATE: Use loop (3) to check. $+20.0 \text{ V} - I_1(2.00 \, \Omega) - 14.0 \text{ V} + 36.0 \text{ V} - I_2(5.00 \, \Omega) = 0$

$(5.21 \text{ A})(2.00 \, \Omega) + (6.32 \text{ A})(5.00 \, \Omega) = 42.0$ V

$10.4 \text{ V} + 31.6 \text{ V} = 42.0 \text{ V},$ so the loop rule is satisfied for this loop.

26.67.   **(a) IDENTIFY:** Break the circuit between points a and b means no current in the middle branch that contains the 3.00-Ω resistor and the 10.0-V battery. The circuit therefore has a single current path. Find the current, so that potential drops across the resistors can be calculated. Calculate V_{ab} by traveling from a to b, keeping track of the potential changes along the path taken.

SET UP: The circuit is sketched in Figure 26.67a.

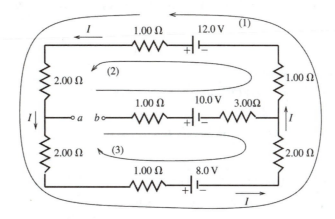

Figure 26.67a

EXECUTE: Apply the loop rule to loop (1).

$+12.0 \text{ V} - I(1.00 \, \Omega + 2.00 \, \Omega + 2.00 \, \Omega + 1.00 \, \Omega) - 8.0 \text{ V} - I(2.00 \, \Omega + 1.00 \, \Omega) = 0$

$I = \dfrac{12.0 \text{ V} - 8.0 \text{ V}}{9.00 \, \Omega} = 0.4444 \text{ A}.$

To find V_{ab} start at point b and travel to a, adding up the potential rises and drops. Travel on path (2) shown on the diagram. The 1.00-Ω and 3.00-Ω resistors in the middle branch have no current through them and hence no voltage across them. Therefore,

$V_b - 10.0 \text{ V} + 12.0 \text{ V} - I(1.00 \, \Omega + 1.00 \, \Omega + 2.00 \, \Omega) = V_a;$ thus

$V_a - V_b = 2.0 \text{ V} - (0.4444 \text{ A})(4.00 \, \Omega) = +0.22 \text{ V}$ (point a is at higher potential)

EVALUATE: As a check on this calculation we also compute V_{ab} by traveling from b to a on path (3).

$V_b - 10.0 \text{ V} + 8.0 \text{ V} + I(2.00 \, \Omega + 1.00 \, \Omega + 2.00 \, \Omega) = V_a$

$V_{ab} = -2.00 \text{ V} + (0.4444 \text{ A})(5.00 \, \Omega) = +0.22 \text{ V},$ which checks.

(b) IDENTIFY and **SET UP:**   With points a and b connected by a wire there are three current branches, as shown in Figure 26.67b.

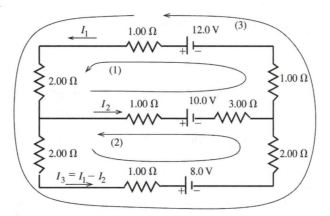

Figure 26.67b

The junction rule has been used to write the third current (in the 8.0-V battery) in terms of the other currents. Apply the loop rule to loops (1) and (2) to obtain two equations for the two unknowns I_1 and I_2.

EXECUTE: Apply the loop rule to loop (1).

$12.0 \text{ V} - I_1(1.00 \,\Omega) - I_1(2.00 \,\Omega) - I_2(1.00 \,\Omega) - 10.0 \text{ V} - I_2(3.00 \,\Omega) - I_1(1.00 \,\Omega) = 0$

$2.0 \text{ V} - I_1(4.00 \,\Omega) - I_2(4.00 \,\Omega) = 0$

$(2.00 \,\Omega)I_1 + (2.00 \,\Omega)I_2 = 1.0 \text{ V}$ eq. (1)

Apply the loop rule to loop (2).

$-(I_1 - I_2)(2.00 \,\Omega) - (I_1 - I_2)(1.00 \,\Omega) - 8.0 \text{ V} - (I_1 - I_2)(2.00 \,\Omega) + I_2(3.00 \,\Omega) + 10.0 \text{ V} + I_2(1.00 \,\Omega) = 0$

$2.0 \text{ V} - (5.00 \,\Omega)I_1 + (9.00 \,\Omega)I_2 = 0$ eq. (2)

Solve eq. (1) for I_2 and use this to replace I_2 in eq. (2).

$I_2 = 0.50 \text{ A} - I_1$

$2.0 \text{ V} - (5.00 \,\Omega)I_1 + (9.00 \,\Omega)(0.50 \text{ A} - I_1) = 0$

$(14.0 \,\Omega)I_1 = 6.50 \text{ V}$ so $I_1 = (6.50 \text{ V})/(14.0 \,\Omega) = 0.464 \text{ A}$.

$I_2 = 0.500 \text{ A} - 0.464 \text{ A} = 0.036 \text{ A}$.

The current in the 12.0-V battery is $I_1 = 0.464 \text{ A}$.

EVALUATE: We can apply the loop rule to loop (3) as a check.

$+12.0 \text{ V} - I_1(1.00 \,\Omega + 2.00 \,\Omega + 1.00 \,\Omega) - (I_1 - I_2)(2.00 \,\Omega + 1.00 \,\Omega + 2.00 \,\Omega) - 8.0 \text{ V} = 4.0 \text{ V} - 1.86 \text{ V} - 2.14 \text{ V} = 0,$ as it should.

26.69.   **IDENTIFY:** In one case, the copper and aluminum lengths are in parallel, while in the other case they are in series.

SET UP: $R = \dfrac{\rho L}{A}$. Table 25.1 in the text gives the resistivities of copper and aluminum to be

$\rho_c = 1.72 \times 10^{-8} \,\Omega \cdot \text{m}$ and $\rho_a = 2.75 \times 10^{-8} \,\Omega \cdot \text{m}$. For the cables in series (end-to-end), $R_{eq} = R_c + R_a$.

For the cables in parallel the equivalent resistance R_{eq} is given by $\dfrac{1}{R_{eq}} = \dfrac{1}{R_c} + \dfrac{1}{R_a}$. Note that in the two

configurations the copper and aluminum sections have different lengths. And, for the parallel cables the cross-sectional area of each cable is half what it is for the end-to-end configuration.

EXECUTE: *End-to-end:* $L = 0.50 \times 10^3$ m for each cable.

$R_c = \dfrac{\rho_c L}{A} = \dfrac{(1.72 \times 10^{-8} \,\Omega \cdot \text{m})(0.50 \times 10^3 \text{ m})}{0.500 \times 10^{-4} \text{ m}^2} = 0.172 \,\Omega.$

$R_a = \dfrac{\rho_a L}{A} = \dfrac{(2.75 \times 10^{-8} \,\Omega \cdot \text{m})(0.50 \times 10^3 \text{ m})}{0.500 \times 10^{-4} \text{ m}^2} = 0.275 \,\Omega.$

$R_{eq} = 0.172 \,\Omega + 0.275 \,\Omega = 0.447 \,\Omega.$

In parallel: Now $L = 1.00 \times 10^3$ m for each cable. L is doubled and A is halved compared to the other configuration, so $R_c = 4(0.172 \,\Omega) = 0.688 \,\Omega$ and $R_a = 4(0.275 \,\Omega) = 1.10 \,\Omega$.

$\dfrac{1}{R_{eq}} = \dfrac{1}{R_c} + \dfrac{1}{R_a} = \dfrac{1}{0.688 \,\Omega} + \dfrac{1}{1.10 \,\Omega}$ and $R_{eq} = 0.423 \,\Omega$. The least resistance is for the cables in parallel.

EVALUATE: The parallel combination has less equivalent resistance even though both cables contain the same volume of each metal.

26.71.   **IDENTIFY** and **SET UP:** Simplify the circuit by replacing the parallel networks of resistors by their equivalents. In this simplified circuit apply the loop and junction rules to find the current in each branch.
EXECUTE: The 20.0-Ω and 30.0-Ω resistors are in parallel and have equivalent resistance 12.0 Ω. The two resistors R are in parallel and have equivalent resistance $R/2$. The circuit is equivalent to the circuit sketched in Figure 26.71.

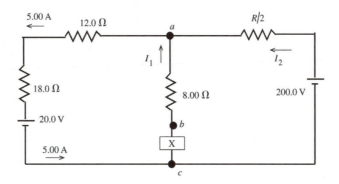

Figure 26.71

(a) Calculate V_{ca} by traveling along the branch that contains the 20.0-V battery, since we know the current in that branch.

$V_a - (5.00 \text{ A})(12.0\,\Omega) - (5.00 \text{ A})(18.0\,\Omega) - 20.0 \text{ V} = V_c$

$V_a - V_c = 20.0 \text{ V} + 90.0 \text{ V} + 60.0 \text{ V} = 170.0 \text{ V}$

$V_b - V_a = V_{ab} = 16.0 \text{ V}$

$X - V_{ba} = 170.0 \text{ V}$ so $X = 186.0 \text{ V}$, with the upper terminal $+$

(b) $I_1 = (16.0 \text{ V}) / (8.0\,\Omega) = 2.00 \text{ A}$

The junction rule applied to point a gives $I_2 + I_1 = 5.00 \text{ A}$, so $I_2 = 3.00 \text{ A}$. The current through the 200.0-V battery is in the direction from the $-$ to the $+$ terminal, as shown in the diagram.

(c) $200.0 \text{ V} - I_2(R/2) = 170.0 \text{ V}$

$(3.00 \text{ A})(R/2) = 30.0 \text{ V}$ so $R = 20.0\,\Omega$

EVALUATE: We can check the loop rule by going clockwise around the outer circuit loop. This gives $+20.0 \text{ V} + (5.00 \text{ A})(18.0\,\Omega + 12.0\,\Omega) + (3.00 \text{ A})(10.0\,\Omega) - 200.0 \text{ V} = 20.0 \text{ V} + 150.0 \text{ V} + 30.0 \text{ V} - 200.0 \text{ V}$, which does equal zero.

26.75. **IDENTIFY:** An initially uncharged capacitor is charged up by an emf source. The current in the circuit and the charge on the capacitor both obey exponential equations.

SET UP: $U_C = \dfrac{q^2}{2C}$, $P_R = i^2 R$, $q = Q_f(1 - e^{-t/RC})$, and $i = I_0 e^{-t/RC}$.

EXECUTE: **(a)** Initially, $q = 0$ so $V_R = \varepsilon$ and $I = \dfrac{\varepsilon}{R} = \dfrac{90.0 \text{ V}}{6.00 \times 10^3\,\Omega} = 0.0150 \text{ A}$. $P_R = I^2 R = 1.35 \text{ W}$.

(b) $U_C = \dfrac{q^2}{2C}$. $P_C = \dfrac{dU_C}{dt} = \dfrac{qi}{C}$. $P_R = i^2 R$. $P_C = P_R$ gives $\dfrac{qi}{C} = i^2 R$. $\dfrac{q}{RC} = i$.

$q = Q_f(1 - e^{-t/RC}) = \varepsilon C(1 - e^{-t/RC})$. $i = I_0 e^{-t/RC} = \dfrac{\varepsilon}{R}e^{-t/RC}$. $i = \dfrac{q}{RC}$ gives

$\dfrac{\varepsilon}{R}e^{-t/RC} = \dfrac{\varepsilon C}{RC}(1 - e^{-t/RC})$. $e^{-t/RC} = 1 - e^{-t/RC}$ and $e^{t/RC} = 2$.

$t = RC \ln 2 = (6.00 \times 10^3\,\Omega)(2.00 \times 10^{-6} \text{ F}) \ln 2 = 8.31 \times 10^{-3} \text{ s} = 8.31 \text{ ms}$.

(c) $i = \dfrac{\varepsilon}{R}e^{-t/RC} = \dfrac{90.0 \text{ V}}{6.00 \times 10^3\,\Omega}e^{-(8.318 \times 10^{-3} \text{ s})/[(6.00 \times 10^3\,\Omega)(2.00 \times 10^{-6} \text{ F})]} = 7.50 \times 10^{-3} \text{ A}$.

$P_R = i^2 R = (7.50 \times 10^{-3} \text{ A})^2 (6.00 \times 10^3\,\Omega) = 0.337 \text{ W}$.

EVALUATE: Initially energy is dissipated in the resistor at a higher rate because the current is high, but as time goes by the current deceases, as does the power dissipated in the resistor.

26.77.   **(a) IDENTIFY** and **SET UP:**   The circuit is sketched in Figure 26.77a.

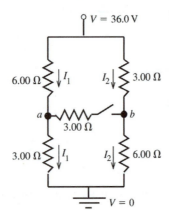

With the switch open there is no current through it and there are only the two currents I_1 and I_2 indicated in the sketch.

Figure 26.77a

The potential drop across each parallel branch is 36.0 V. Use this fact to calculate I_1 and I_2. Then travel from point a to point b and keep track of the potential rises and drops in order to calculate V_{ab}.

EXECUTE: $-I_1(6.00\,\Omega + 3.00\,\Omega) + 36.0\text{ V} = 0$

$$I_1 = \frac{36.0\text{ V}}{6.00\,\Omega + 3.00\,\Omega} = 4.00\text{ A}$$

$-I_2(3.00\,\Omega + 6.00\,\Omega) + 36.0\text{ V} = 0$

$$I_2 = \frac{36.0\text{ V}}{3.00\,\Omega + 6.00\,\Omega} = 4.00\text{ A}$$

To calculate $V_{ab} = V_a - V_b$ start at point b and travel to point a, adding up all the potential rises and drops along the way. We can do this by going from b up through the 3.00-Ω resistor:

$V_b + I_2(3.00\,\Omega) - I_1(6.00\,\Omega) = V_a$

$V_a - V_b = (4.00\text{ A})(3.00\,\Omega) - (4.00\text{ A})(6.00\,\Omega) = 12.0\text{ V} - 24.0\text{ V} = -12.0\text{ V}$

$V_{ab} = -12.0\text{ V}$ (point a is 12.0 V lower in potential than point b)

EVALUATE: Alternatively, we can go from point b down through the 6.00-Ω resistor.

$V_b - I_2(6.00\,\Omega) + I_1(3.00\,\Omega) = V_a$

$V_a - V_b = -(4.00\text{ A})(6.00\,\Omega) + (4.00\text{ A})(3.00\,\Omega) = -24.0\text{ V} + 12.0\text{ V} = -12.0\text{ V}$, which checks.

(b) IDENTIFY: Now there are multiple current paths, as shown in Figure 26.77b. Use the junction rule to write the current in each branch in terms of three unknown currents I_1, I_2 and I_3. Apply the loop rule to three loops to get three equations for the three unknowns. The target variable is I_3, the current through the switch. R_{eq} is calculated from $V = IR_{eq}$, where I is the total current that passes through the network.

SET UP:

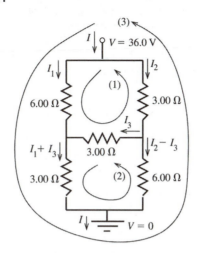

The three unknown currents I_1, I_2 and I_3 are labeled on Figure 26.77b.

Figure 26.77b

EXECUTE: Apply the loop rule to loops (1), (2) and (3).

loop (1): $-I_1(6.00\,\Omega) + I_3(3.00\,\Omega) + I_2(3.00\,\Omega) = 0$

$I_2 = 2I_1 - I_3$ eq. (1)

loop (2): $-(I_1 + I_3)(3.00\,\Omega) + (I_2 - I_3)(6.00\,\Omega) - I_3(3.00\,\Omega) = 0$

$6I_2 - 12I_3 - 3I_1 = 0$ so $2I_2 - 4I_3 - I_1 = 0$

Use eq (1) to replace I_2:

$4I_1 - 2I_3 - 4I_3 - I_1 = 0$

$3I_1 = 6I_3$ and $I_1 = 2I_3$ eq. (2)

loop (3) (This loop is completed through the battery [not shown], in the direction from the $-$ to the $+$ terminal.):

$-I_1(6.00\,\Omega) - (I_1 + I_3)(3.00\,\Omega) + 36.0\text{ V} = 0$

$9I_1 + 3I_3 = 36.0$ A and $3I_1 + I_3 = 12.0$ A   eq. (3)

Use eq. (2) in eq. (3) to replace I_1:

$3(2I_3) + I_3 = 12.0$ A

$I_3 = 12.0$ A$/7 = 1.71$ A

$I_1 = 2I_3 = 3.42$ A

$I_2 = 2I_1 - I_3 = 2(3.42\text{ A}) - 1.71\text{ A} = 5.13$ A

The current through the switch is $I_3 = 1.71$ A.

(c) From the results in part (a) the current through the battery is $I = I_1 + I_2 = 3.42$ A $+ 5.13$ A $= 8.55$ A.

The equivalent circuit is a single resistor that produces the same current through the 36.0-V battery, as shown in Figure 26.77c.

$-IR + 36.0\text{ V} = 0$

$R = \dfrac{36.0\text{ V}}{I} = \dfrac{36.0\text{ V}}{8.55\text{ A}} = 4.21\,\Omega$

$I = 8.55$ A   36.0 V R

Figure 26.77c

EVALUATE: With the switch open (part a), point b is at higher potential than point a, so when the switch is closed the current flows in the direction from b to a. With the switch closed the circuit cannot be simplified using series and parallel combinations but there is still an equivalent resistance that represents the network.

26.79.   **(a) IDENTIFY:** Connecting the voltmeter between point b and ground gives a resistor network and we can solve for the current through each resistor. The voltmeter reading equals the potential drop across the 200-kΩ resistor.

SET UP: For resistors in parallel, $\dfrac{1}{R_{\mathrm{eq}}} = \dfrac{1}{R_1} + \dfrac{1}{R_2}$. For resistors in series, $R_{\mathrm{eq}} = R_1 + R_2$.

EXECUTE: (a) $R_{\mathrm{eq}} = 100 \text{ k}\Omega + \left(\dfrac{1}{200 \text{ k}\Omega} + \dfrac{1}{50 \text{ k}\Omega} \right)^{-1} = 140 \text{ k}\Omega$. The total current is

$I = \dfrac{0.400 \text{ kV}}{140 \text{ k}\Omega} = 2.86 \times 10^{-3} \text{ A}$. The voltage across the 200-kΩ resistor is

$V_{200\text{k}\Omega} = IR = (2.86 \times 10^{-3} \text{ A}) \left(\dfrac{1}{200 \text{ k}\Omega} + \dfrac{1}{50 \text{ k}\Omega} \right)^{-1} = 114.4 \text{ V}$.

(b) If $V_R = 5.00 \times 10^6 \ \Omega$, then we carry out the same calculations as above to find $R_{\mathrm{eq}} = 292 \text{ k}\Omega$,

$I = 1.37 \times 10^{-3} \text{ A}$ and $V_{200\text{k}\Omega} = 263 \text{ V}$.

(c) If $V_R = \infty$, then we find $R_{\mathrm{eq}} = 300 \text{ k}\Omega$, $I = 1.33 \times 10^{-3} \text{ A}$ and $V_{200\text{k}\Omega} = 266 \text{ V}$.

EVALUATE: When a voltmeter of finite resistance is connected to a circuit, current flows through the voltmeter and the presence of the voltmeter alters the currents and voltages in the original circuit. The effect of the voltmeter on the circuit decreases as the resistance of the voltmeter increases.

26.83.   **IDENTIFY** and **SET UP:** Without the meter, the circuit consists of the two resistors in series. When the meter is connected, its resistance is added to the circuit in parallel with the resistor it is connected across.

(a) EXECUTE: $I = I_1 = I_2$

$I = \dfrac{90.0 \text{ V}}{R_1 + R_2} = \dfrac{90.0 \text{ V}}{224 \ \Omega + 589 \ \Omega} = 0.1107 \text{ A}$

$V_1 = I_1 R_1 = (0.1107 \text{ A})(224 \ \Omega) = 24.8 \text{ V}; \ V_2 = I_2 R_2 = (0.1107 \text{ A})(589 \ \Omega) = 65.2 \text{ V}$

(b) SET UP: The resistor network is sketched in Figure 26.83a.

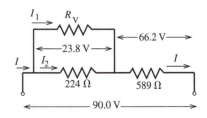

The voltmeter reads the potential difference across its terminals, which is 23.8 V. If we can find the current I_1 through the voltmeter then we can use Ohm's law to find its resistance.

Figure 26.83a

EXECUTE: The voltage drop across the 589-Ω resistor is $90.0 \text{ V} - 23.8 \text{ V} = 66.2 \text{ V}$, so

$I = \dfrac{V}{R} = \dfrac{66.2 \text{ V}}{589 \ \Omega} = 0.1124 \text{ A}$. The voltage drop across the 224-Ω resistor is 23.8 V, so

$I_2 = \dfrac{V}{R} = \dfrac{23.8 \text{ V}}{224 \ \Omega} = 0.1062 \text{ A}$. Then $I = I_1 + I_2$ gives $I_1 = I - I_2 = 0.1124 \text{ A} - 0.1062 \text{ A} = 0.0062 \text{ A}$.

$R_V = \dfrac{V}{I_1} = \dfrac{23.8 \text{ V}}{0.0062 \ A} = 3840 \ \Omega$

(c) SET UP: The circuit with the voltmeter connected is sketched in Figure 26.83b.

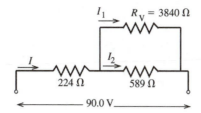

Figure 26.83b

EXECUTE: Replace the two resistors in parallel by their equivalent, as shown in Figure 26.83c.

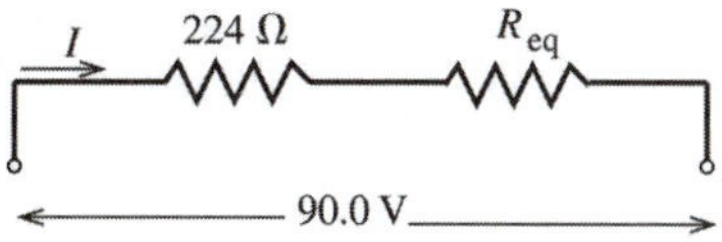

$$\frac{1}{R_{eq}} = \frac{1}{3840\,\Omega} + \frac{1}{589\,\Omega};$$

$$R_{eq} = \frac{(3840\,\Omega)(589\,\Omega)}{3840\,\Omega + 589\,\Omega} = 510.7\,\Omega$$

Figure 26.83c

$$I = \frac{90.0\,\text{V}}{224\,\Omega + 510.7\,\Omega} = 0.1225\,\text{A}$$

The potential drop across the 224-Ω resistor then is $IR = (0.1225\,\text{A})(224\,\Omega) = 27.4\,\text{V}$, so the potential drop across the 589-Ω resistor and across the voltmeter (what the voltmeter reads) is 90.0 V − 27.4 V = 62.6 V.

EVALUATE: **(d)** No, any real voltmeter will draw some current and thereby reduce the current through the resistance whose voltage is being measured. Thus the presence of the voltmeter connected in parallel with the resistance lowers the voltage drop across that resistance. The resistance of the voltmeter in this problem is only about a factor of ten larger than the resistances in the circuit, so the voltmeter has a noticeable effect on the circuit.

MAGNETIC FIELD AND MAGNETIC FORCES

27.7. **IDENTIFY:** Apply $\vec{F} = q\vec{v} \times \vec{B}$.

SET UP: $\vec{v} = v_y \hat{j}$, with $v_y = -3.80 \times 10^3$ m/s. $F_x = +7.60 \times 10^{-3}$ N, $F_y = 0$, and $F_z = -5.20 \times 10^{-3}$ N.

EXECUTE: **(a)** $F_x = q(v_y B_z - v_z B_y) = qv_y B_z$.

$B_z = F_x/qv_y = (7.60 \times 10^{-3} \text{ N})/[(7.80 \times 10^{-6} \text{ C})(-3.80 \times 10^3 \text{ m/s})] = -0.256$ T

$F_y = q(v_z B_x - v_x B_z) = 0$, which is consistent with $\vec{F}$ as given in the problem. There is no force component along the direction of the velocity.

$F_z = q(v_x B_y - v_y B_x) = -qv_y B_x$. $B_x = -F_z/qv_y = -0.175$ T.

(b) B_y is not determined. No force due to this component of $\vec{B}$ along $\vec{v}$; measurement of the force tells us nothing about B_y.

(c) $\vec{B} \cdot \vec{F} = B_x F_x + B_y F_y + B_z F_z = (-0.175 \text{ T})(+7.60 \times 10^{-3} \text{ N}) + (-0.256 \text{ T})(-5.20 \times 10^{-3} \text{ N})$

$\vec{B} \cdot \vec{F} = 0$. $\vec{B}$ and $\vec{F}$ are perpendicular (angle is $90°$).

EVALUATE: The force is perpendicular to both $\vec{v}$ and $\vec{B}$, so $\vec{v} \cdot \vec{F}$ is also zero.

27.9. **IDENTIFY:** Apply $\vec{F} = q\vec{v} \times \vec{B}$ to the force on the proton and to the force on the electron. Solve for the components of $\vec{B}$ and use them to find its magnitude and direction.

SET UP: $\vec{F}$ is perpendicular to both $\vec{v}$ and $\vec{B}$. Since the force on the proton is in the $+y$-direction, $B_y = 0$ and $\vec{B} = B_x \hat{i} + B_z \hat{k}$. For the proton, $\vec{v}_p = (1.50 \text{ km/s}) \hat{i} = v_p \hat{i}$ and $\vec{F}_p = (2.25 \times 10^{-16} \text{ N}) \hat{j} = F_p \hat{j}$. For the electron, $\vec{v}_e = -(4.75 \text{ km/s}) \hat{k} = -v_e \hat{k}$ and $\vec{F}_e = (8.50 \times 10^{-16} \text{ N}) \hat{j} = F_e \hat{j}$. The magnetic force is $\vec{F} = q\vec{v} \times \vec{B}$.

EXECUTE: **(a)** For the proton, $\vec{F}_p = q\vec{v}_p \times \vec{B}$ gives $F_p \hat{j} = ev_p \hat{i} \times (B_x \hat{i} + B_z \hat{k}) = -ev_p B_z \hat{j}$. Solving for B_z

gives $B_z = -\dfrac{F_p}{ev_p} = -\dfrac{2.25 \times 10^{-16} \text{ N}}{(1.60 \times 10^{-19} \text{ C})(1500 \text{ m/s})} = -0.9375$ T. For the electron, $\vec{F}_e = -e\vec{v}_e \times \vec{B}$, which gives

$F_e \hat{j} = (-e)(-v_e \hat{k}) \times (B_x \hat{i} + B_z \hat{k}) = ev_e B_x \hat{j}$. Solving for B_x gives

$B_x = \dfrac{F_e}{ev_e} = \dfrac{8.50 \times 10^{-16} \text{ N}}{(1.60 \times 10^{-19} \text{ C})(4750 \text{ m/s})} = 1.118$ T. Therefore $\vec{B} = 1.118 \text{ T} \hat{i} - 0.9375 \text{ T} \hat{k}$. The magnitude of

the field is $B = \sqrt{B_x^2 + B_z^2} = \sqrt{(1.118 \text{ T})^2 + (-0.9375 \text{ T})^2} = 1.46$ T. Calling θ the angle that the magnetic

field makes with the $+x$-axis, we have $\tan \theta = \dfrac{B_z}{B_x} = \dfrac{-0.9375 \text{ T}}{1.118 \text{ T}} = -0.8386$, so $\theta = -40.0°$. Therefore the

magnetic field is in the xz-plane directed at $40.0°$ from the $+x$-axis toward the $-z$-axis, having a magnitude of 1.46 T.

(b) $\vec{B} = B_x\hat{i} + B_z\hat{k}$ and $\vec{v} = (3.2 \text{ km/s})(-\hat{j})$.

$\vec{F} = q\vec{v}\times\vec{B} = (-e)(3.2 \text{ km/s})(-\hat{j})\times(B_x\hat{i} + B_z\hat{k}) = e(3.2\times10^3 \text{ m/s})[B_x(-\hat{k}) + B_z\hat{i}]$.

$\vec{F} = e(3.2\times10^3 \text{ m/s})(-1.118 \text{ T}\hat{k} - 0.9375 \text{ T}\hat{i}) = -4.80\times10^{-16} \text{ N}\hat{i} - 5.724\times10^{-16} \text{ N}\hat{k}$.

$F = \sqrt{F_x^2 + F_z^2} = 7.47\times10^{-16}$ N. Calling θ the angle that the force makes with the $-x$-axis, we have

$\tan\theta = \dfrac{F_z}{F_x} = \dfrac{-5.724\times10^{-16} \text{ N}}{-4.800\times10^{-16} \text{ N}}$, which gives $\theta = 50.0°$. The force is in the xz-plane and is directed at

50.0° from the $-x$-axis toward either the $-z$-axis.

EVALUATE: The force on the electrons in parts (a) and (b) are comparable in magnitude because the electron speeds are comparable in both cases.

27.13. **IDENTIFY:** The total flux through the bottle is zero because it is a closed surface.

SET UP: The total flux through the bottle is the flux through the plastic plus the flux through the open cap, so the sum of these must be zero. $\Phi_{\text{plastic}} + \Phi_{\text{cap}} = 0$.

$\Phi_{\text{plastic}} = -\Phi_{\text{cap}} = -B\,A\cos\phi = -B(\pi r^2)\cos\phi$

EXECUTE: Substituting the numbers gives $\Phi_{\text{plastic}} = -(1.75 \text{ T})\pi(0.0125 \text{ m})^2 \cos 25° = -7.8\times10^{-4}$ Wb

EVALUATE: It would be very difficult to calculate the flux through the plastic directly because of the complicated shape of the bottle, but with a little thought we can find this flux through a simple calculation.

27.15. **(a) IDENTIFY:** Apply Eq. (27.2) to relate the magnetic force $\vec{F}$ to the directions of $\vec{v}$ and $\vec{B}$. The electron has negative charge so $\vec{F}$ is opposite to the direction of $\vec{v}\times\vec{B}$. For motion in an arc of a circle the acceleration is toward the center of the arc so $\vec{F}$ must be in this direction. $a = v^2/R$.

SET UP:

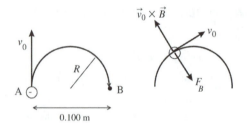

As the electron moves in the semicircle, its velocity is tangent to the circular path. The direction of $\vec{v}_0\times\vec{B}$ at a point along the path is shown in Figure 27.15.

Figure 27.15

EXECUTE: For circular motion the acceleration of the electron $\vec{a}_{\text{rad}}$ is directed in toward the center of the circle. Thus the force $\vec{F}_B$ exerted by the magnetic field, since it is the only force on the electron, must be radially inward. Since q is negative, $\vec{F}_B$ is opposite to the direction given by the right-hand rule for $\vec{v}_0\times\vec{B}$. Thus $\vec{B}$ is directed into the page. Apply Newton's second law to calculate the magnitude of $\vec{B}$:

$\Sigma\vec{F} = m\vec{a}$ gives $\Sigma F_{\text{rad}} = ma$ $F_B = m(v^2/R)$

$F_B = |q|vB\sin\phi = |q|vB$, so $|q|vB = m(v^2/R)$

$B = \dfrac{mv}{|q|R} = \dfrac{(9.109\times10^{-31} \text{ kg})(1.41\times10^6 \text{ m/s})}{(1.602\times10^{-19} \text{ C})(0.050 \text{ m})} = 1.60\times10^{-4}$ T

(b) IDENTIFY and SET UP: The speed of the electron as it moves along the path is constant. ($\vec{F}_B$ changes the direction of $\vec{v}$ but not its magnitude.) The time is given by the distance divided by v_0.

EXECUTE: The distance along the semicircular path is πR, so $t = \dfrac{\pi R}{v_0} = \dfrac{\pi(0.050 \text{ m})}{1.41\times10^6 \text{ m/s}} = 1.11\times10^{-7}$ s.

EVALUATE: The magnetic field required increases when v increases or R decreases and also depends on the mass to charge ratio of the particle.

27.17.   **IDENTIFY** and **SET UP:** Use conservation of energy to find the speed of the ball when it reaches the bottom of the shaft. The right-hand rule gives the direction of $\vec{F}$ and Eq. (27.1) gives its magnitude. The number of excess electrons determines the charge of the ball.

EXECUTE: $q = (4.00 \times 10^8)(-1.602 \times 10^{-19} \text{ C}) = -6.408 \times 10^{-11} \text{ C}$

speed at bottom of shaft: $\frac{1}{2}mv^2 = mgy$; $v = \sqrt{2gy} = 49.5 \text{ m/s}$

$\vec{v}$ is downward and $\vec{B}$ is west, so $\vec{v} \times \vec{B}$ is north. Since $q < 0$, $\vec{F}$ is south.

$F = |q|vB\sin\theta = (6.408 \times 10^{-11} \text{ C})(49.5 \text{ m/s})(0.250 \text{ T})\sin 90° = 7.93 \times 10^{-10} \text{ N}$

EVALUATE: Both the charge and speed of the ball are relatively small so the magnetic force is small, much less than the gravity force of 1.5 N.

27.21.   **(a) IDENTIFY** and **SET UP:** Apply Newton's second law, with $a = v^2/R$ since the path of the particle is circular.

EXECUTE: $\sum\vec{F} = m\vec{a}$ says $|q|vB = m(v^2/R)$

$v = \dfrac{|q|BR}{m} = \dfrac{(1.602 \times 10^{-19} \text{ C})(2.50 \text{ T})(6.96 \times 10^{-3} \text{ m})}{3.34 \times 10^{-27} \text{ kg}} = 8.35 \times 10^5 \text{ m/s}$

(b) IDENTIFY and **SET UP:** The speed is constant so $t = \text{distance}/v$.

EXECUTE: $t = \dfrac{\pi R}{v} = \dfrac{\pi(6.96 \times 10^{-3} \text{ m})}{8.35 \times 10^5 \text{ m/s}} = 2.62 \times 10^{-8} \text{ s}$

(c) IDENTIFY and **SET UP:** kinetic energy gained = electric potential energy lost

EXECUTE: $\frac{1}{2}mv^2 = |q|V$

$V = \dfrac{mv^2}{2|q|} = \dfrac{(3.34 \times 10^{-27} \text{ kg})(8.35 \times 10^5 \text{ m/s})^2}{2(1.602 \times 10^{-19} \text{ C})} = 7.27 \times 10^3 \text{ V} = 7.27 \text{ kV}$

EVALUATE: The deutron has a much larger mass to charge ratio than an electron so a much larger B is required for the same v and R. The deutron has positive charge so gains kinetic energy when it goes from high potential to low potential.

27.25.   **IDENTIFY:** When a particle of charge $-e$ is accelerated through a potential difference of magnitude V, it gains kinetic energy eV. When it moves in a circular path of radius R, its acceleration is $\dfrac{v^2}{R}$.

SET UP: An electron has charge $q = -e = -1.60 \times 10^{-19} \text{ C}$ and mass $9.11 \times 10^{-31} \text{ kg}$.

EXECUTE: $\frac{1}{2}mv^2 = eV$ and $v = \sqrt{\dfrac{2eV}{m}} = \sqrt{\dfrac{2(1.60 \times 10^{-19} \text{ C})(2.00 \times 10^3 \text{ V})}{9.11 \times 10^{-31} \text{ kg}}} = 2.65 \times 10^7 \text{ m/s}$. $\vec{F} = m\vec{a}$

gives $|q|vB\sin\phi = m\dfrac{v^2}{R}$. $\phi = 90°$ and $B = \dfrac{mv}{|q|R} = \dfrac{(9.11 \times 10^{-31} \text{ kg})(2.65 \times 10^7 \text{ m/s})}{(1.60 \times 10^{-19} \text{ C})(0.180 \text{ m})} = 8.38 \times 10^{-4} \text{ T}$.

EVALUATE: The smaller the radius of the circular path, the larger the magnitude of the magnetic field that is required.

27.27.   **(a) IDENTIFY** and **SET UP:** Eq. (27.4) gives the total force on the proton. At $t = 0$,

$\vec{F} = q\vec{v} \times \vec{B} = q(v_x\hat{i} + v_z\hat{k}) \times B_x\hat{i} = qv_zB_x\hat{j}$.

$\vec{F} = (1.60 \times 10^{-19} \text{ C})(2.00 \times 10^5 \text{ m/s})(0.500 \text{ T})\hat{j} = (1.60 \times 10^{-14} \text{ N})\hat{j}$.

(b) Yes. The electric field exerts a force in the direction of the electric field, since the charge of the proton is positive, and there is a component of acceleration in this direction.

(c) EXECUTE: In the plane perpendicular to $\vec{B}$ (the yz-plane) the motion is circular. But there is a velocity component in the direction of $\vec{B}$, so the motion is a helix. The electric field in the $+\hat{i}$ direction exerts a force in the $+\hat{i}$ direction. This force produces an acceleration in the $+\hat{i}$ direction and this causes the pitch of the helix to vary. The force does not affect the circular motion in the yz-plane, so the electric field does not affect the radius of the helix.

(d) IDENTIFY and SET UP: Eq. (27.12) and $T = 2\pi/\omega$ to calculate the period of the motion. Calculate a_x produced by the electric force and use a constant acceleration equation to calculate the displacement in the x-direction in time $T/2$.

EXECUTE: Calculate the period T: $\omega = |q| B/m$

$$T = \frac{2\pi}{\omega} = \frac{2\pi m}{|q| B} = \frac{2\pi(1.67 \times 10^{-27} \text{ kg})}{(1.60 \times 10^{-19} \text{ C})(0.500 \text{ T})} = 1.312 \times 10^{-7} \text{ s. Then } t = T/2 = 6.56 \times 10^{-8} \text{ s.}$$

$$v_{0x} = 1.50 \times 10^5 \text{ m/s}$$

$$a_x = \frac{F_x}{m} = \frac{(1.60 \times 10^{-19} \text{ C})(2.00 \times 10^4 \text{ V/m})}{1.67 \times 10^{-27} \text{ kg}} = +1.916 \times 10^{12} \text{ m/s}^2$$

$$x - x_0 = v_{0x}t + \tfrac{1}{2}a_x t^2$$

$$x - x_0 = (1.50 \times 10^5 \text{ m/s})(6.56 \times 10^{-8} \text{ s}) + \tfrac{1}{2}(1.916 \times 10^{12} \text{ m/s}^2)(6.56 \times 10^{-8} \text{ s})^2 = 1.40 \text{ cm}$$

EVALUATE: The electric and magnetic fields are in the same direction but produce forces that are in perpendicular directions to each other.

27.31.   **IDENTIFY:** For the alpha particles to emerge from the plates undeflected, the magnetic force on them must exactly cancel the electric force. The battery produces an electric field between the plates, which acts on the alpha particles.

SET UP: First use energy conservation to find the speed of the alpha particles as they enter the region between the plates: $qV = 1/2 \ mv^2$. The electric field between the plates due to the battery is $E = V_b d$. For the alpha particles not to be deflected, the magnetic force must cancel the electric force, so $qvB = qE$, giving $B = E/v$.

EXECUTE: Solve for the speed of the alpha particles just as they enter the region between the plates. Their charge is $2e$.

$$v_\alpha = \sqrt{\frac{2(2e)V}{m}} = \sqrt{\frac{4(1.60 \times 10^{-19} \text{ C})(1750\text{V})}{6.64 \times 10^{-27} \text{ kg}}} = 4.11 \times 10^5 \text{ m/s}$$

The electric field between the plates, produced by the battery, is

$$E = V_b/d = (150 \text{ V})/(0.00820 \text{ m}) = 18,300 \text{ V/m}$$

The magnetic force must cancel the electric force:

$$B = E/v_\alpha = (18,300 \text{ V/m})/(4.11 \times 10^5 \text{ m/s}) = 0.0445 \text{ T}$$

The magnetic field is perpendicular to the electric field. If the charges are moving to the right and the electric field points upward, the magnetic field is out of the page.

EVALUATE: The sign of the charge of the alpha particle does not enter the problem, so negative charges of the same magnitude would also not be deflected.

27.33.   **IDENTIFY:** The velocity selector eliminates all ions not having the desired velocity. Then the magnetic field bends the ions in a circular arc.

SET UP: In a velocity selector, $E = vB$. For motion in a circular arc in a magnetic field of magnitude B,

$R = \dfrac{mv}{|q| B}$. The ion has charge $+e$.

EXECUTE: (a) $v = \dfrac{E}{B} = \dfrac{155 \text{ V/m}}{0.0315 \text{ T}} = 4.92 \times 10^3 \text{ m/s.}$

(b) $m = \dfrac{R|q| B}{v} = \dfrac{(0.175 \text{ m})(1.60 \times 10^{-19} \text{ C})(0.0175 \text{ T})}{4.92 \times 10^3 \text{ m/s}} = 9.96 \times 10^{-26} \text{ kg.}$

EVALUATE: Ions with larger ratio $\dfrac{m}{|q|}$ will move in a path of larger radius.

27.35.   **IDENTIFY:** A mass spectrometer separates ions by mass. Since ^{14}N and ^{15}N have different masses they will be separated and the relative amounts of these isotopes can be determined.

SET UP: $R = \dfrac{mv}{|q|B}$. For $m = 1.99 \times 10^{-26}$ kg (^{12}C), $R_{12} = 12.5$ cm. The separation of the isotopes at the detector is $2(R_{15} - R_{14})$.

EXECUTE: Since $R = \dfrac{mv}{|q|B}$, $\dfrac{R}{m} = \dfrac{v}{|q|B} = $ constant. Therefore $\dfrac{R_{14}}{m_{14}} = \dfrac{R_{12}}{m_{12}}$ which gives

$R_{14} = R_{12}\left(\dfrac{m_{14}}{m_{12}}\right) = (12.5 \text{ cm})\left(\dfrac{2.32 \times 10^{-26} \text{ kg}}{1.99 \times 10^{-26} \text{ kg}}\right) = 14.6$ cm and

$R_{15} = R_{12}\left(\dfrac{m_{15}}{m_{12}}\right) = (12.5 \text{ cm})\left(\dfrac{2.49 \times 10^{-26} \text{ kg}}{1.99 \times 10^{-26} \text{ kg}}\right) = 15.6$ cm. The separation of the isotopes at the detector is

$2(R_{15} - R_{14}) = 2(15.6 \text{ cm} - 14.6 \text{ cm}) = 2.0$ cm.

EVALUATE: The separation is large enough to be easily detectable. Since the diameter of the ion path is large, about 30 cm, the uniform magnetic field within the instrument must extend over a large area.

27.39.   **IDENTIFY:**   Apply $F = IlB\sin\phi$.

SET UP: Label the three segments in the field as a, b, and c. Let x be the length of segment a. Segment b has length 0.300 m and segment c has length $0.600 \text{ m} - x$. Figure 27.39a shows the direction of the force on each segment. For each segment, $\phi = 90°$. The total force on the wire is the vector sum of the forces on each segment.

EXECUTE: $F_a = IlB = (4.50 \text{ A})x(0.240 \text{ T})$. $F_c = (4.50 \text{ A})(0.600 \text{ m} - x)(0.240 \text{ T})$. Since $\vec{F}_a$ and $\vec{F}_c$ are in the same direction their vector sum has magnitude

$F_{ac} = F_a + F_c = (4.50 \text{ A})(0.600 \text{ m})(0.240 \text{ T}) = 0.648$ N and is directed toward the bottom of the page in Figure 27.39a. $F_b = (4.50 \text{ A})(0.300 \text{ m})(0.240 \text{ T}) = 0.324$ N and is directed to the right. The vector addition diagram for $\vec{F}_{ac}$ and $\vec{F}_b$ is given in Figure 27.39b.

$F = \sqrt{F_{ac}^2 + F_b^2} = \sqrt{(0.648 \text{ N})^2 + (0.324 \text{ N})^2} = 0.724$ N. $\tan\theta = \dfrac{F_{ac}}{F_b} = \dfrac{0.648 \text{ N}}{0.324 \text{ N}}$ and $\theta = 63.4°$. The net force has magnitude 0.724 N and its direction is specified by $\theta = 63.4°$ in Figure 27.39b.

EVALUATE: All three current segments are perpendicular to the magnetic field, so $\phi = 90°$ for each in the force equation. The direction of the force on a segment depends on the direction of the current for that segment.

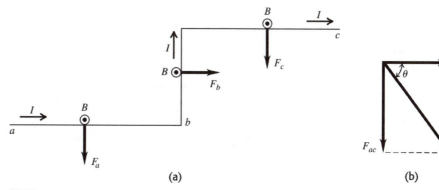

(a)   (b)

Figure 27.39

27.41.   **IDENTIFY** and **SET UP:**   The magnetic force is given by Eq. (27.19). $F_I = mg$ when the bar is just ready to levitate. When I becomes larger, $F_I > mg$ and $F_I - mg$ is the net force that accelerates the bar upward. Use Newton's second law to find the acceleration.

(a) EXECUTE: $IlB = mg$, $I = \dfrac{mg}{lB} = \dfrac{(0.750 \text{ kg})(9.80 \text{ m/s}^2)}{(0.500 \text{ m})(0.450 \text{ T})} = 32.67$ A

$\mathcal{E} = IR = (32.67 \text{ A})(25.0 \ \Omega) = 817$ V

(b) $R = 2.0\,\Omega, I = \mathcal{E}/R = (816.7\text{ V})/(2.0\,\Omega) = 408$ A

$F_I = IlB = 92$ N

$a = (F_I - mg)/m = 113$ m/s^2

EVALUATE: I increases by over an order of magnitude when R changes to $F_I \gg mg$ and a is an order of magnitude larger than g.

27.47. **IDENTIFY:** The magnetic field exerts a torque on the current-carrying coil, which causes it to turn. We can use the rotational form of Newton's second law to find the angular acceleration of the coil.

SET UP: The magnetic torque is given by $\vec{\tau} = \vec{\mu} \times \vec{B}$, and the rotational form of Newton's second law is $\sum \tau = I\alpha$. The magnetic field is parallel to the plane of the loop.

EXECUTE: **(a)** The coil rotates about axis A_2 because the only torque is along top and bottom sides of the coil.

(b) To find the moment of inertia of the coil, treat the two 1.00-m segments as point-masses (since all the points in them are 0.250 m from the rotation axis) and the two 0.500-m segments as thin uniform bars rotated about their centers. Since the coil is uniform, the mass of each segment is proportional to its fraction of the total perimeter of the coil. Each 1.00-m segment is 1/3 of the total perimeter, so its mass is $(1/3)(210$ g$) = 70$ g $= 0.070$ kg. The mass of each 0.500-m segment is half this amount, or 0.035 kg. The result is

$$I = 2(0.070\text{ kg})(0.250\text{ m})^2 + 2\tfrac{1}{12}(0.035\text{ kg})(0.500\text{ m})^2 = 0.0102\text{ kg}\cdot\text{m}^2$$

The torque is

$$|\vec{\tau}| = |\vec{\mu} \times \vec{B}| = IAB\sin 90° = (2.00\text{A})(0.500\text{m})(1.00\text{m})(3.00\text{T}) = 3.00\text{ N}\cdot\text{m}$$

Using the above values, the rotational form of Newton's second law gives

$$\alpha = \frac{\tau}{I} = 290\text{ rad/s}^2$$

EVALUATE: This angular acceleration will not continue because the torque changes as the coil turns.

27.49. **IDENTIFY and SET UP:** The potential energy is given by Eq. (27.27): $U = -\vec{\mu} \cdot \vec{B}$. The scalar product depends on the angle between $\vec{\mu}$ and $\vec{B}$.

EXECUTE: For $\vec{\mu}$ and $\vec{B}$ parallel, $\phi = 0°$ and $\vec{\mu} \cdot \vec{B} = \mu B\cos\phi = \mu B$. For $\vec{\mu}$ and $\vec{B}$ antiparallel, $\phi = 180°$ and $\vec{\mu} \cdot \vec{B} = \mu B\cos\phi = -\mu B$.

$U_1 = +\mu B,\ U_2 = -\mu B$

$\Delta U = U_2 - U_1 = -2\mu B = -2(1.45\text{ A}\cdot\text{m}^2)(0.835\text{ T}) = -2.42$ J

EVALUATE: U is maximum when $\vec{\mu}$ and $\vec{B}$ are antiparallel and minimum when they are parallel. When the coil is rotated as specified its magnetic potential energy decreases.

27.51. **IDENTIFY:** The circuit consists of two parallel branches with the potential difference of 120 V applied across each. One branch is the rotor, represented by a resistance R_r and an induced emf that opposes the applied potential. Apply the loop rule to each parallel branch and use the junction rule to relate the currents through the field coil and through the rotor to the 4.82 A supplied to the motor.

SET UP: The circuit is sketched in Figure 27.51.

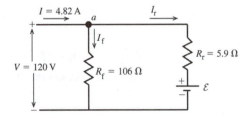

$\mathcal{E}$ is the induced emf developed by the motor. It is directed so as to oppose the current through the rotor.

Figure 27.51

EXECUTE: (a) The field coils and the rotor are in parallel with the applied potential difference

V, so $V = I_f R_f$. $I_f = \dfrac{V}{R_f} = \dfrac{120 \text{ V}}{106 \text{ }\Omega} = 1.13 \text{ A}.$

(b) Applying the junction rule to point a in the circuit diagram gives $I - I_f - I_r = 0$.

$I_r = I - I_f = 4.82 \text{ A} - 1.13 \text{ A} = 3.69 \text{ A}.$

(c) The potential drop across the rotor, $I_r R_r + \mathcal{E}$, must equal the applied potential difference

$V: V = I_r R_r + \mathcal{E}$

$\mathcal{E} = V - I_r R_r = 120 \text{ V} - (3.69 \text{ A})(5.9 \text{ }\Omega) = 98.2 \text{ V}$

(d) The mechanical power output is the electrical power input minus the rate of dissipation of electrical energy in the resistance of the motor:

electrical power input to the motor

$P_{in} = IV = (4.82 \text{ A})(120 \text{ V}) = 578 \text{ W}$

electrical power loss in the two resistances

$P_{loss} = I_f^2 R_f + I_r^2 R_r = (1.13 \text{ A})^2 (106 \text{ }\Omega) + (3.69 \text{ A})^2 (5.9 \text{ }\Omega) = 216 \text{ W}$

mechanical power output

$P_{out} = P_{in} - P_{loss} = 578 \text{ W} - 216 \text{ W} = 362 \text{ W}$

The mechanical power output is the power associated with the induced emf $\mathcal{E}$.

$P_{out} = P_{\mathcal{E}} = \mathcal{E} I_r = (98.2 \text{ V})(3.69 \text{ A}) = 362 \text{ W}$, which agrees with the above calculation.

EVALUATE: The induced emf reduces the amount of current that flows through the rotor. This motor differs from the one described in Example 27.11. In that example the rotor and field coils are connected in series and in this problem they are in parallel.

27.53.   IDENTIFY: The drift velocity is related to the current density by Eq. (25.4). The electric field is determined by the requirement that the electric and magnetic forces on the current-carrying charges are equal in magnitude and opposite in direction.

(a) SET UP: The section of the silver ribbon is sketched in Figure 27.53a.

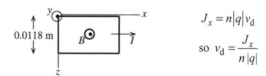

Figure 27.53a

EXECUTE: $J_x = \dfrac{I}{A} = \dfrac{I}{y_1 z_1} = \dfrac{120 \text{ A}}{(0.23 \times 10^{-3} \text{ m})(0.0118 \text{ m})} = 4.42 \times 10^7 \text{ A/m}^2$

$v_d = \dfrac{J_x}{n|q|} = \dfrac{4.42 \times 10^7 \text{ A/m}^2}{(5.85 \times 10^{28}/\text{m}^3)(1.602 \times 10^{-19} \text{ C})} = 4.7 \times 10^{-3} \text{ m/s} = 4.7 \text{ mm/s}$

(b) <u>magnitude of $\vec{E}$</u>

$|q| E_z = |q| v_d B_y$

$E_z = v_d B_y = (4.7 \times 10^{-3} \text{ m/s})(0.95 \text{ T}) = 4.5 \times 10^{-3} \text{ V/m}$

<u>direction of $\vec{E}$</u>

The drift velocity of the electrons is in the opposite direction to the current, as shown in Figure 27.53b.

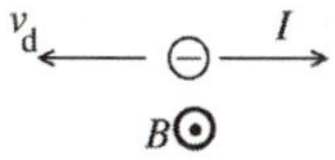

$v_d \overset{\longleftarrow}{} \; \ominus \; \overset{I}{\longrightarrow} \qquad\qquad \vec{v} \times \vec{B} \uparrow$

$B \odot \qquad\qquad\qquad \vec{F}_B = q\vec{v} \times \vec{B} = -e\vec{v} \times \vec{B} \downarrow$

Figure 27.53b

The directions of the electric and magnetic forces on an electron in the ribbon are shown in Figure 27.53c.

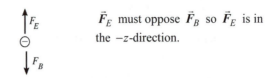

$\vec{F}_E$ must oppose $\vec{F}_B$ so $\vec{F}_E$ is in the $-z$-direction.

Figure 27.53c

$\vec{F}_E = q\vec{E} = -e\vec{E}$ so $\vec{E}$ is opposite to the direction of $\vec{F}_E$ and thus $\vec{E}$ is in the $+z$-direction.

(c) The Hall emf is the potential difference between the two edges of the strip (at $z = 0$ and $z = z_1$) that results from the electric field calculated in part (b). $\varepsilon_{\text{Hall}} = Ez_1 = (4.5 \times 10^{-3} \text{ V/m})(0.0118 \text{ m}) = 53 \ \mu\text{V}$.

EVALUATE: Even though the current is quite large the Hall emf is very small. Our calculated Hall emf is more than an order of magnitude larger than in Example 27.12. In this problem the magnetic field and current density are larger than in the example, and this leads to a larger Hall emf.

27.55.   **(a) IDENTIFY:** Use Eq. (27.2) to relate $\vec{v}$, $\vec{B}$ and $\vec{F}$.

SET UP: The directions of $\vec{v}_1$ and $\vec{F}_1$ are shown in Figure 27.55a.

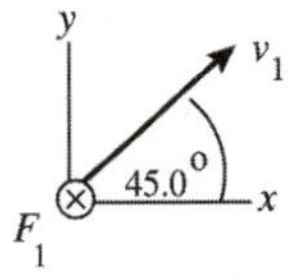

$\vec{F} = q\vec{v} \times \vec{B}$ says that $\vec{F}$ is perpendicular to $\vec{v}$ and $\vec{B}$. The information given here means that $\vec{B}$ can have no z-component.

Figure 27.55a

The directions of $\vec{v}_2$ and $\vec{F}_2$ are shown in Figure 27.55b.

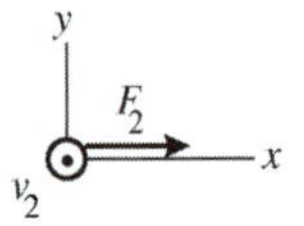

$\vec{F}$ is perpendicular to $\vec{v}$ and $\vec{B}$, so $\vec{B}$ can have no x-component.

Figure 27.55b

Both pieces of information taken together say that $\vec{B}$ is in the y-direction; $\vec{B} = B_y\hat{j}$.

EXECUTE: Use the information given about $\vec{F}_2$ to calculate B_y: $\vec{F}_2 = F_2\hat{i}, \vec{v}_2 = v_2\hat{k}, \vec{B} = B_y\hat{j}$.

$\vec{F}_2 = q\vec{v}_2 \times \vec{B}$ says $F_2\hat{i} = qv_2B_y\hat{k} \times \hat{j} = qv_2B_y(-\hat{i})$ and $F_2 = -qv_2B_y$

$B_y = -F_2/(qv_2) = -F_2/(qv_1)$. $\vec{B}$ has the magnitude $F_2/(qv_1)$ and is in the $-y$-direction.

(b) $F_1 = qvB\sin\phi = qv_1|B_y|/\sqrt{2} = F_2/\sqrt{2}$

EVALUATE: $v_1 = v_2$. $\vec{v}_2$ is perpendicular to $\vec{B}$ whereas only the component of $\vec{v}_1$ perpendicular to $\vec{B}$ contributes to the force, so it is expected that $F_2 > F_1$, as we found.

27.57.   **IDENTIFY:** In part (a), apply conservation of energy to the motion of the two nuclei. In part (b) apply $|q|vB = mv^2/R$.

SET UP: In part (a), let point 1 be when the two nuclei are far apart and let point 2 be when they are at their closest separation.

EXECUTE: **(a)** $K_1 + U_1 = K_2 + U_2$. $U_1 = K_2 = 0$, so $K_1 = U_2$. There are two nuclei having equal kinetic energy, so $\frac{1}{2}mv^2 + \frac{1}{2}mv^2 = ke^2/r$. Solving for v gives

$$v = e\sqrt{\frac{k}{mr}} = (1.602\times10^{-19}\ \text{C})\sqrt{\frac{8.99\times10^9\ \text{N}\cdot\text{m}^2/\text{C}^2}{(3.34\times10^{-27}\ \text{kg})(1.0\times10^{-15}\ \text{m})}} = 8.3\times10^6\ \text{m/s}.$$

(b) $\sum \vec{F} = m\vec{a}$ gives $qvB = mv^2/r$. $B = \dfrac{mv}{qr} = \dfrac{(3.34\times10^{-27}\ \text{kg})(8.3\times10^6\ \text{m/s})}{(1.602\times10^{-19}\ \text{C})(1.25\ \text{m})} = 0.14\ \text{T}.$

EVALUATE: The speed calculated in part (a) is large, nearly 3% of the speed of light.

27.61. **(a) IDENTIFY and SET UP:** The maximum radius of the orbit determines the maximum speed v of the protons. Use Newton's second law and $a_{\text{rad}} = v^2/R$ for circular motion to relate the variables. The energy of the particle is the kinetic energy $K = \frac{1}{2}mv^2$.

EXECUTE: $\sum \vec{F} = m\vec{a}$ gives $|q|vB = m(v^2/R)$

$v = \dfrac{|q|BR}{m} = \dfrac{(1.60\times10^{-19}\ \text{C})(0.85\ \text{T})(0.40\ \text{m})}{1.67\times10^{-27}\ \text{kg}} = 3.257\times10^7\ \text{m/s}.$ The kinetic energy of a proton moving

with this speed is $K = \frac{1}{2}mv^2 = \frac{1}{2}(1.67\times10^{-27}\ \text{kg})(3.257\times10^7\ \text{m/s})^2 = 8.9\times10^{-13}\ \text{J} = 5.5\ \text{MeV}$

(b) The time for one revolution is the period $T = \dfrac{2\pi R}{v} = \dfrac{2\pi(0.40\ \text{m})}{3.257\times10^7\ \text{m/s}} = 7.7\times10^{-8}\ \text{s}.$

(c) $K = \frac{1}{2}mv^2 = \frac{1}{2}m\left(\dfrac{|q|BR}{m}\right)^2 = \frac{1}{2}\dfrac{|q|^2 B^2 R^2}{m}$. Or, $B = \dfrac{\sqrt{2Km}}{|q|R}$. B is proportional to $\sqrt{K}$, so if K is increased

by a factor of 2 then B must be increased by a factor of $\sqrt{2}$. $B = \sqrt{2}(0.85\ \text{T}) = 1.2\ \text{T}.$

(d) $v = \dfrac{|q|BR}{m} = \dfrac{(3.20\times10^{-19}\ \text{C})(0.85\ \text{T})(0.40\ \text{m})}{6.65\times10^{-27}\ \text{kg}} = 1.636\times10^7\ \text{m/s}$

$K = \frac{1}{2}mv^2 = \frac{1}{2}(6.65\times10^{-27}\ \text{kg})(1.636\times10^7\ \text{m/s})^2 = 8.9\times10^{-13}\ \text{J} = 5.5\ \text{MeV}$, the same as the maximum energy for protons.

EVALUATE: We can see that the maximum energy must be approximately the same as follows: From part

(c), $K = \frac{1}{2}m\left(\dfrac{|q|BR}{m}\right)^2$. For alpha particles $|q|$ is larger by a factor of 2 and m is larger by a factor of 4

(approximately). Thus $|q|^2/m$ is unchanged and K is the same.

27.63. **IDENTIFY and SET UP:** Use Eq. (27.2) to relate q, $\vec{v}$, $\vec{B}$ and $\vec{F}$. The force $\vec{F}$ and $\vec{a}$ are related by Newton's second law. $\vec{B} = -(0.120\ \text{T})\hat{k}$, $\vec{v} = (1.05\times10^6\ \text{m/s})(-3\hat{i} + 4\hat{j} + 12\hat{k})$, $F = 2.45\ \text{N}$.

(a) EXECUTE: $\vec{F} = q\vec{v}\times\vec{B}$. $\vec{F} = q(-0.120\ \text{T})(1.05\times10^6\ \text{m/s})(-3\hat{i}\times\hat{k} + 4\hat{j}\times\hat{k} + 12\hat{k}\times\hat{k})$

$\hat{i}\times\hat{k} = -\hat{j}$, $\hat{j}\times\hat{k} = \hat{i}$, $\hat{k}\times\hat{k} = 0$. $\vec{F} = -q(1.26\times10^5\ \text{N/C})(+3\hat{j} + 4\hat{i}) = -q(1.26\times10^5\ \text{N/C})(+4\hat{i} + 3\hat{j})$. The

magnitude of the vector $+4\hat{i} + 3\hat{j}$ is $\sqrt{3^2 + 4^2} = 5$. Thus $F = -q(1.26\times10^5\ \text{N/C})(5)$.

$q = -\dfrac{F}{5(1.26\times10^5\ \text{N/C})} = -\dfrac{2.45\ \text{N}}{5(1.26\times10^5\ \text{N/C})} = -3.89\times10^{-6}\ \text{C}.$

(b) $\sum \vec{F} = m\vec{a}$ so $\vec{a} = \vec{F}/m$.

$\vec{F} = -q(1.26\times10^5\ \text{N/C})(+4\hat{i} + 3\hat{j}) = -(-3.89\times10^{-6}\ \text{C})(1.26\times10^5\ \text{N/C})(+4\hat{i} + 3\hat{j}) = +0.490\ \text{N}(+4\hat{i} + 3\hat{j})$.

Then

$\vec{a} = \vec{F}/m = \left(\dfrac{0.490\ \text{N}}{2.58\times10^{-15}\ \text{kg}}\right)(+4\hat{i} + 3\hat{j}) = (1.90\times10^{14}\ \text{m/s}^2)(+4\hat{i} + 3\hat{j}) = 7.60\times10^{14}\ \text{m/s}^2\hat{i} + 5.70\times10^{14}\ \text{m/s}^2\hat{j}.$

(c) IDENTIFY and SET UP: $\vec{F}$ is in the xy-plane, so in the z-direction the particle moves with constant speed 12.6×10^6 m/s. In the xy-plane the force $\vec{F}$ causes the particle to move in a circle, with $\vec{F}$ directed in toward the center of the circle.

EXECUTE: $\sum \vec{F} = m\vec{a}$ gives $F = m(v^2/R)$ and $R = mv^2/F$.

$v^2 = v_x^2 + v_y^2 = (-3.15 \times 10^6 \text{ m/s})^2 + (+4.20 \times 10^6 \text{ m/s})^2 = 2.756 \times 10^{13} \text{ m}^2/\text{s}^2$.

$F = \sqrt{F_x^2 + F_y^2} = (0.490 \text{ N})\sqrt{4^2 + 3^2} = 2.45 \text{ N}$.

$R = \dfrac{mv^2}{F} = \dfrac{(2.58 \times 10^{-15} \text{ kg})(2.756 \times 10^{13} \text{ m}^2/\text{s}^2)}{2.45 \text{ N}} = 0.0290 \text{ m} = 2.90 \text{ cm}$.

(d) IDENTIFY and SET UP: By Eq. (27.12) the cyclotron frequency is $f = \omega/2\pi = v/2\pi R$.

EXECUTE: The circular motion is in the xy-plane, so $v = \sqrt{v_x^2 + v_y^2} = 5.25 \times 10^6$ m/s.

$f = \dfrac{v}{2\pi R} = \dfrac{5.25 \times 10^6 \text{ m/s}}{2\pi(0.0290 \text{ m})} = 2.88 \times 10^7 \text{ Hz}$, and $\omega = 2\pi f = 1.81 \times 10^8$ rad/s.

(e) IDENTIFY and SET UP: Compare t to the period T of the circular motion in the xy-plane to find the x and y coordinates at this t. In the z-direction the particle moves with constant speed, so $z = z_0 + v_z t$.

EXECUTE: The period of the motion in the xy-plane is given by $T = \dfrac{1}{f} = \dfrac{1}{2.88 \times 10^7 \text{ Hz}} = 3.47 \times 10^{-8}$ s. In $t = 2T$ the particle has returned to the same x and y coordinates. The z-component of the motion is motion with a constant velocity of $v_z = +12.6 \times 10^6$ m/s. Thus

$z = z_0 + v_z t = 0 + (12.6 \times 10^6 \text{ m/s})(2)(3.47 \times 10^{-8} \text{ s}) = +0.874$ m. The coordinates at $t = 2T$ are $x = R = 0.0290$ m, $y = 0$, $z = +0.874$ m.

EVALUATE: The circular motion is in the plane perpendicular to $\vec{B}$. The radius of this motion gets smaller when B increases and it gets larger when v increases. There is no magnetic force in the direction of $\vec{B}$ so the particle moves with constant velocity in that direction. The superposition of circular motion in the xy-plane and constant speed motion in the z-direction is a helical path.

27.67.   **IDENTIFY:** For the velocity selector, $E = vB$. For circular motion in the field B', $R = \dfrac{mv}{|q|B'}$.

SET UP: $B = B' = 0.682$ T.

EXECUTE: $v = \dfrac{E}{B} = \dfrac{1.88 \times 10^4 \text{ N/C}}{0.682 \text{ T}} = 2.757 \times 10^4$ m/s. $R = \dfrac{mv}{qB'}$, so

$R_{82} = \dfrac{82(1.66 \times 10^{-27} \text{ kg})(2.757 \times 10^4 \text{ m/s})}{(1.60 \times 10^{-19} \text{ C})(0.682 \text{ T})} = 0.0344 \text{ m} = 3.44 \text{ cm}$.

$R_{84} = \dfrac{84(1.66 \times 10^{-27} \text{ kg})(2.757 \times 10^4 \text{ m/s})}{(1.60 \times 10^{-19} \text{ C})(0.682 \text{ T})} = 0.0352 \text{ m} = 3.52 \text{ cm}$.

$R_{86} = \dfrac{86(1.66 \times 10^{-27} \text{ kg})(2.757 \times 10^4 \text{ m/s})}{(1.60 \times 10^{-19} \text{ C})(0.682 \text{ T})} = 0.0361 \text{ m} = 3.61 \text{ cm}$.

The distance between two adjacent lines is $2\Delta R = 2(3.52 \text{ cm} - 3.44 \text{ cm}) = 0.16 \text{ cm} = 1.6$ mm.

EVALUATE: The distance between the ^{82}Kr line and the ^{84}Kr line is 1.6 mm and the distance between the ^{84}Kr line and the ^{86}Kr line is 1.6 mm. Adjacent lines are equally spaced since the ^{82}Kr versus ^{84}Kr and ^{84}Kr versus ^{86}Kr mass differences are the same.

27.69.   **IDENTIFY:** The force exerted by the magnetic field is given by Eq. (27.19). The net force on the wire must be zero.

SET UP: For the wire to remain at rest the force exerted on it by the magnetic field must have a component directed up the incline. To produce a force in this direction, the current in the wire must be directed from right to left in Figure P27.69 in the textbook. Or, viewing the wire from its left-hand end the directions are shown in Figure 27.69a.

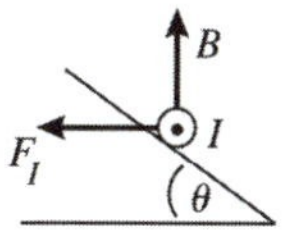

Figure 27.69a

The free-body diagram for the wire is given in Figure 27.69b.

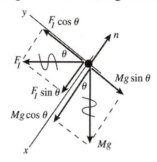

EXECUTE: $\sum F_y = 0$

$F_I \cos\theta - Mg\sin\theta = 0$

$F_I = ILB\sin\phi$

$\phi = 90°$ since $\vec{B}$ is perpendicular to the current direction.

Figure 27.69b

Thus $(ILB)\cos\theta - Mg\sin\theta = 0$ and $I = \dfrac{Mg\tan\theta}{LB}$.

EVALUATE: The magnetic and gravitational forces are in perpendicular directions so their components parallel to the incline involve different trig functions. As the tilt angle θ increases there is a larger component of Mg down the incline and the component of F_I up the incline is smaller; I must increase with θ to compensate. As $\theta \to 0$, $I \to 0$ and as $\theta \to 90°$, $I \to \infty$.

27.73. **IDENTIFY:** $R = \dfrac{mv}{|q|B}$.

SET UP: After completing one semicircle the separation between the ions is the difference in the diameters of their paths, or $2(R_{13} - R_{12})$. A singly ionized ion has charge $+e$.

EXECUTE: **(a)** $B = \dfrac{mv}{|q|R} = \dfrac{(1.99 \times 10^{-26}\text{ kg})(8.50 \times 10^3\text{ m/s})}{(1.60 \times 10^{-19}\text{ C})(0.125\text{ m})} = 8.46 \times 10^{-3}$ T.

(b) The only difference between the two isotopes is their masses. $\dfrac{R}{m} = \dfrac{v}{|q|B} = $ constant and $\dfrac{R_{12}}{m_{12}} = \dfrac{R_{13}}{m_{13}}$.

$R_{13} = R_{12}\left(\dfrac{m_{13}}{m_{12}}\right) = (12.5\text{ cm})\left(\dfrac{2.16 \times 10^{-26}\text{ kg}}{1.99 \times 10^{-26}\text{ kg}}\right) = 13.6$ cm. The diameter is 27.2 cm.

(c) The separation is $2(R_{13} - R_{12}) = 2(13.6\text{ cm} - 12.5\text{ cm}) = 2.2$ cm. This distance can be easily observed.

EVALUATE: Decreasing the magnetic field increases the separation between the two isotopes at the detector.

27.77. **IDENTIFY:** For the loop to be in equilibrium the net torque on it must be zero. Use Eq. (27.26) to calculate the torque due to the magnetic field and use Eq. (10.3) for the torque due to the gravity force.

SET UP: See Figure 27.77a.

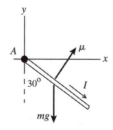

Use $\sum \tau_A = 0$, where point A is at the origin.

Figure 27.77a

EXECUTE: See Figure 27.77b.

$\tau_{mg} = mgr\sin\phi = mg(0.400 \text{ m})\sin 30.0°$

The torque is clockwise; $\vec{\tau}_{mg}$ is directed into the paper.

Figure 27.77b

For the loop to be in equilibrium the torque due to $\vec{B}$ must be counterclockwise (opposite to $\vec{\tau}_{mg}$) and it must be that $\tau_B = \tau_{mg}$. See Figure 27.77c.

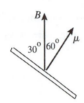

$\vec{\tau}_B = \vec{\mu} \times \vec{B}$. For this torque to be counter-clockwise ($\vec{\tau}_B$ directed out of the paper), $\vec{B}$ must be in the $+y$-direction.

Figure 27.77c

$\tau_B = \mu B \sin\phi = IAB \sin 60.0°$

$\tau_B = \tau_{mg}$ gives $IAB\sin 60.0° = mg(0.0400 \text{ m})\sin 30.0°$

$m = (0.15 \text{ g/cm})2(8.00 \text{ cm} + 6.00 \text{ cm}) = 4.2 \text{ g} = 4.2\times 10^{-3} \text{ kg}$

$A = (0.0800 \text{ m})(0.0600 \text{ m}) = 4.80\times 10^{-3} \text{ m}^2$

$B = \dfrac{mg(0.0400 \text{ m})(\sin 30.0°)}{IA\sin 60.0°}$

$B = \dfrac{(4.2\times 10^{-3} \text{ kg})(9.80 \text{ m/s}^2)(0.0400 \text{ m})\sin 30.0°}{(8.2 \text{ A})(4.80\times 10^{-3} \text{ m}^2)\sin 60.0°} = 0.024 \text{ T}$

EVALUATE: As the loop swings up the torque due to $\vec{B}$ decreases to zero and the torque due to mg increases from zero, so there must be an orientation of the loop where the net torque is zero.

27.79.   **IDENTIFY:** Use Eq. (27.20) to calculate the force and then the torque on each small section of the rod and integrate to find the total magnetic torque. At equilibrium the torques from the spring force and from the magnetic force cancel. The spring force depends on the amount x the spring is stretched and then $U = \frac{1}{2}kx^2$ gives the energy stored in the spring.

(a) SET UP:

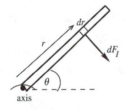

Divide the rod into infinitesimal sections of length dr, as shown in Figure 27.79.

Figure 27.79

EXECUTE: The magnetic force on this section is $dF_I = IBdr$ and is perpendicular to the rod. The torque $d\tau$ due to the force on this section is $d\tau = r dF_I = IBr\,dr$. The total torque is

$$\int d\tau = IB \int_0^l r\,dr = \frac{1}{2}Il^2B = 0.0442 \text{ N}\cdot\text{m, clockwise.}$$

(b) SET UP: F_I produces a clockwise torque so the spring force must produce a counterclockwise torque. The spring force must be to the left; the spring is stretched.

EXECUTE: Find x, the amount the spring is stretched:

$\sum \tau = 0$, axis at hinge, counterclockwise torques positive

$(kx)l\sin 53° - \frac{1}{2}Il^2 B = 0$

$x = \dfrac{IlB}{2k\sin 53.0°} = \dfrac{(6.50 \text{ A})(0.200 \text{ m})(0.340 \text{ T})}{2(4.80 \text{ N/m})\sin 53.0°} = 0.05765 \text{ m}$

(c) $U = \frac{1}{2}kx^2 = 7.98 \times 10^{-3}$ J

EVALUATE: The magnetic torque calculated in part (a) is the same torque calculated from a force diagram in which the total magnetic force $F_I = IlB$ acts at the center of the rod. We didn't include a gravity torque since the problem said the rod had negligible mass.

27.81. **IDENTIFY:** The contact at a will break if the bar rotates about b. The magnetic field is directed into the page, so the magnetic torque is counterclockwise, whereas the gravity torque is clockwise in the figure in the problem. The maximum current corresponds to zero net torque, in which case the torque due to gravity is just equal to the torque due to the magnetic field.

SET UP: The magnetic force is perpendicular to the bar and has moment arm $l/2$, where $l = 0.750 \text{ m}$ is the length of the bar. The gravity torque is $mg\left(\dfrac{l}{2}\cos 60.0°\right)$ and $F_B = IlB\sin\phi = IlB$. The results of Problem 27.79 show that we can take F_B to act at the center of the bar. F_B is perpendicular to the bar. Apply $\sum\tau_z = 0$ with the axis at b and counterclockwise torques positive.

EXECUTE: $F_B\dfrac{l}{2} - mg\left(\dfrac{l}{2}\cos 60.0°\right) = 0$. $IlB = mg\cos 60.0°$.

$I = \dfrac{mg\cos 60.0°}{lB} = \dfrac{(0.458 \text{ kg})(9.80 \text{ m/s}^2)\cos 60.0°}{(0.750 \text{ m})(1.25 \text{ T})} = 2.39 \text{ A}$.

EVALUATE: Once contact is broken, the magnetic torque ceases.

27.83. **IDENTIFY:** Use Eq. (27.20) to calculate the force on a short segment of the coil and integrate over the entire coil to find the total force.

SET UP: See Figure 27.83a.

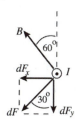

Consider the force $d\vec{F}$ on a short segment dl at the left-hand side of the coil, as viewed in Figure P27.83 in the textbook. The current at this point is directed out of the page. $d\vec{F}$ is perpendicular both to $\vec{B}$ and to the direction of I.

Figure 27.83a

See Figure 27.83b.

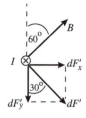

Consider also the force $d\vec{F}'$ on a short segment on the opposite side of the coil, at the right-hand side of the coil in Figure P27.83 in the textbook. The current at this point is directed into the page.

Figure 27.83b

The two sketches show that the x-components cancel and that the y-components add. This is true for all pairs of short segments on opposite sides of the coil. The net magnetic force on the coil is in the y-direction and its magnitude is given by $F = \int dF_y$.

EXECUTE: $dF = Idl\,B\sin\phi$. But $\vec{B}$ is perpendicular to the current direction so $\phi = 90°$.

$dF_y = dF\cos 30.0 = IB\cos 30.0° dl$

$F = \int dF_y = IB\cos 30.0° \int dl$

But $\int dl = N(2\pi r)$, the total length of wire in the coil.

$F = IB\cos 30.0° N(2\pi r) = (0.950\text{ A})(0.220\text{ T})(\cos 30.0°)(50)2\pi(0.0078\text{ m}) = 0.444\text{ N}$ and $\vec{F} = -(0.444\text{ N})\hat{j}$

EVALUATE: The magnetic field makes a constant angle with the plane of the coil but has a different direction at different points around the circumference of the coil so is not uniform. The net force is proportional to the magnitude of the current and reverses direction when the current reverses direction.

27.87. **IDENTIFY:** While the ends of the wire are in contact with the mercury and current flows in the wire, the magnetic field exerts an upward force and the wire has an upward acceleration. After the ends leave the mercury the electrical connection is broken and the wire is in free-fall.

(a) SET UP: After the wire leaves the mercury its acceleration is g, downward. The wire travels upward a total distance of 0.350 m from its initial position. Its ends lose contact with the mercury after the wire has traveled 0.025 m, so the wire travels upward 0.325 m after it leaves the mercury. Consider the motion of the wire after it leaves the mercury. Take $+y$ to be upward and take the origin at the position of the wire as it leaves the mercury.

$a_y = -9.80\text{ m/s}^2$, $y - y_0 = +0.325\text{ m}$, $v_y = 0$ (at maximum height), $v_{0y} = ?$

$v_y^2 = v_{0y}^2 + 2a_y(y - y_0)$

EXECUTE: $v_{0y} = \sqrt{-2a_y(y - y_0)} = \sqrt{-2(-9.80\text{ m/s}^2)(0.325\text{ m})} = 2.52\text{ m/s}$

(b) SET UP: Now consider the motion of the wire while it is in contact with the mercury. Take $+y$ to be upward and the origin at the initial position of the wire. Calculate the acceleration:

$y - y_0 = +0.025\text{ m}$, $v_{0y} = 0$ (starts from rest), $v_y = +2.52\text{ m/s}$ (from part (a)), $a_y = ?$

$v_y^2 = v_{0y}^2 + 2a_y(y - y_0)$

EXECUTE: $a_y = \dfrac{v_y^2}{2(y - y_0)} = \dfrac{(2.52\text{ m/s})^2}{2(0.025\text{ m})} = 127\text{ m/s}^2$

SET UP: The free-body diagram for the wire is given in Figure 27.87.

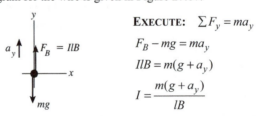

EXECUTE: $\sum F_y = ma_y$

$F_B - mg = ma_y$

$IlB = m(g + a_y)$

$I = \dfrac{m(g + a_y)}{lB}$

Figure 27.87

l is the length of the horizontal section of the wire; $l = 0.150\text{ m}$

$I = \dfrac{(5.40\times 10^{-5}\text{ kg})(9.80\text{ m/s}^2 + 127\text{ m/s}^2)}{(0.150\text{ m})(0.00650\text{ T})} = 7.58\text{ A}$

(c) IDENTIFY and **SET UP:** Use Ohm's law.

EXECUTE: $V = IR$ so $R = \dfrac{V}{I} = \dfrac{1.50\text{ V}}{7.58\text{ A}} = 0.198\ \Omega$

EVALUATE: The current is large and the magnetic force provides a large upward acceleration. During this upward acceleration the wire moves a much shorter distance as it gains speed than the distance it moves while in free-fall with a much smaller acceleration, as it loses the speed it gained. The large current means the resistance of the wire must be small.

28

SOURCES OF MAGNETIC FIELD

28.1. **IDENTIFY** and **SET UP:** Use Eq. (28.2) to calculate $\vec{B}$ at each point.

$\vec{B} = \dfrac{\mu_0}{4\pi} \dfrac{q\vec{v} \times \hat{r}}{r^2} = \dfrac{\mu_0}{4\pi} \dfrac{q\vec{v} \times \vec{r}}{r^3}$, since $\hat{r} = \dfrac{\vec{r}}{r}$.

$\vec{v} = (8.00 \times 10^6 \text{ m/s})\hat{j}$ and $\vec{r}$ is the vector from the charge to the point where the field is calculated.

EXECUTE: (a) $\vec{r} = (0.500 \text{ m})\hat{i}, r = 0.500 \text{ m}$

$\vec{v} \times \vec{r} = vr\hat{j} \times \hat{i} = -vr\hat{k}$

$\vec{B} = -\dfrac{\mu_0}{4\pi} \dfrac{qv}{r^2}\hat{k} = -(1 \times 10^{-7} \text{ T} \cdot \text{m/A})\dfrac{(6.00 \times 10^{-6} \text{ C})(8.00 \times 10^6 \text{ m/s})}{(0.500 \text{ m})^2}\hat{k}$

$\vec{B} = -(1.92 \times 10^{-5} \text{ T})\hat{k}$

(b) $\vec{r} = -(0.500 \text{ m})\hat{j}, r = 0.500 \text{ m}$

$\vec{v} \times \vec{r} = -vr\hat{j} \times \hat{j} = 0$ and $\vec{B} = 0$.

(c) $\vec{r} = (0.500 \text{ m})\hat{k}, r = 0.500 \text{ m}$

$\vec{v} \times \vec{r} = vr\hat{j} \times \hat{k} = vr\hat{i}$

$\vec{B} = (1 \times 10^{-7} \text{ T} \cdot \text{m/A})\dfrac{(6.00 \times 10^{-6} \text{ C})(8.00 \times 10^6 \text{ m/s})}{(0.500 \text{ m})^2}\hat{i} = +(1.92 \times 10^{-5} \text{ T})\hat{i}$

(d) $\vec{r} = -(0.500 \text{ m})\hat{j} + (0.500 \text{ m})\hat{k}, r = \sqrt{(0.500 \text{ m})^2 + (0.500 \text{ m})^2} = 0.7071 \text{ m}$

$\vec{v} \times \vec{r} = v(0.500 \text{ m})(-\hat{j} \times \hat{j} + \hat{j} \times \hat{k}) = (4.00 \times 10^6 \text{ m}^2/\text{s})\hat{i}$

$\vec{B} = (1 \times 10^{-7} \text{ T} \cdot \text{m/A})\dfrac{(6.00 \times 10^{-6} \text{ C})(4.00 \times 10^6 \text{ m}^2/\text{s})}{(0.7071 \text{ m})^3}\hat{i} = +(6.79 \times 10^{-6} \text{ T})\hat{i}$

EVALUATE: At each point $\vec{B}$ is perpendicular to both $\vec{v}$ and $\vec{r}$. $B = 0$ along the direction of $\vec{v}$.

28.3. **IDENTIFY:** A moving charge creates a magnetic field.

SET UP: The magnetic field due to a moving charge is $B = \dfrac{\mu_0}{4\pi} \dfrac{qv \sin\phi}{r^2}$.

EXECUTE: Substituting numbers into the above equation gives

(a) $B = \dfrac{\mu_0}{4\pi} \dfrac{qv \sin\phi}{r^2} = \dfrac{4\pi \times 10^{-7} \text{ T} \cdot \text{m/A}}{4\pi} \dfrac{(1.6 \times 10^{-19}\text{C})(3.0 \times 10^7 \text{m/s})\sin 30°}{(2.00 \times 10^{-6}\text{m})^2}$.

$B = 6.00 \times 10^{-8} \text{ T}$, out of the paper, and it is the same at point B.

(b) $B = (1.00 \times 10^{-7} \text{ T} \cdot \text{m/A})(1.60 \times 10^{-19} \text{ C})(3.00 \times 10^7 \text{ m/s})/(2.00 \times 10^{-6} \text{ m})^2$

$B = 1.20 \times 10^{-7} \text{ T}$, out of the page.

(c) $B = 0$ T since $\sin(180°) = 0$.

EVALUATE: Even at high speeds, these charges produce magnetic fields much less than the earth's magnetic field.

28.7.   **IDENTIFY:**   Apply $\vec{B} = \dfrac{\mu_0}{4\pi}\dfrac{q\vec{v}\times\vec{r}}{r^3}$. For the magnetic force on q', use $\vec{F}_B = q'\vec{v}'\times\vec{B}_q$ and for the magnetic

force on q use $\vec{F}_B = q\vec{v}\times\vec{B}_{q'}$.

SET UP: In part (a), $r = d$ and $\dfrac{|\vec{v}\times\vec{r}|}{r^3} = \dfrac{v}{d^2}$.

EXECUTE: (a) $q' = -q$; $B_q = \dfrac{\mu_0 qv}{4\pi d^2}$, into the page; $B_{q'} = \dfrac{\mu_0 qv'}{4\pi d^2}$, out of the page.

(i) $v' = \dfrac{v}{2}$ gives $B = \dfrac{\mu_0 qv}{4\pi d^2}\left(1-\tfrac{1}{2}\right) = \dfrac{\mu_0 qv}{4\pi(2d^2)}$, into the page. (ii) $v' = v$ gives $B = 0$.

(iii) $v' = 2v$ gives $B = \dfrac{\mu_0 qv}{4\pi d^2}$, out of the page.

(b) The force that q exerts on q' is given by $\vec{F} = q'\vec{v}'\times\vec{B}_q$, so $F = \dfrac{\mu_0 q^2 v' v}{4\pi(2d)^2}$. $\vec{B}_q$ is into the page, so the force

on q' is toward q. The force that q' exerts on q is toward q'. The force between the two charges is attractive.

(c) $F_B = \dfrac{\mu_0 q^2 vv'}{4\pi(2d)^2}$, $F_C = \dfrac{q^2}{4\pi\epsilon_0(2d)^2}$ so $\dfrac{F_B}{F_C} = \mu_0\epsilon_0 vv' = \mu_0\epsilon_0(3.00\times10^5 \text{ m/s})^2 = 1.00\times10^{-6}$.

EVALUATE: When charges of opposite sign move in opposite directions, the force between them is attractive. For the values specified in part (c), the magnetic force between the two charges is much smaller in magnitude than the Coulomb force between them.

28.11.   **IDENTIFY:**   A current segment creates a magnetic field.

SET UP: The law of Biot and Savart gives $dB = \dfrac{\mu_0}{4\pi}\dfrac{Idl\sin\phi}{r^2}$.

EXECUTE: Applying the law of Biot and Savart gives

(a) $dB = \dfrac{4\pi\times10^{-7} \text{ T}\cdot\text{m/A}}{4\pi}\dfrac{(10.0 \text{ A})(0.00110 \text{ m}) \sin 90°}{(0.0500 \text{ m})^2} = 4.40\times10^{-7}$ T, out of the paper.

(b) The same as above, except $r = \sqrt{(5.00 \text{ cm})^2 + (14.0 \text{ cm})^2}$ and $\phi = \arctan(5/14) = 19.65°$, giving

$dB = 1.67\times10^{-8}$ T, out of the page.

(c) $dB = 0$ since $\phi = 0°$.

EVALUATE: This is a very small field, but it comes from a very small segment of current.

28.15.   **IDENTIFY:**   A current segment creates a magnetic field.

SET UP: The law of Biot and Savart gives $dB = \dfrac{\mu_0}{4\pi}\dfrac{Idl\sin\phi}{r^2}$. Both fields are into the page, so their

magnitudes add.

EXECUTE: Applying the Biot and Savart law, where $r = \tfrac{1}{2}\sqrt{(3.00 \text{ cm})^2 + (3.00 \text{ cm})^2} = 2.121$ cm, we have

$$dB = 2\dfrac{4\pi\times10^{-7} \text{ T}\cdot\text{m/A}}{4\pi}\dfrac{(28.0 \text{ A})(0.00200 \text{ m})\sin 45.0°}{(0.02121 \text{ m})^2} = 1.76\times10^{-5} \text{ T, into the paper.}$$

EVALUATE: Even though the two wire segments are at right angles, the magnetic fields they create are in the same direction.

28.21.   **IDENTIFY:**   The long current-carrying wire produces a magnetic field.

SET UP: The magnetic field due to a long wire is $B = \dfrac{\mu_0 I}{2\pi r}$.

EXECUTE: First solve for the current, then substitute the numbers using the above equation.

(a) Solving for the current gives

$$I = 2\pi rB/\mu_0 = 2\pi(0.0200 \text{ m})(1.00\times10^{-4} \text{ T})/(4\pi\times10^{-7} \text{ T}\cdot\text{m/A}) = 10.0 \text{ A}$$

(b) The earth's horizontal field points northward, so at all points directly above the wire the field of the wire would point northward.

(c) At all points directly east of the wire, its field would point northward.

EVALUATE: Even though the earth's magnetic field is rather weak, it requires a fairly large current to cancel this field.

28.23. **IDENTIFY:**   The total magnetic field is the vector sum of the constant magnetic field and the wire's magnetic field.

SET UP: For the wire, $B_{\text{wire}} = \dfrac{\mu_0 I}{2\pi r}$ and the direction of B_{wire} is given by the right-hand rule that is illustrated in Figure 28.6 in the textbook. $\vec{B}_0 = (1.50 \times 10^{-6}\text{ T})\hat{\imath}$.

EXECUTE: **(a)** At $(0, 0, 1\text{ m})$, $\vec{B} = \vec{B}_0 - \dfrac{\mu_0 I}{2\pi r}\hat{\imath} = (1.50 \times 10^{-6}\text{ T})\hat{\imath} - \dfrac{\mu_0 (8.00\text{ A})}{2\pi(1.00\text{ m})}\hat{\imath} = -(1.0 \times 10^{-7}\text{ T})\hat{\imath}$.

(b) At $(1\text{ m}, 0, 0)$, $\vec{B} = \vec{B}_0 + \dfrac{\mu_0 I}{2\pi r}\hat{k} = (1.50 \times 10^{-6}\text{ T})\hat{\imath} + \dfrac{\mu_0 (8.00\text{ A})}{2\pi(1.00\text{ m})}\hat{k}$.

$\vec{B} = (1.50 \times 10^{-6}\text{ T})\hat{\imath} + (1.6 \times 10^{-6}\text{ T})\hat{k} = 2.19 \times 10^{-6}\text{ T}$, at $\theta = 46.8°$ from x to z.

(c) At $(0, 0, -0.25\text{ m})$, $\vec{B} = \vec{B}_0 + \dfrac{\mu_0 I}{2\pi r}\hat{\imath} = (1.50 \times 10^{-6}\text{ T})\hat{\imath} + \dfrac{\mu_0 (8.00\text{ A})}{2\pi(0.25\text{ m})}\hat{\imath} = (7.9 \times 10^{-6}\text{ T})\hat{\imath}$.

EVALUATE: At point c the two fields are in the same direction and their magnitudes add. At point a they are in opposite directions and their magnitudes subtract. At point b the two fields are perpendicular.

28.25. **IDENTIFY:** $B = \dfrac{\mu_0 I}{2\pi r}$. The direction of $\vec{B}$ is given by the right-hand rule.

SET UP: Call the wires a and b, as indicated in Figure 28.25. The magnetic fields of each wire at points P_1 and P_2 are shown in Figure 28.25a. The fields at point 3 are shown in Figure 28.25b.

EXECUTE: **(a)** At P_1, $B_a = B_b$ and the two fields are in opposite directions, so the net field is zero.

(b) $B_a = \dfrac{\mu_0 I}{2\pi r_a}$. $B_b = \dfrac{\mu_0 I}{2\pi r_b}$. $\vec{B}_a$ and $\vec{B}_b$ are in the same direction so

$B = B_a + B_b = \dfrac{\mu_0 I}{2\pi}\left(\dfrac{1}{r_a} + \dfrac{1}{r_b}\right) = \dfrac{(4\pi \times 10^{-7}\text{ T} \cdot \text{m/A})(4.00\text{ A})}{2\pi}\left[\dfrac{1}{0.300\text{ m}} + \dfrac{1}{0.200\text{ m}}\right] = 6.67 \times 10^{-6}\text{ T}$

$\vec{B}$ has magnitude $6.67\ \mu\text{T}$ and is directed toward the top of the page.

(c) In Figure 28.25b, $\vec{B}_a$ is perpendicular to $\vec{r}_a$ and $\vec{B}_b$ is perpendicular to $\vec{r}_b$. $\tan\theta = \dfrac{5\text{ cm}}{20\text{ cm}}$ and $\theta = 14.04°$. $r_a = r_b = \sqrt{(0.200\text{ m})^2 + (0.050\text{ m})^2} = 0.206\text{ m}$ and $B_a = B_b$.

$B = B_a \cos\theta + B_b \cos\theta = 2B_a \cos\theta = 2\left(\dfrac{\mu_0 I}{2\pi r_a}\right)\cos\theta = \dfrac{2(4\pi \times 10^{-7}\text{ T} \cdot \text{m/A})(4.0\text{ A})\cos 14.04°}{2\pi(0.206\text{ m})} = 7.54\ \mu\text{T}$

B has magnitude $7.53\ \mu\text{T}$ and is directed to the left.

EVALUATE: At points directly to the left of both wires the net field is directed toward the bottom of the page.

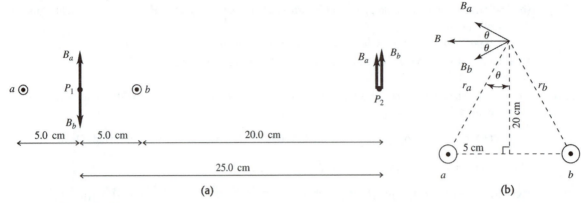

Figure 28.25

28.27.   **IDENTIFY:** The net magnetic field at the center of the square is the vector sum of the fields due to each wire.

SET UP: For each wire, $B = \dfrac{\mu_0 I}{2\pi r}$ and the direction of $\vec{B}$ is given by the right-hand rule that is illustrated in Figure 28.6 in the textbook.

EXECUTE: **(a)** and **(b)** $B = 0$ since the magnetic fields due to currents at opposite corners of the square cancel.

(c) The fields due to each wire are sketched in Figure 28.27.

$$B = B_a \cos 45° + B_b \cos 45° + B_c \cos 45° + B_d \cos 45° = 4B_a \cos 45° = 4\left(\dfrac{\mu_0 I}{2\pi r}\right) \cos 45°.$$

$$r = \sqrt{(10 \text{ cm})^2 + (10 \text{ cm})^2} = 10\sqrt{2} \text{ cm} = 0.10\sqrt{2} \text{ m, so}$$

$$B = 4\dfrac{(4\pi \times 10^{-7} \text{ T} \cdot \text{m/A})(100 \text{ A})}{2\pi(0.10\sqrt{2} \text{ m})} \cos 45° = 4.0 \times 10^{-4} \text{ T, to the left.}$$

EVALUATE: In part (c), if all four currents are reversed in direction, the net field at the center of the square would be to the right.

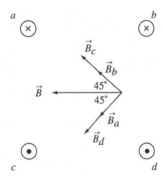

Figure 28.27

28.29.   **IDENTIFY:** The net magnetic field at any point is the vector sum of the magnetic fields of the two wires.

SET UP: For each wire $B = \dfrac{\mu_0 I}{2\pi r}$ and the direction of $\vec{B}$ is determined by the right-hand rule described in the text. Let the wire with 12.0 A be wire 1 and the wire with 10.0 A be wire 2.

EXECUTE: **(a)** *Point Q:* $B_1 = \dfrac{\mu_0 I_1}{2\pi r_1} = \dfrac{(4\pi \times 10^{-7} \text{ T} \cdot \text{m/A})(12.0 \text{ A})}{2\pi(0.15 \text{ m})} = 1.6 \times 10^{-5} \text{ T.}$

The direction of $\vec{B}_1$ is out of the page. $B_2 = \dfrac{\mu_0 I_2}{2\pi r_2} = \dfrac{(4\pi \times 10^{-7} \text{ T} \cdot \text{m/A})(10.0 \text{ A})}{2\pi(0.80 \text{ m})} = 2.5 \times 10^{-5} \text{ T.}$

The direction of $\vec{B}_2$ is out of the page. Since $\vec{B}_1$ and $\vec{B}_2$ are in the same direction,

$B = B_1 + B_2 = 4.1\times10^{-5}$ T and $\vec{B}$ is directed out of the page.

Point P: $B_1 = 1.6\times10^{-5}$ T, directed into the page. $B_2 = 2.5\times10^{-5}$ T, directed into the page.

$B = B_1 + B_2 = 4.1\times10^{-5}$ T and $\vec{B}$ is directed into the page.

(b) $\vec{B}_1$ is the same as in part (a), out of the page at Q and into the page at P. The direction of $\vec{B}_2$ is reversed from what it was in (a) so is into the page at Q and out of the page at P.

Point Q: $\vec{B}_1$ and $\vec{B}_2$ are in opposite directions so $B = B_2 - B_1 = 2.5\times10^{-5}$ T -1.6×10^{-5} T $= 9.0\times10^{-6}$ T and $\vec{B}$ is directed into the page.

Point P: $\vec{B}_1$ and $\vec{B}_2$ are in opposite directions so $B = B_2 - B_1 = 9.0\times10^{-6}$ T and $\vec{B}$ is directed out of the page.

EVALUATE: Points P and Q are the same distances from the two wires. The only difference is that the fields point in either the same direction or in opposite directions.

28.31.   **IDENTIFY:** Apply Eq. (28.11).

SET UP: Two parallel conductors carrying current in the same direction attract each other. Parallel conductors carrying currents in opposite directions repel each other.

EXECUTE: **(a)** $F = \dfrac{\mu_0 I_1 I_2 L}{2\pi r} = \dfrac{\mu_0 (5.00 \text{ A})(2.00 \text{ A})(1.20 \text{ m})}{2\pi(0.400 \text{ m})} = 6.00\times10^{-6}$ N, and the force is repulsive since the currents are in opposite directions.

(b) Doubling the currents makes the force increase by a factor of four to $F = 2.40\times10^{-5}$ N.

EVALUATE: Doubling the current in a wire doubles the magnetic field of that wire. For fixed magnetic field, doubling the current in a wire doubles the force that the magnetic field exerts on the wire.

28.37.   **IDENTIFY:** Calculate the magnetic field vector produced by each wire and add these fields to get the total field.

SET UP: First consider the field at P produced by the current I_1 in the upper semicircle of wire. See Figure 28.37a.

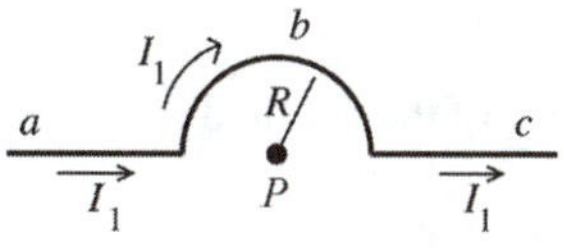

Consider the three parts of this wire
a: long straight section
b: semicircle
c: long, straight section

Figure 28.37a

Apply the Biot-Savart law $d\vec{B} = \dfrac{\mu_0}{4\pi}\dfrac{I d\vec{l} \times \hat{r}}{r^2} = \dfrac{\mu_0}{4\pi}\dfrac{I d\vec{l} \times \vec{r}}{r^3}$ to each piece.

EXECUTE: part a See Figure 28.37b.

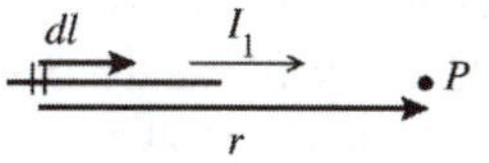

$d\vec{l} \times \vec{r} = 0$,
so $dB = 0$

Figure 28.37b

The same is true for all the infinitesimal segments that make up this piece of the wire, so $B = 0$ for this piece.

part c See Figure 28.37c.

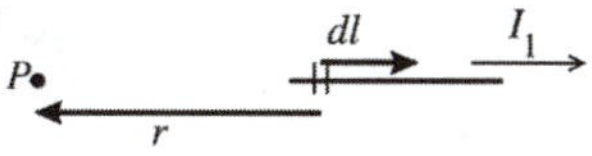

$d\vec{l} \times \vec{r} = 0$,
so $dB = 0$ and $B = 0$ for this piece.

Figure 28.37c

part b See Figure 28.37d.

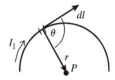

$d\vec{l} \times \vec{r}$ is directed into the paper for all infinitesimal segments that make up this semicircular piece, so $\vec{B}$ is directed into the paper and $B = \int dB$ (the vector sum of the $d\vec{B}$ is obtained by adding their magnitudes since they are in the same direction).

Figure 28.37d

$\left| d\vec{l} \times \vec{r} \right| = rdl \sin\theta$. The angle θ between $d\vec{l}$ and $\vec{r}$ is $90°$ and $r = R$, the radius of the semicircle. Thus $\left| d\vec{l} \times \vec{r} \right| = R\,dl$

$$dB = \frac{\mu_0}{4\pi} \frac{I \left| d\vec{l} \times \vec{r} \right|}{r^3} = \frac{\mu_0 I_1}{4\pi} \frac{R}{R^3} dl = \left(\frac{\mu_0 I_1}{4\pi R^2} \right) dl$$

$$B = \int dB = \left(\frac{\mu_0 I_1}{4\pi R^2} \right) \int dl = \left(\frac{\mu_0 I_1}{4\pi R^2} \right) (\pi R) = \frac{\mu_0 I_1}{4R}$$

(We used that $\int dl$ is equal to πR, the length of wire in the semicircle.) We have shown that the two straight sections make zero contribution to $\vec{B}$, so $B_1 = \mu_0 I_1 / 4R$ and is directed into the page.

For current in the direction shown in Figure 28.37e, a similar analysis gives $B_2 = \mu_0 I_2 / 4R$, out of the paper.

Figure 28.37e

$\vec{B}_1$ and $\vec{B}_2$ are in opposite directions, so the magnitude of the net field at P is $B = \left| B_1 - B_2 \right| = \frac{\mu_0 \left| I_1 - I_2 \right|}{4R}$.

EVALUATE: When $I_1 = I_2$, $B = 0$.

28.39. **IDENTIFY:** Apply Eq. (28.16).
SET UP: At the center of the coil, $x = 0$. a is the radius of the coil, 0.020 m.

EXECUTE: **(a)** $B_{\text{center}} = \frac{\mu_0 NI}{2a} = \frac{\mu_0 (600)(0.500\ \text{A})}{2(0.020\ \text{m})} = 9.42 \times 10^{-3}\ \text{T}$.

(b) $B(x) = \frac{\mu_0 NIa^2}{2(x^2 + a^2)^{3/2}}$. $B(0.08\ \text{m}) = \frac{\mu_0 (600)(0.500\ \text{A})(0.020\ \text{m})^2}{2((0.080\ \text{m})^2 + (0.020\ \text{m})^2)^{3/2}} = 1.34 \times 10^{-4}\ \text{T}$.

EVALUATE: As shown in Figure 28.14 in the textbook, the field has its largest magnitude at the center of the coil and decreases with distance along the axis from the center.

28.41. **IDENTIFY:** The field at the center of the loops is the vector sum of the field due to each loop. They must be in opposite directions in order to add to zero.
SET UP: Let wire 1 be the inner wire with diameter 20.0 cm and let wire 2 be the outer wire with diameter 30.0 cm. To produce zero net field, the fields $\vec{B}_1$ and $\vec{B}_2$ of the two wires must have equal magnitudes and opposite directions. At the center of a wire loop $B = \frac{\mu_0 I}{2R}$. The direction of $\vec{B}$ is given by the right-hand rule applied to the current direction.

EXECUTE: $B_1 = \dfrac{\mu_0 I}{2R_1}$, $B_2 = \dfrac{\mu_0 I}{2R_2}$. $B_1 = B_2$ gives $\dfrac{\mu_0 I_1}{2R_1} = \dfrac{\mu_0 I_2}{2R_2}$. Solving for I_2 gives

$I_2 = \left(\dfrac{R_2}{R_1}\right)I_1 = \left(\dfrac{15.0\text{ cm}}{10.0\text{ cm}}\right)(12.0\text{ A}) = 18.0\text{ A}$. The directions of I_1 and of its field are shown in Figure 28.41.

Since $\vec{B}_1$ is directed into the page, $\vec{B}_2$ must be directed out of the page and I_2 is counterclockwise.

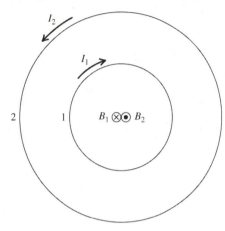

Figure 28.41

EVALUATE: The outer current, I_2, must be larger than the inner current, I_1, because the outer ring is larger than the inner ring, which makes the outer current farther from the center than the inner current is.

28.45.   **IDENTIFY:**   Apply Ampere's law.

SET UP: To calculate the magnetic field at a distance r from the center of the cable, apply Ampere's law to a circular path of radius r. By symmetry, $\oint \vec{B} \cdot d\vec{l} = B(2\pi r)$ for such a path.

EXECUTE: **(a)** For $a < r < b$, $I_{\text{encl}} = I \Rightarrow \oint \vec{B} \cdot d\vec{l} = \mu_0 I \Rightarrow B2\pi r = \mu_0 I \Rightarrow B = \dfrac{\mu_0 I}{2\pi r}$.

(b) For $r > c$, the enclosed current is zero, so the magnetic field is also zero.

EVALUATE: A useful property of coaxial cables for many applications is that the current carried by the cable doesn't produce a magnetic field outside the cable.

28.47.   **IDENTIFY:**   The largest value of the field occurs at the surface of the cylinder. Inside the cylinder, the field increases linearly from zero at the center, and outside the field decreases inversely with distance from the central axis of the cylinder.

SET UP: At the surface of the cylinder, $B = \dfrac{\mu_0 I}{2\pi R}$, inside the cylinder, Eq. 28.21 gives $B = \dfrac{\mu_0 I}{2\pi}\dfrac{r}{R^2}$ and

outside the field is $B = \dfrac{\mu_0 I}{2\pi r}$.

EXECUTE: For points inside the cylinder, the field is half its maximum value when $\dfrac{\mu_0 I}{2\pi}\dfrac{r}{R^2} = \dfrac{1}{2}\left(\dfrac{\mu_0 I}{2\pi R}\right)$,

which gives $r = R/2$. Outside the cylinder, we have $\dfrac{\mu_0 I}{2\pi r} = \dfrac{1}{2}\left(\dfrac{\mu_0 I}{2\pi R}\right)$, which gives $r = 2R$.

EVALUATE: The field has half its maximum value at all points on cylinders coaxial with the wire but of radius $R/2$ and of radius $2R$.

28.49.   **(a) IDENTIFY** and **SET UP:**   The magnetic field near the center of a long solenoid is given by Eq. (28.23), $B = \mu_0 n I$.

EXECUTE: Turns per unit length $n = \dfrac{B}{\mu_0 I} = \dfrac{0.0270\text{ T}}{(4\pi \times 10^{-7}\text{ T}\cdot\text{m/A})(12.0\text{ A})} = 1790$ turns/m

(b) $N = nL = (1790 \text{ turns/m})(0.400 \text{ m}) = 716 \text{ turns}$

Each turn of radius R has a length $2\pi R$ of wire. The total length of wire required is

$N(2\pi R) = (716)(2\pi)(1.40 \times 10^{-2} \text{ m}) = 63.0 \text{ m}.$

EVALUATE: A large length of wire is required. Due to the length of wire the solenoid will have appreciable resistance.

28.53. **IDENTIFY:** Example 28.10 shows that inside a toroidal solenoid, $B = \dfrac{\mu_0 NI}{2\pi r}$.

SET UP: $r = 0.070 \text{ m}$

EXECUTE: $B = \dfrac{\mu_0 NI}{2\pi r} = \dfrac{\mu_0 (600)(0.650 \text{ A})}{2\pi (0.070 \text{ m})} = 1.11 \times 10^{-3} \text{ T}.$

EVALUATE: If the radial thickness of the torus is small compared to its mean diameter, B is approximately uniform inside its windings.

28.55. **IDENTIFY and SET UP:** $B = \dfrac{K_m \mu_0 NI}{2\pi r}$ (Eq. 28.24, with μ_0 replaced by $K_m \mu_0$)

EXECUTE: **(a)** $K_m = 1400$

$I = \dfrac{2\pi rB}{\mu_0 K_m N} = \dfrac{(2.90 \times 10^{-2} \text{ m})(0.350 \text{ T})}{(2 \times 10^{-7} \text{ T} \cdot \text{m/A})(1400)(500)} = 0.0725 \text{ A}$

(b) $K_m = 5200$

$I = \dfrac{2\pi rB}{\mu_0 K_m N} = \dfrac{(2.90 \times 10^{-2} \text{ m})(0.350 \text{ T})}{(2 \times 10^{-7} \text{ T} \cdot \text{m/A})(5200)(500)} = 0.0195 \text{ A}$

EVALUATE: If the solenoid were air-filled instead, a much larger current would be required to produce the same magnetic field.

28.57. **IDENTIFY:** The magnetic field from the solenoid alone is $B_0 = \mu_0 nI$. The total magnetic field is $B = K_m B_0$. M is given by Eq. (28.29).

SET UP: $n = 6000 \text{ turns/m}$

EXECUTE: **(a)** (i) $B_0 = \mu_0 nI = \mu_0 (6000 \text{ m}^{-1})(0.15 \text{ A}) = 1.13 \times 10^{-3} \text{ T}.$

(ii) $M = \dfrac{K_m - 1}{\mu_0} B_0 = \dfrac{5199}{\mu_0}(1.13 \times 10^{-3} \text{ T}) = 4.68 \times 10^6 \text{ A/m}.$

(iii) $B = K_m B_0 = (5200)(1.13 \times 10^{-3} \text{ T}) = 5.88 \text{ T}.$

(b) The directions of $\vec{B}$, $\vec{B}_0$ and $\vec{M}$ are shown in Figure 28.57. Silicon steel is paramagnetic and $\vec{B}_0$ and $\vec{M}$ are in the same direction.

EVALUATE: The total magnetic field is much larger than the field due to the solenoid current alone.

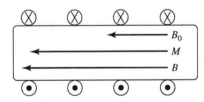

Figure 28.57

28.59. **IDENTIFY:** Moving charges create magnetic fields. The net field is the vector sum of the two fields. A charge moving in an external magnetic field feels a force.

(a) SET UP: The magnetic field due to a moving charge is $B = \dfrac{\mu_0}{4\pi} \dfrac{qv\sin\phi}{r^2}$. Both fields are into the paper,

so their magnitudes add, giving $B_{\text{net}} = B + B' = \dfrac{\mu_0}{4\pi}\left(\dfrac{qv\sin\phi}{r^2} + \dfrac{q'v'\sin\phi'}{r'^2}\right).$

EXECUTE: Substituting numbers gives

$$B_{net} = \frac{\mu_0}{4\pi}\left[\frac{(8.00\ \mu C)(9.00\times10^4\ m/s)\sin90°}{(0.300\ m)^2} + \frac{(5.00\ \mu C)(6.50\times10^4\ m/s)\sin90°}{(0.400\ m)^2} \right]$$

$B_{net} = 1.00\times10^{-6}\ T = 1.00\mu T$, into the paper.

(b) SET UP: The magnetic force on a moving charge is $\vec{F} = q\vec{v}\times\vec{B}$, and the magnetic field of charge q' at the location of charge q is into the page. The force on q is

$$\vec{F} = q\vec{v}\times\vec{B}' = (qv)\hat{i}\times\frac{\mu_0}{4\pi}\frac{q\vec{v}'\times\hat{r}}{r^2} = (qv)\hat{i}\times\left(\frac{\mu_0}{4\pi}\frac{qv'\sin\phi}{r^2}\right)(-\hat{k}) = \left(\frac{\mu_0}{4\pi}\frac{qq'vv'\sin\phi}{r^2}\right)\hat{j}$$

where ϕ is the angle between $\vec{v}'$ and $\hat{r}'$.

EXECUTE: Substituting numbers gives

$$\vec{F} = \frac{\mu_0}{4\pi}\left[\frac{(8.00\times10^{-6}\ C)(5.00\times10^{-6}\ C)(9.00\times10^4\ m/s)(6.50\times10^4\ m/s)}{(0.500\ m)^2}\left(\frac{0.400}{0.500}\right) \right]\hat{j}$$

$$\vec{F} = (7.49\times10^{-8}\ N)\hat{j}.$$

EVALUATE: These are small fields and small forces, but if the charge has small mass, the force can affect its motion.

28.61. **IDENTIFY:** Use Eq. (28.9) and the right-hand rule to determine points where the fields of the two wires cancel.
(a) SET UP: The only place where the magnetic fields of the two wires are in opposite directions is between the wires, in the plane of the wires. Consider a point a distance x from the wire carrying $I_2 = 75.0\ A$. B_{tot} will be zero where $B_1 = B_2$.

EXECUTE: $\dfrac{\mu_0 I_1}{2\pi(0.400\ m - x)} = \dfrac{\mu_0 I_2}{2\pi x}$

$I_2(0.400\ m - x) = I_1 x$; $I_1 = 25.0\ A$, $I_2 = 75.0\ A$

$x = 0.300\ m$; $B_{tot} = 0$ along a line 0.300 m from the wire carrying 75.0 A and 0.100 m from the wire carrying current 25.0 A.

(b) SET UP: Let the wire with $I_1 = 25.0\ A$ be 0.400 m above the wire with $I_2 = 75.0\ A$. The magnetic fields of the two wires are in opposite directions in the plane of the wires and at points above both wires or below both wires. But to have $B_1 = B_2$ must be closer to wire #1 since $I_1 < I_2$, so can have $B_{tot} = 0$ only at points above both wires. Consider a point a distance x from the wire carrying $I_1 = 25.0\ A$. B_{tot} will be zero where $B_1 = B_2$.

EXECUTE: $\dfrac{\mu_0 I_1}{2\pi x} = \dfrac{\mu_0 I_2}{2\pi(0.400\ m + x)}$

$I_2 x = I_1(0.400\ m + x)$; $x = 0.200\ m$

$B_{tot} = 0$ along a line 0.200 m from the wire carrying current 25.0 A and 0.600 m from the wire carrying current $I_2 = 75.0\ A$.

EVALUATE: For parts (a) and (b) the locations of zero field are in different regions. In each case the points of zero field are closer to the wire that has the smaller current.

28.63. **IDENTIFY:** Find the force that the magnetic field of the wire exerts on the electron.
SET UP: The force on a moving charge has magnitude $F = |q|vB\sin\phi$ and direction given by the right-hand rule. For a long straight wire, $B = \dfrac{\mu_0 I}{2\pi r}$ and the direction of $\vec{B}$ is given by the right-hand rule.

EXECUTE: **(a)** $a = \dfrac{F}{m} = \dfrac{|q|vB\sin\phi}{m} = \dfrac{ev}{m}\left(\dfrac{\mu_0 I}{2\pi r}\right)$. Substituting numbers gives

$$a = \frac{(1.6\times10^{-19}\ C)(2.50\times10^5\ m/s)(4\pi\times10^{-7}\ T\cdot m/A)(13.0\ A)}{(9.11\times10^{-31}\ kg)(2\pi)(0.0200\ m)} = 5.7\times10^{12}\ m/s^2,\ \text{away from the wire.}$$

(b) The electric force must balance the magnetic force. $eE = evB$, and

$$E = vB = v\frac{\mu_0 I}{2\pi r} = \frac{(250{,}000 \text{ m/s})(4\pi \times 10^{-7} \text{ T} \cdot \text{m/A})(13.0 \text{ A})}{2\pi(0.0200 \text{ m})} = 32.5 \text{ N/C}.$$ The magnetic force is directed

away from the wire so the force from the electric field must be toward the wire. Since the charge of the electron is negative, the electric field must be directed away from the wire to produce a force in the desired direction.

EVALUATE: (c) $mg = (9.11 \times 10^{-31} \text{ kg})(9.8 \text{ m/s}^2) \approx 10^{-29}$ N.

$F_{el} = eE = (1.6 \times 10^{-19} \text{ C})(32.5 \text{ N/C}) \approx 5 \times 10^{-18}$ N. $F_{el} \approx 5 \times 10^{11} F_{grav}$, so we can neglect gravity.

28.65.   **IDENTIFY:**   Find the net magnetic field due to the two loops at the location of the proton and then find the force these fields exert on the proton.

SET UP: For a circular loop, the field on the axis, a distance x from the center of the loop is

$$B = \frac{\mu_0 I R^2}{2(R^2 + x^2)^{3/2}}.$$ $R = 0.200$ m and $x = 0.125$ m.

EXECUTE: The fields add, so $B = B_1 + B_2 = 2B_1 = 2\left[\frac{\mu_0 I R^2}{2(R^2 + x^2)^{3/2}}\right]$. Putting in the numbers gives

$$B = \frac{(4\pi \times 10^{-7} \text{ T} \cdot \text{m/A})(3.80 \text{ A})(0.200 \text{ m})^2}{[(0.200 \text{ m})^2 + (0.125 \text{ m})^2]^{3/2}} = 1.46 \times 10^{-5} \text{ T}.$$ The magnetic force is

$F = |q|vB \sin\phi = (1.6 \times 10^{-19} \text{ C})(2{,}400{,}000 \text{ m/s})(1.46 \times 10^{-5} \text{ T})\sin 90° = 5.59 \times 10^{-18}$ N.

EVALUATE: The weight of a proton is $w = mg = 1.6 \times 10^{-24}$ N, so the force from the loops is much greater than the gravity force on the proton.

28.69.   **IDENTIFY:**   Apply $F = lB \sin\phi$, with the magnetic field at point P that is calculated in Problem 28.68.

SET UP: The net field of the first two wires at the location of the third wire is $B = \dfrac{\mu_0 Ia}{\pi(x^2 + a^2)}$, in the

+x-direction.

EXECUTE: (a) Wire is carrying current into the page, so it feels a force in the $-y$-direction.

$$\frac{F}{L} = IB = I\left(\frac{\mu_0 Ia}{\pi(x^2 + a^2)}\right) = \frac{\mu_0(6.00 \text{ A})^2(0.400 \text{ m})}{\pi((0.600 \text{ m})^2 + (0.400 \text{ m})^2)} = 1.11 \times 10^{-5} \text{ N/m}.$$

(b) If the wire carries current out of the page then the force felt will be in the opposite direction as in part (a). Thus the force will be 1.11×10^{-5} N/m, in the $+y$-direction.

EVALUATE: We could also calculate the force exerted by each of the first two wires and find the vector sum of the two forces.

28.71.   **IDENTIFY:**   Apply $\sum \vec{F} = 0$ to one of the wires. The force one wire exerts on the other depends on I so $\sum \vec{F} = 0$ gives two equations for the two unknowns T and I.

SET UP: The force diagram for one of the wires is given in Figure 28.71.

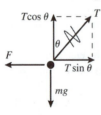

The force one wire exerts on the other is $F = \left(\dfrac{\mu_0 I^2}{2\pi r}\right) L$, where

$r = 2(0.040 \text{ m})\sin\theta = 8.362 \times 10^{-3}$ m is the distance between the two wires.

Figure 28.71

EXECUTE: $\sum F_y = 0$ gives $T \cos\theta = mg$ and $T = mg/\cos\theta$

$\sum F_x = 0$ gives $F = T \sin\theta = (mg/\cos\theta)\sin\theta = mg \tan\theta$

And $m = \lambda L$, so $F = \lambda Lg \tan\theta$

$$\left(\frac{\mu_0 I^2}{2\pi r}\right)L = \lambda Lg \tan\theta$$

$$I = \sqrt{\frac{\lambda gr \tan\theta}{(\mu_0/2\pi)}}$$

$$I = \sqrt{\frac{(0.0125 \text{ kg/m})(9.80 \text{ m/s}^2)(\tan 6.00°)(8.362 \times 10^{-3} \text{ m})}{2 \times 10^{-7} \text{ T} \cdot \text{m/A}}} = 23.2 \text{ A}$$

EVALUATE: Since the currents are in opposite directions the wires repel. When I is increased, the angle θ from the vertical increases; a large current is required even for the small displacement specified in this problem.

28.73.   **IDENTIFY:**   Knowing the magnetic field at the center of the ring, we can calculate the current running through it. We can then use this current to calculate the torque that the external magnetic field exerts on the ring.

SET UP: The torque on a current loop is $\tau = IAB \sin\phi$. We can use the magnetic field of the ring,

$B = \dfrac{\mu_0 I}{2R}$, to calculate the current in the ring.

EXECUTE: $I = \dfrac{2RB_{\text{ring}}}{\mu_0} = \dfrac{2(2.50 \times 10^{-2} \text{ m})(75.4 \times 10^{-6} \text{ T})}{4\pi \times 10^{-7} \text{ T} \cdot \text{m/A}} = 3.00 \text{ A}$. The torque is a maximum when

$\phi = 90°$ and the plane of the ring is parallel to the field.

$\tau_{\text{max}} = IAB = (3.00 \text{ A})(0.375 \text{ T})\pi(2.50 \times 10^{-2} \text{ m})^2 = 2.21 \times 10^{-3} \text{ N} \cdot \text{m}.$

EVALUATE: When the external field is perpendicular to the plane of the ring the torque on the ring is zero.

ELECTROMAGNETIC INDUCTION

29.3. **IDENTIFY and SET UP:** Use Faraday's law to calculate the average induced emf and apply Ohm's law to the coil to calculate the average induced current and charge that flows.

(a) EXECUTE: The magnitude of the average emf induced in the coil is $\left|\varepsilon_{\mathrm{av}}\right| = N\left|\dfrac{\Delta\Phi_B}{\Delta t}\right|$. Initially,

$\Phi_{Bi} = BA\cos\phi = BA$. The final flux is zero, so $\left|\varepsilon_{\mathrm{av}}\right| = N\dfrac{\left|\Phi_{Bf} - \Phi_{Bi}\right|}{\Delta t} = \dfrac{NBA}{\Delta t}$. The average induced current

is $I = \dfrac{\left|\varepsilon_{\mathrm{av}}\right|}{R} = \dfrac{NBA}{R\Delta t}$. The total charge that flows through the coil is $Q = I\Delta t = \left(\dfrac{NBA}{R\Delta t}\right)\Delta t = \dfrac{NBA}{R}$.

EVALUATE: The charge that flows is proportional to the magnetic field but does not depend on the time Δt.

(b) The magnetic stripe consists of a pattern of magnetic fields. The pattern of charges that flow in the reader coil tells the card reader the magnetic field pattern and hence the digital information coded onto the card.

(c) According to the result in part (a) the charge that flows depends only on the change in the magnetic flux and it does not depend on the rate at which this flux changes.

29.7. **IDENTIFY:** Calculate the flux through the loop and apply Faraday's law.

SET UP: To find the total flux integrate $d\Phi_B$ over the width of the loop. The magnetic field of a long

straight wire, at distance r from the wire, is $B = \dfrac{\mu_0 I}{2\pi r}$. The direction of $\vec{B}$ is given by the right-hand rule.

EXECUTE: **(a)** $B = \dfrac{\mu_0 i}{2\pi r}$, into the page.

(b) $d\Phi_B = BdA = \dfrac{\mu_0 i}{2\pi r}Ldr$.

(c) $\Phi_B = \displaystyle\int_a^b d\Phi_B = \dfrac{\mu_0 iL}{2\pi}\int_a^b \dfrac{dr}{r} = \dfrac{\mu_0 iL}{2\pi}\ln(b/a)$.

(d) $\left|\varepsilon\right| = \dfrac{d\Phi_B}{dt} = \dfrac{\mu_0 L}{2\pi}\ln(b/a)\dfrac{di}{dt}$.

(e) $\left|\varepsilon\right| = \dfrac{\mu_0(0.240\text{ m})}{2\pi}\ln(0.360/0.120)(9.60\text{ A/s}) = 5.06\times 10^{-7}\text{ V}$.

EVALUATE: The induced emf is proportional to the rate at which the current in the long straight wire is changing

29.9. **IDENTIFY and SET UP:** Use Faraday's law to calculate the emf (magnitude and direction). The direction of the induced current is the same as the direction of the emf. The flux changes because the area of the loop is changing; relate dA/dt to dc/dt, where c is the circumference of the loop.

(a) EXECUTE: $c = 2\pi r$ and $A = \pi r^2$ so $A = c^2/4\pi$

$\Phi_B = BA = (B/4\pi)c^2$

$$|\varepsilon| = \left|\frac{d\Phi_B}{dt}\right| = \left(\frac{B}{2\pi}\right)c\left|\frac{dc}{dt}\right|$$

At $t = 9.0$ s, $c = 1.650$ m $- (9.0$ s$)(0.120$ m/s$) = 0.570$ m

$|\varepsilon| = (0.500$ T$)(1/2\pi)(0.570$ m$)(0.120$ m/s$) = 5.44$ mV

(b) **SET UP:** The loop and magnetic field are sketched in Figure 29.9.

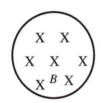

Take into the page to be the positive direction for $\vec{A}$. Then the magnetic flux is positive.

Figure 29.9

EXECUTE: The positive flux is decreasing in magnitude; $d\Phi_B/dt$ is negative and ε is positive. By the right-hand rule, for $\vec{A}$ into the page, positive ε is clockwise.

EVALUATE: Even though the circumference is changing at a constant rate, dA/dt is not constant and $|\varepsilon|$ is not constant. Flux $\otimes$ is decreasing so the flux of the induced current is $\otimes$ and this means that I is clockwise, which checks.

29.11.   **IDENTIFY:** A change in magnetic flux through a coil induces an emf in the coil.
SET UP: The flux through a coil is $\Phi_B = NBA\cos\phi$ and the induced emf is $\varepsilon = -d\Phi_B/dt$.

EXECUTE: **(a)** $|\varepsilon| = d\Phi_B/dt = d[A(B_0 + bx)]/dt = bA\,dx/dt = bAv$

(b) clockwise

(c) Same answers except the current is counterclockwise.

EVALUATE: Even though the coil remains within the magnetic field, the flux through it changes because the strength of the field is changing.

29.13.   **IDENTIFY:** Apply the results of Example 29.3.
SET UP: $\varepsilon_{max} = NBA\omega$

EXECUTE: $\omega = \dfrac{\varepsilon_{max}}{NBA} = \dfrac{2.40\times10^{-2}\text{ V}}{(120)(0.0750\text{ T})(0.016\text{ m})^2} = 10.4$ rad/s

EVALUATE: We may also express ω as 99.3 rev/min or 1.66 rev/s.

29.17.   **IDENTIFY and SET UP:** Apply Lenz's law, in the form that states that the flux of the induced current tends to oppose the change in flux.
EXECUTE: **(a)** With the switch closed the magnetic field of coil A is to the right at the location of coil B. When the switch is opened the magnetic field of coil A goes away. Hence by Lenz's law the field of the current induced in coil B is to the right, to oppose the decrease in the flux in this direction. To produce magnetic field that is to the right the current in the circuit with coil B must flow through the resistor in the direction a to b.

(b) With the switch closed the magnetic field of coil A is to the right at the location of coil B. This field is stronger at points closer to coil A so when coil B is brought closer the flux through coil B increases. By Lenz's law the field of the induced current in coil B is to the left, to oppose the increase in flux to the right. To produce magnetic field that is to the left the current in the circuit with coil B must flow through the resistor in the direction b to a.

(c) With the switch closed the magnetic field of coil A is to the right at the location of coil B. The current in the circuit that includes coil A increases when R is decreased and the magnetic field of coil A increases when the current through the coil increases. By Lenz's law the field of the induced current in coil B is to the left, to oppose the increase in flux to the right. To produce magnetic field that is to the left the current in the circuit with coil B must flow through the resistor in the direction b to a.

EVALUATE: In parts (b) and (c) the change in the circuit causes the flux through circuit B to increase and in part (a) it causes the flux to decrease. Therefore, the direction of the induced current is the same in parts (b) and (c) and opposite in part (a).

29.21.   **IDENTIFY:** The changing flux through the loop due to the changing magnetic field induces a current in the wire.

SET UP: The magnitude of the induced emf is $\left|\varepsilon\right| = \left|\dfrac{d\Phi_B}{dt}\right| = \pi r^2 \left|\dfrac{dB}{dt}\right|$, $I = \varepsilon/R$.

EXECUTE: $\vec{B}$ is into the page and Φ_B is increasing, so the field of the induced current is directed out of the page inside the loop and the induced current is counterclockwise.

$\left|\varepsilon\right| = \left|\dfrac{d\Phi_B}{dt}\right| = \pi r^2 \left|\dfrac{dB}{dt}\right| = \pi(0.0250\ \text{m})^2 (0.380\ \text{T/s}^3)(3t^2) = (2.238 \times 10^{-3}\ \text{V/s}^2)t^2.$

$I = \dfrac{\left|\varepsilon\right|}{R} = (5.739 \times 10^{-3}\ \text{A/s}^2)t^2.$ When $B = 1.33\ \text{T}$, we have $1.33\ \text{T} = (0.380\ \text{T/s}^3)t^3$, which gives

$t = 1.518\ \text{s}.$ At this t, $I = (5.739 \times 10^{-3}\ \text{A/s}^2)(1.518\ \text{s})^2 = 0.0132\ \text{A}.$

EVALUATE: As the field changes, the current will also change.

29.27.   **IDENTIFY and SET UP:** $\varepsilon = vBL.$ Use Lenz's law to determine the direction of the induced current. The force F_{ext} required to maintain constant speed is equal and opposite to the force F_I that the magnetic field exerts on the rod because of the current in the rod.

EXECUTE: **(a)** $\varepsilon = vBL = (7.50\ \text{m/s})(0.800\ \text{T})(0.500\ \text{m}) = 3.00\ \text{V}$

(b) $\vec{B}$ is into the page. The flux increases as the bar moves to the right, so the magnetic field of the induced current is out of the page inside the circuit. To produce magnetic field in this direction the induced current must be counterclockwise, so from b to a in the rod.

(c) $I = \dfrac{\varepsilon}{R} = \dfrac{3.00\ \text{V}}{1.50\ \Omega} = 2.00\ \text{A}.$ $F_I = ILB \sin\phi = (2.00\ \text{A})(0.500\ \text{m})(0.800\ \text{T})\sin 90° = 0.800\ \text{N}.$ $\vec{F}_I$ is to the

left. To keep the bar moving to the right at constant speed an external force with magnitude $F_{\text{ext}} = 0.800\ \text{N}$ and directed to the right must be applied to the bar.

(d) The rate at which work is done by the force F_{ext} is $F_{\text{ext}}v = (0.800\ \text{N})(7.50\ \text{m/s}) = 6.00\ \text{W}.$ The rate at

which thermal energy is developed in the circuit is $I^2 R = (2.00\ \text{A})^2 (1.50\ \Omega) = 6.00\ \text{W}.$ These two rates are equal, as is required by conservation of energy.

EVALUATE: The force on the rod due to the induced current is directed to oppose the motion of the rod. This agrees with Lenz's law.

29.29.   **IDENTIFY:** The motion of the bar due to the applied force causes a motional emf to be induced across the ends of the bar, which induces a current through the bar. The magnetic field exerts a force on the bar due to this current.

SET UP: The applied force is to the left and equal to $F_{\text{applied}} = F_B = ILB.$ $\varepsilon = BvL$ and $I = \dfrac{\varepsilon}{R} = \dfrac{BvL}{R}.$

EXECUTE: **(a)** $\vec{B}$ out of page and Φ_B decreasing, so the field of the induced current is out of the page inside the loop and the induced current is counterclockwise.

(b) Combining $F_{\text{applied}} = F_B = ILB$ and $\varepsilon = BvL$, we have $I = \dfrac{\varepsilon}{R} = \dfrac{BvL}{R}.$ $F_{\text{applied}} = \dfrac{vB^2 L^2}{R}.$ The rate at

which this force does work is $P_{\text{applied}} = F_{\text{applied}}v = \dfrac{(vBL)^2}{R} = \dfrac{[(5.90\ \text{m/s})(0.650\ \text{T})(0.360\ \text{m})]^2}{45.0\ \Omega} = 0.0424\ \text{W}.$

EVALUATE: The power is small because the magnetic force is usually small compared to everyday forces.

29.33.   **IDENTIFY:** A bar moving in a magnetic field has an emf induced across its ends.

SET UP: The induced potential is $\varepsilon = vBL \sin\phi.$

EXECUTE: Note that $\phi = 90°$ in all these cases because the bar moved perpendicular to the magnetic field. But the effective length of the bar, $L \sin\theta$, is different in each case.

(a) $\varepsilon = vBL \sin \theta = (2.50 \text{ m/s})(1.20 \text{ T})(1.41 \text{ m}) \sin (37.0°) = 2.55$ V, with a at the higher potential because positive charges are pushed toward that end.

(b) Same as (a) except $\theta = 53.0°$, giving 3.38 V, with a at the higher potential.

(c) Zero, since the velocity is parallel to the magnetic field.

(d) The bar must move perpendicular to its length, for which the emf is 4.23 V. For $V_b > V_a$, it must move upward and to the left (toward the second quadrant) perpendicular to its length.

EVALUATE: The orientation of the bar affects the potential induced across its ends.

29.37. **IDENTIFY:** Apply Eq. (29.11) with $\Phi_B = \mu_0 niA$.

SET UP: $A = \pi r^2$, where $r = 0.0110$ m. In Eq. (29.11), $r = 0.0350$ m.

EXECUTE: $|\varepsilon| = \left|\dfrac{d\Phi_B}{dt}\right| = \left|\dfrac{d}{dt}(BA)\right| = \left|\dfrac{d}{dt}(\mu_0 niA)\right| = \mu_0 nA \left|\dfrac{di}{dt}\right|$ and $|\varepsilon| = E(2\pi r)$. Therefore, $\left|\dfrac{di}{dt}\right| = \dfrac{E2\pi r}{\mu_0 nA}$.

$\left|\dfrac{di}{dt}\right| = \dfrac{(8.00\times10^{-6} \text{ V/m})2\pi(0.0350 \text{ m})}{\mu_0 (400 \text{ m}^{-1})\pi(0.0110 \text{ m})^2} = 9.21$ A/s.

EVALUATE: Outside the solenoid the induced electric field decreases with increasing distance from the axis of the solenoid.

29.39. **IDENTIFY:** Apply Faraday's law in the form $|\varepsilon_{\text{av}}| = N\left|\dfrac{\Delta\Phi_B}{\Delta t}\right|$.

SET UP: The magnetic field of a large straight solenoid is $B = \mu_0 nI$ inside the solenoid and zero outside. $\Phi_B = BA$, where A is 8.00 cm², the cross-sectional area of the long straight solenoid.

EXECUTE: $|\varepsilon_{\text{av}}| = N\left|\dfrac{\Delta\Phi_B}{\Delta t}\right| = \left|\dfrac{NA(B_{\text{f}} - B_{\text{i}})}{\Delta t}\right| = \dfrac{NA\mu_0 nI}{\Delta t}$.

$\varepsilon_{\text{av}} = \dfrac{\mu_0 (12)(8.00\times10^{-4} \text{ m}^2)(9000 \text{ m}^{-1})(0.350 \text{ A})}{0.0400 \text{ s}} = 9.50\times10^{-4}$ V.

EVALUATE: An emf is induced in the second winding even though the magnetic field of the solenoid is zero at the location of the second winding. The changing magnetic field induces an electric field outside the solenoid and that induced electric field produces the emf.

29.41. **IDENTIFY:** Apply Eq. (29.14), where $\epsilon = K\epsilon_0$.

SET UP: $d\Phi_E/dt = 4(8.76\times10^3 \text{ V} \cdot \text{m/s}^4)t^3$. $\epsilon_0 = 8.854\times10^{-12}$ F/m.

EXECUTE: $\epsilon = \dfrac{i_D}{(d\Phi_E/dt)} = \dfrac{12.9\times10^{-12} \text{ A}}{4(8.76\times10^3 \text{ V} \cdot \text{m/s}^4)(26.1\times10^{-3} \text{ s})^3} = 2.07\times10^{-11}$ F/m. The dielectric

constant is $K = \dfrac{\epsilon}{\epsilon_0} = 2.34$.

EVALUATE: The larger the dielectric constant, the larger is the displacement current for a given $d\Phi_E/dt$.

29.43. **IDENTIFY:** $q = CV$. For a parallel-plate capacitor, $C = \dfrac{\epsilon A}{d}$, where $\epsilon = K\epsilon_0$. $i_C = dq/dt$. $j_D = \epsilon\dfrac{E}{dt}$.

SET UP: $E = q/\epsilon A$ so $dE/dt = i_C/\epsilon A$.

EXECUTE: **(a)** $q = CV = \left(\dfrac{\epsilon A}{d}\right)V = \dfrac{(4.70)\epsilon_0 (3.00\times10^{-4} \text{ m}^2)(120 \text{ V})}{2.50\times10^{-3} \text{ m}} = 5.99\times10^{-10}$ C.

(b) $\dfrac{dq}{dt} = i_C = 6.00\times10^{-3}$ A.

(c) $j_D = \epsilon\dfrac{dE}{dt} = K\epsilon_0\dfrac{i_C}{K\epsilon_0 A} = \dfrac{i_C}{A} = j_C$, so $i_D = i_C = 6.00\times10^{-3}$ A.

EVALUATE: $i_D = i_C$, so Kirchhoff's junction rule is satisfied where the wire connects to each capacitor plate.

29.47. **IDENTIFY:** $\vec{B} = \vec{B}_0 + \mu_0 \vec{M}$.

SET UP: When the magnetic flux is expelled from the material the magnetic field $\vec{B}$ in the material is zero. When the material is completely normal, the magnetization is close to zero.

EXECUTE: **(a)** When $\vec{B}_0$ is just under $\vec{B}_{c1}$ (threshold of superconducting phase), the magnetic field in the material must be zero, and $\vec{M} = -\dfrac{\vec{B}_{c1}}{\mu_0} = -\dfrac{(55 \times 10^{-3}\text{ T})\hat{i}}{\mu_0} = -(4.38 \times 10^4\text{ A/m})\hat{i}$.

(b) When $\vec{B}_0$ is just over $\vec{B}_{c2}$ (threshold of normal phase), there is zero magnetization, and $\vec{B} = \vec{B}_{c2} = (15.0\text{ T})\hat{i}$.

EVALUATE: Between B_{c1} and B_{c2} there are filaments of normal phase material and there is magnetic field along these filaments.

29.49. **IDENTIFY:** Apply Faraday's law and Lenz's law.

SET UP: For a discharging RC circuit, $i(t) = \dfrac{V_0}{R} e^{-t/RC}$, where V_0 is the initial voltage across the capacitor. The resistance of the small loop is $(25)(0.600\text{ m})(1.0\ \Omega/\text{m}) = 15.0\ \Omega$.

EXECUTE: **(a)** The large circuit is an RC circuit with a time constant of $\tau = RC = (10\ \Omega)(20 \times 10^{-6}\text{ F}) = 200\ \mu\text{s}$. Thus, the current as a function of time is $i = ((100\text{ V})/(10\ \Omega)) e^{-t/200\ \mu\text{s}}$. At $t = 200\ \mu\text{s}$, we obtain $i = (10\text{ A})(e^{-1}) = 3.7\text{ A}$.

(b) Assuming that only the long wire nearest the small loop produces an appreciable magnetic flux through the small loop and referring to the solution of Exercise 29.7 we obtain $\Phi_B = \displaystyle\int_c^{c+a} \dfrac{\mu_0 ib}{2\pi r}\,dr = \dfrac{\mu_0 ib}{2\pi}\ln\left(1 + \dfrac{a}{c}\right)$.

Therefore, the emf induced in the small loop at $t = 200\ \mu\text{s}$ is $\varepsilon = -N\dfrac{d\Phi_B}{dt} = -\dfrac{N\mu_0 b}{2\pi}\ln\left(1 + \dfrac{a}{c}\right)\dfrac{di}{dt}$.

$\varepsilon = -\dfrac{(25)(4\pi \times 10^{-7}\text{ Wb/A}\cdot\text{m}^2)(0.200\text{ m})}{2\pi}\ln(3.0)\left(-\dfrac{3.7\text{ A}}{200 \times 10^{-6}\text{ s}}\right) = +20.0\text{ mV}$. Thus, the induced current in the small loop is $i' = \dfrac{\varepsilon}{R} = \dfrac{20.0\text{ mV}}{15.0\ \Omega} = 1.33\text{ mA}$.

(c) The magnetic field from the large loop is directed out of the page within the small loop. The induced current will act to oppose the decrease in flux from the large loop. Thus, the induced current flows counterclockwise.

EVALUATE: **(d)** Three of the wires in the large loop are too far away to make a significant contribution to the flux in the small loop—as can be seen by comparing the distance c to the dimensions of the large loop.

29.51. **IDENTIFY:** The changing current in the solenoid will cause a changing magnetic field (and hence changing flux) through the secondary winding, which will induce an emf in the secondary coil.

SET UP: The magnetic field of the solenoid is $B = \mu_0 ni$, and the induced emf is $|\varepsilon| = N\left|\dfrac{d\Phi_B}{dt}\right|$.

EXECUTE: $B = \mu_0 ni = (4\pi \times 10^{-7}\text{ T}\cdot\text{m/A})(90.0 \times 10^2\text{ m}^{-1})(0.160\text{ A/s}^2)t^2 = (1.810 \times 10^{-3}\text{ T/s}^2)t^2$. The total flux through secondary winding is $(5.0)B(2.00 \times 10^{-4}\text{ m}^2) = (1.810 \times 10^{-6}\text{ Wb/s}^2)t^2$.

$|\varepsilon| = N\left|\dfrac{d\Phi_B}{dt}\right| = (3.619 \times 10^{-6}\text{ V/s})t$. $i = 3.20\text{ A}$ says $3.20\text{ A} = (0.160\text{ A/s}^2)t^2$ and $t = 4.472\text{ s}$. This gives $|\varepsilon| = (3.619 \times 10^6\text{ V/s})(4.472\text{ s}) = 1.62 \times 10^{-5}\text{ V}$.

EVALUATE: This a very small voltage, about $16\ \mu\text{V}$.

29.55. **IDENTIFY:** Apply the results of Example 29.3, so $\varepsilon_{\max} = N\omega BA$ for N loops.

SET UP: For the minimum ω, let the rotating loop have an area equal to the area of the uniform magnetic field, so $A = (0.100\text{ m})^2$.

EXECUTE: $N = 400$, $B = 1.5$ T, $A = (0.100\text{ m})^2$ and $\varepsilon_{max} = 120$ V gives

$\omega = \varepsilon_{max}/NBA = (20\text{ rad/s})(1\text{ rev}/2\pi\text{ rad})(60\text{ s/1 min}) = 190$ rpm.

EVALUATE: In $\varepsilon_{max} = \omega BA$, ω is in rad/s.

29.57.   **IDENTIFY:**   Apply Faraday's law in the form $\varepsilon_{av} = -N\dfrac{\Delta\Phi_B}{\Delta t}$ to calculate the average emf. Apply Lenz's

law to calculate the direction of the induced current.

SET UP: $\Phi_B = BA$. The flux changes because the area of the loop changes.

EXECUTE: (a) $\varepsilon_{av} = \left|\dfrac{\Delta\Phi_B}{\Delta t}\right| = B\left|\dfrac{\Delta A}{\Delta t}\right| = B\dfrac{\pi r^2}{\Delta t} = (1.35\text{ T})\dfrac{\pi(0.0650/2\text{ m})^2}{0.250\text{ s}} = 0.0179$ V $= 17.9$ mV.

(b) Since the magnetic field is directed into the page and the magnitude of the flux through the loop is decreasing, the induced current must produce a field that goes into the page. Therefore the current flows from point a through the resistor to point b.

EVALUATE: Faraday's law can be used to find the direction of the induced current. Let $\vec{A}$ be into the page. Then Φ_B is positive and decreasing in magnitude, so $d\Phi_B/dt < 0$. Therefore $\varepsilon > 0$ and the induced current is clockwise around the loop.

29.61.   **IDENTIFY:**   Apply $\varepsilon = BvL$. Use $\sum\vec{F} = m\vec{a}$ applied to the satellite motion to find the speed v of the satellite.

SET UP: The gravitational force on the satellite is $F_g = G\dfrac{mm_E}{r^2}$, where m is the mass of the satellite and r is the radius of its orbit.

EXECUTE: $B = 8.0\times10^{-5}$ T, $L = 2.0$ m. $G\dfrac{mm_E}{r^2} = m\dfrac{v^2}{r}$ and $r = 400\times10^3$ m $+ R_E$ gives $v = \sqrt{\dfrac{Gm_E}{r}} =$

7.665×10^3 m/s. Using this v in $\varepsilon = vBL$ gives $\varepsilon = (8.0\times10^{-5}\text{ T})(7.665\times10^3\text{ m/s})(2.0\text{ m}) = 1.2$ V.

EVALUATE: The induced emf is large enough to be measured easily.

29.63.   **IDENTIFY:**   Find the magnetic field at a distance r from the center of the wire. Divide the rectangle into narrow strips of width dr, find the flux through each strip and integrate to find the total flux.

SET UP: Example 28.8 uses Ampere's law to show that the magnetic field inside the wire, a distance r from the axis, is $B(r) = \mu_0 Ir/2\pi R^2$.

EXECUTE: Consider a small strip of length W and width dr that is a distance r from the axis of the wire, as shown in Figure 29.63. The flux through the strip is $d\Phi_B = B(r)W\,dr = \dfrac{\mu_0 IW}{2\pi R^2}r\,dr$. The total flux through

the rectangle is $\Phi_B = \int d\Phi_B = \left(\dfrac{\mu_0 IW}{2\pi R^2}\right)\displaystyle\int_0^R r\,dr = \dfrac{\mu_0 IW}{4\pi}$.

EVALUATE: Note that the result is independent of the radius R of the wire.

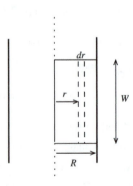

Figure 29.63

29.65. **(a)** and **(b)** IDENTIFY and SET UP:

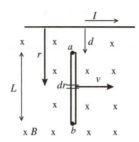

Figure 29.65a

The magnetic field of the wire is given by $B = \dfrac{\mu_0 I}{2\pi r}$ and varies along the length of the bar. At every point along the bar $\vec{B}$ has direction into the page. Divide the bar up into thin slices, as shown in Figure 29.65a.

EXECUTE: The emf $d\varepsilon$ induced in each slice is given by $d\varepsilon = \vec{v} \times \vec{B} \cdot d\vec{l}$. $\vec{v} \times \vec{B}$ is directed toward the wire, so $d\varepsilon = -vB\,dr = -v\left(\dfrac{\mu_0 I}{2\pi r}\right)dr$. The total emf induced in the bar is

$$V_{ba} = \int_a^b d\varepsilon = -\int_d^{d+L}\left(\frac{\mu_0 I v}{2\pi r}\right)dr = -\frac{\mu_0 I v}{2\pi}\int_d^{d+L}\frac{dr}{r} = -\frac{\mu_0 I v}{2\pi}\big[\ln(r)\big]_d^{d+L}$$

$$V_{ba} = -\frac{\mu_0 I v}{2\pi}(\ln(d+L) - \ln(d)) = -\frac{\mu_0 I v}{2\pi}\ln(1 + L/d)$$

EVALUATE: The minus sign means that V_{ba} is negative, point a is at higher potential than point b. (The force $\vec{F} = q\vec{v} \times \vec{B}$ on positive charge carriers in the bar is towards a, so a is at higher potential.) The potential difference increases when I or v increase, or d decreases.

(c) IDENTIFY: Use Faraday's law to calculate the induced emf.

SET UP: The wire and loop are sketched in Figure 29.65b.

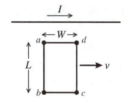

Figure 29.65b

EXECUTE: As the loop moves to the right the magnetic flux through it doesn't change. Thus $\varepsilon = -\dfrac{d\Phi_B}{dt} = 0$ and $I = 0$.

EVALUATE: This result can also be understood as follows. The induced emf in section ab puts point a at higher potential; the induced emf in section dc puts point d at higher potential. If you travel around the loop then these two induced emf's sum to zero. There is no emf in the loop and hence no current.

29.67. **(a)** IDENTIFY: Use the expression for motional emf to calculate the emf induced in the rod.

SET UP: The rotating rod is shown in Figure 29.67a.

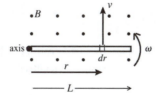

Figure 29.67a

The emf induced in a thin slice is $d\varepsilon = \vec{v} \times \vec{B} \cdot d\vec{l}$.

EXECUTE: Assume that $\vec{B}$ is directed out of the page. Then $\vec{v} \times \vec{B}$ is directed radially outward and $dl = dr$, so $\vec{v} \times \vec{B} \cdot d\vec{l} = vB\,dr$

$v = r\omega$ so $d\varepsilon = \omega Br\,dr$.

The $d\varepsilon$ for all the thin slices that make up the rod are in series so they add:

$$\varepsilon = \int d\varepsilon = \int_0^L \omega Br\, dr = \tfrac{1}{2}\omega B L^2 = \tfrac{1}{2}(8.80\text{ rad/s})(0.650\text{ T})(0.240\text{ m})^2 = 0.165\text{ V}$$

EVALUATE: ε increases with ω, B or L^2.

(b) No current flows so there is no IR drop in potential. Thus the potential difference between the ends equals the emf of 0.165 V calculated in part (a).

(c) SET UP: The rotating rod is shown in Figure 29.67b.

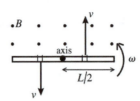

Figure 29.67b

EXECUTE: The emf between the center of the rod and each end is

$\varepsilon = \tfrac{1}{2}\omega B (L/2)^2 = \tfrac{1}{4}(0.165\text{ V}) = 0.0412\text{ V}$, with the direction of the emf from the center of the rod toward

each end. The emfs in each half of the rod thus oppose each other and there is no net emf between the ends of the rod.

EVALUATE: ω and B are the same as in part (a) but L of each half is $\tfrac{1}{2}L$ for the whole rod. ε is

proportional to L^2, so is smaller by a factor of $\tfrac{1}{4}$.

29.73. **IDENTIFY:** Apply Eq. (29.14).

SET UP: $\epsilon = 3.5 \times 10^{-11}$ F/m

EXECUTE: $i_D = \epsilon \dfrac{d\Phi_E}{dt} = (3.5 \times 10^{-11}\text{ F/m})(24.0 \times 10^3\text{ V}\cdot\text{m/s}^3)t^2$. $i_D = 21 \times 10^{-6}$ A gives $t = 5.0$ s.

EVALUATE: i_D depends on the rate at which Φ_E is changing.

29.75. **IDENTIFY:** The conduction current density is related to the electric field by Ohm's law. The displacement current density is related to the rate of change of the electric field by Eq. (29.16).

SET UP: $dE/dt = \omega E_0 \cos \omega t$

EXECUTE: **(a)** $j_C(\text{max}) = \dfrac{E_0}{\rho} = \dfrac{0.450\text{ V/m}}{2300\ \Omega\cdot\text{m}} = 1.96 \times 10^{-4}\text{ A/m}^2$

(b) $j_D(\text{max}) = \epsilon_0 \left(\dfrac{dE}{dt}\right)_{\text{max}} = \epsilon_0 \omega E_0 = 2\pi\epsilon_0 f E_0 = 2\pi\epsilon_0 (120\text{ Hz})(0.450\text{ V/m}) = 3.00 \times 10^{-9}\text{ A/m}^2$

(c) If $j_C = j_D$ then $\dfrac{E_0}{\rho} = \omega\epsilon_0 E_0$ and $\omega = \dfrac{1}{\rho\epsilon_0} = 4.91 \times 10^7$ rad/s

$$f = \frac{\omega}{2\pi} = \frac{4.91 \times 10^7\text{ rad/s}}{2\pi} = 7.82 \times 10^6\text{ Hz}.$$

EVALUATE: **(d)** The two current densities are out of phase by $90°$ because one has a sine function and the other has a cosine, so the displacement current leads the conduction current by $90°$.

30

30.3. **IDENTIFY:** A coil is wound around a solenoid, so magnetic flux from the solenoid passes through the coil.

SET UP: Example 30.1 shows that the mutual inductance for this configuration of coils is

$M = \dfrac{\mu_0 N_1 N_2 A}{l}$, where l is the length of coil 1.

EXECUTE: Using the formula for M gives

$$M = \frac{(4\pi \times 10^{-7} \text{ Wb/m} \cdot \text{A})(800)(50)\pi(0.200 \times 10^{-2} \text{ m})^2}{0.100 \text{ m}} = 6.32 \times 10^{-6} \text{ H} = 6.32 \ \mu\text{H}.$$

EVALUATE: This result is a physically reasonable mutual inductance.

30.5. **IDENTIFY and SET UP:** Apply Eq. (30.5).

EXECUTE: **(a)** $M = \dfrac{N_2 \Phi_{B2}}{i_1} = \dfrac{400(0.0320 \text{ Wb})}{6.52 \text{ A}} = 1.96 \text{ H}$

(b) $M = \dfrac{N_1 \Phi_{B1}}{i_2}$ so $\Phi_{B1} = \dfrac{M i_2}{N_1} = \dfrac{(1.96 \text{ H})(2.54 \text{ A})}{700} = 7.11 \times 10^{-3} \text{ Wb}$

EVALUATE: M relates the current in one coil to the flux through the other coil. Eq. (30.5) shows that M is the same for a pair of coils, no matter which one has the current and which one has the flux.

30.7. **IDENTIFY:** We can relate the known self-inductance of the toroidal solenoid to its geometry to calculate the number of coils it has. Knowing the induced emf, we can find the rate of change of the current.

SET UP: Example 30.3 shows that the self-inductance of a toroidal solenoid is $L = \dfrac{\mu_0 N^2 A}{2\pi r}$. The voltage

across the coil is related to the rate at which the current in it is changing by $\varepsilon = L \left| \dfrac{di}{dt} \right|$.

EXECUTE: **(a)** Solving $L = \dfrac{\mu_0 N^2 A}{2\pi r}$ for N gives

$$N = \sqrt{\frac{2\pi r L}{\mu_0 A}} = \sqrt{\frac{2\pi(0.0600 \text{ m})(2.50 \times 10^{-3} \text{ H})}{(4\pi \times 10^{-7} \text{ T} \cdot \text{m/A})(2.00 \times 10^{-4} \text{ m}^2)}} = 1940 \text{ turns.}$$

(b) $\left| \dfrac{di}{dt} \right| = \dfrac{\varepsilon}{L} = \dfrac{2.00 \text{ V}}{2.50 \times 10^{-3} \text{ H}} = 800 \text{ A/s.}$

EVALUATE: The inductance is determined solely by how the coil is constructed. The induced emf depends on the rate at which the current through the coil is changing.

30.9. **IDENTIFY:** $\varepsilon = L \left| \dfrac{\Delta i}{\Delta t} \right|$ and $L = \dfrac{N \Phi_B}{i}$.

SET UP: $\dfrac{\Delta i}{\Delta t} = 0.0640 \text{ A/s}$

EXECUTE: **(a)** $L = \dfrac{\varepsilon}{|\Delta i / \Delta t|} = \dfrac{0.0160 \text{ V}}{0.0640 \text{ A/s}} = 0.250 \text{ H}$

(b) The average flux through each turn is $\Phi_B = \dfrac{Li}{N} = \dfrac{(0.250 \text{ H})(0.720 \text{ A})}{400} = 4.50 \times 10^{-4} \text{ Wb}$.

EVALUATE: The self-induced emf depends on the rate of change of flux and therefore on the rate of change of the current, not on the value of the current.

30.13.   **IDENTIFY:** The inductance depends only on the geometry of the object, and the resistance of the wire depends on its length.

SET UP: $L = \dfrac{\mu_0 N^2 A}{2\pi r}$.

EXECUTE: **(a)** $N = \sqrt{\dfrac{2\pi r L}{\mu_0 A}} = \sqrt{\dfrac{(0.120 \text{ m})(0.100 \times 10^{-3} \text{ H})}{(2\times 10^{-7} \text{ T}\cdot\text{m/A})(0.600\times 10^{-4} \text{ m}^2)}} = 1.00 \times 10^3 \text{ turns}.$

(b) $A = \pi d^2/4$ and $c = \pi d$, so $c = \sqrt{4\pi A} = \sqrt{4\pi(0.600\times 10^{-4} \text{ m}^2)} = 0.02746 \text{ m}$. The total length of the wire is $(1000)(0.02746 \text{ m}) = 27.46 \text{ m}$. Therefore $R = (0.0760 \ \Omega/\text{m})(27.46 \text{ m}) = 2.09 \ \Omega$.

EVALUATE: A resistance of $2 \ \Omega$ is large enough to be significant in a circuit.

30.19.   **IDENTIFY:** A current-carrying inductor has a magnetic field inside of itself and hence stores magnetic energy.
(a) SET UP: The magnetic field inside a solenoid is $B = \mu_0 n I$.

EXECUTE: $B = \dfrac{(4\pi\times 10^{-7} \text{ T}\cdot\text{m/A})(400)(80.0 \text{ A})}{0.250 \text{ m}} = 0.161 \text{ T}$

(b) SET UP: The energy density in a magnetic field is $u = \dfrac{B^2}{2\mu_0}$.

EXECUTE: $u = \dfrac{(0.161 \text{ T})^2}{2(4\pi\times 10^{-7} \text{ T}\cdot\text{m/A})} = 1.03 \times 10^4 \text{ J/m}^3$

(c) SET UP: The total stored energy is $U = uV$.

EXECUTE: $U = uV = u(lA) = (1.03\times 10^4 \text{ J/m}^3)(0.250 \text{ m})(0.500\times 10^{-4} \text{ m}^2) = 0.129 \text{ J}$

(d) SET UP: The energy stored in an inductor is $U = \frac{1}{2}LI^2$.

EXECUTE: Solving for L and putting in the numbers gives

$$L = \dfrac{2U}{I^2} = \dfrac{2(0.129 \text{ J})}{(80.0 \text{ A})^2} = 4.02 \times 10^{-5} \text{ H}$$

EVALUATE: An inductor stores its energy in the magnetic field inside of it.

30.23.   **IDENTIFY:** Apply Kirchhoff's loop rule to the circuit. $i(t)$ is given by Eq. (30.14).
SET UP: The circuit is sketched in Figure 30.23.

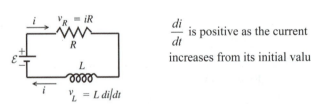

$\dfrac{di}{dt}$ is positive as the current increases from its initial value of zero.

Figure 30.23

EXECUTE: $\mathcal{E} - v_R - v_L = 0$

$\mathcal{E} - iR - L\dfrac{di}{dt} = 0$ so $i = \dfrac{\mathcal{E}}{R}(1 - e^{-(R/L)t})$

(a) Initially $(t = 0)$, $i = 0$ so $\mathcal{E} - L\dfrac{di}{dt} = 0$

$\dfrac{di}{dt} = \dfrac{\mathcal{E}}{L} = \dfrac{6.00 \text{ V}}{2.50 \text{ H}} = 2.40 \text{ A/s}$

(b) $\mathcal{E} - iR - L\dfrac{di}{dt} = 0$ (Use this equation rather than Eq. (30.15) since i rather than t is given.)

Thus $\dfrac{di}{dt} = \dfrac{\mathcal{E} - iR}{L} = \dfrac{6.00\text{ V} - (0.500\text{ A})(8.00\ \Omega)}{2.50\text{ H}} = 0.800\text{ A/s}$

(c) $i = \dfrac{\mathcal{E}}{R}(1 - e^{-(R/L)t}) = \left(\dfrac{6.00\text{ V}}{8.00\ \Omega}\right)(1 - e^{-(8.00\ \Omega/2.50\text{ H})(0.250\text{ s})}) = 0.750\text{ A}(1 - e^{-0.800}) = 0.413\text{ A}$

(d) Final steady state means $t \rightarrow \infty$ and $\dfrac{di}{dt} \rightarrow 0$, so $\mathcal{E} - iR = 0$.

$i = \dfrac{\mathcal{E}}{R} = \dfrac{6.00\text{ V}}{8.00\ \Omega} = 0.750\text{ A}$

EVALUATE: Our results agree with Figure 30.12 in the textbook. The current is initially zero and increases to its final value of $\mathcal{E}/R$. The slope of the current in the figure, which is di/dt, decreases with t.

30.25.   **IDENTIFY:** $i = \mathcal{E}/R(1 - e^{-t/\tau})$, with $\tau = L/R$. The energy stored in the inductor is $U = \frac{1}{2}Li^2$.

       SET UP: The maximum current occurs after a long time and is equal to $\mathcal{E}/R$.

       EXECUTE: **(a)** $i_{max} = \mathcal{E}/R$ so $i = i_{max}/2$ when $(1 - e^{-t/\tau}) = \frac{1}{2}$ and $e^{-t/\tau} = \frac{1}{2}$. $-t/\tau = \ln\left(\frac{1}{2}\right)$.

       $t = \dfrac{L\ln 2}{R} = \dfrac{(\ln 2)(1.25 \times 10^{-3}\text{ H})}{50.0\ \Omega} = 17.3\ \mu s$

       (b) $U = \frac{1}{2}U_{max}$ when $i = i_{max}/\sqrt{2}$. $1 - e^{-t/\tau} = 1/\sqrt{2}$, so $e^{-t/\tau} = 1 - 1/\sqrt{2} = 0.2929$.

       $t = -L\ln(0.2929)/R = 30.7\ \mu s$.

       EVALUATE: $\tau = L/R = 2.50 \times 10^{-5}\text{ s} = 25.0\ \mu s$. The time in part (a) is 0.692τ and the time in part (b) is 1.23τ.

30.29.   **IDENTIFY:** $i(t)$ is given by Eq. (30.14).

       SET UP: The power input from the battery is $\mathcal{E}i$. The rate of dissipation of energy in the resistance is i^2R. The voltage across the inductor has magnitude Ldi/dt, so the rate at which energy is being stored in the inductor is $iLdi/dt$.

       EXECUTE: **(a)** $P = \mathcal{E}i = \mathcal{E}I_0(1 - e^{-(R/L)t}) = \dfrac{\mathcal{E}^2}{R}(1 - e^{-(R/L)t}) = \dfrac{(6.00\text{ V})^2}{8.00\ \Omega}(1 - e^{-(8.00\ \Omega/2.50\text{ H})t})$.

       $P = (4.50\text{ W})(1 - e^{-(3.20\text{ s}^{-1})t})$.

       (b) $P_R = i^2R = \dfrac{\mathcal{E}^2}{R}(1 - e^{-(R/L)t})^2 = \dfrac{(6.00\text{ V})^2}{8.00\ \Omega}(1 - e^{-(8.00\ \Omega/2.50\text{ H})t})^2 = (4.50\text{ W})(1 - e^{-(3.20\text{ s}^{-1})t})^2$

       (c) $P_L = iL\dfrac{di}{dt} = \dfrac{\mathcal{E}}{R}(1 - e^{-(R/L)t})L\left(\dfrac{\mathcal{E}}{L}e^{-(R/L)t}\right) = \dfrac{\mathcal{E}^2}{R}(e^{-(R/L)t} - e^{-2(R/L)t})$

       $P_L = (4.50\text{ W})(e^{-(3.20\text{ s}^{-1})t} - e^{-(6.40\text{ s}^{-1})t})$.

       EVALUATE: **(d)** Note that if we expand the square in part (b), then parts (b) and (c) add to give part (a), and the total power delivered is dissipated in the resistor and inductor. Conservation of energy requires that this be so.

30.33.   **IDENTIFY:** The energy moves back and forth between the inductor and capacitor.

       (a) SET UP: The period is $T = \dfrac{1}{f} = \dfrac{1}{\omega/2\pi} = \dfrac{2\pi}{\omega} = 2\pi\sqrt{LC}$.

       EXECUTE: Solving for L gives

              $L = \dfrac{T^2}{4\pi^2 C} = \dfrac{(8.60 \times 10^{-5}\text{ s})^2}{4\pi^2(7.50 \times 10^{-9}\text{ C})} = 2.50 \times 10^{-2}\text{ H} = 25.0\text{ mH}$

       (b) SET UP: The charge on a capacitor is $Q = CV$.

       EXECUTE: $Q = CV = (7.50 \times 10^{-9}\text{ F})(12.0\text{ V}) = 9.00 \times 10^{-8}\text{ C}$

(c) SET UP: The stored energy is $U = Q^2/2C$.

EXECUTE: $U = \dfrac{(9.00 \times 10^{-8} \text{ C})^2}{2(7.50 \times 10^{-9} \text{ F})} = 5.40 \times 10^{-7}$ J

(d) SET UP: The maximum current occurs when the capacitor is discharged, so the inductor has all the initial energy. $U_L + U_C = U_{\text{Total}}$. $\frac{1}{2} L I^2 + 0 = U_{\text{Total}}$.

EXECUTE: Solve for the current:

$$I = \sqrt{\frac{2 U_{\text{Total}}}{L}} = \sqrt{\frac{2(5.40 \times 10^{-7} \text{ J})}{2.50 \times 10^{-2} \text{ H}}} = 6.58 \times 10^{-3} \text{ A} = 6.58 \text{ mA}$$

EVALUATE: The energy oscillates back and forth forever. However, if there is any resistance in the circuit, no matter how small, all this energy will eventually be dissipated as heat in the resistor.

30.39. **IDENTIFY:** Evaluate Eq. (30.29).

SET UP: The angular frequency of the circuit is ω'.

EXECUTE: **(a)** When $R = 0$, $\omega_0 = \dfrac{1}{\sqrt{LC}} = \dfrac{1}{\sqrt{(0.450 \text{ H})(2.50 \times 10^{-5} \text{ F})}} = 298$ rad/s.

(b) We want $\dfrac{\omega'}{\omega_0} = 0.95$, so $\dfrac{(1/LC - R^2/4L^2)}{1/LC} = 1 - \dfrac{R^2 C}{4L} = (0.95)^2$. This gives

$R = \sqrt{\dfrac{4L}{C}(1 - (0.95)^2)} = \sqrt{\dfrac{4(0.450 \text{ H})(0.0975)}{(2.50 \times 10^{-5} \text{ F})}} = 83.8 \ \Omega$.

EVALUATE: When R increases, the angular frequency decreases and approaches zero as $R \rightarrow 2\sqrt{L/C}$.

30.47. **IDENTIFY:** Set $U_B = K$, where $K = \frac{1}{2} m v^2$.

SET UP: The energy density in the magnetic field is $u_B = B^2/2\mu_0$. Consider volume $V = 1 \text{ m}^3$ of sunspot material.

EXECUTE: The energy density in the sunspot is $u_B = B^2/2\mu_0 = 6.366 \times 10^4 \text{ J/m}^3$. The total energy stored in volume V of the sunspot is $U_B = u_B V$. The mass of the material in volume V of the sunspot is $m = \rho V$. $K = U_B$ so $\frac{1}{2} m v^2 = U_B$. $\frac{1}{2} \rho V v^2 = u_B V$. The volume divides out, and $v = \sqrt{2 u_B/\rho} = 2 \times 10^4$ m/s.

EVALUATE: The speed we calculated is about 30 times smaller than the escape speed.

30.49. **(a) IDENTIFY and SET UP:** An end view is shown in Figure 30.49.

Apply Ampere's law to a circular path of radius r.
$$\oint \vec{B} \cdot d\vec{l} = \mu_0 I_{\text{encl}}$$

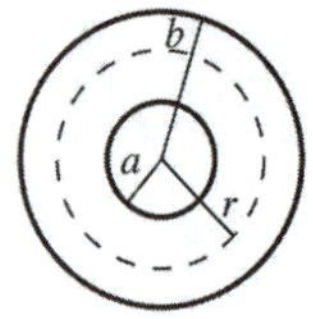

Figure 30.49

EXECUTE: $\oint \vec{B} \cdot d\vec{l} = B(2\pi r)$

$I_{\text{encl}} = i$, the current in the inner conductor

Thus $B(2\pi r) = \mu_0 i$ and $B = \dfrac{\mu_0 i}{2\pi r}$.

(b) IDENTIFY and SET UP: Follow the procedure specified in the problem.

EXECUTE: $u = \dfrac{B^2}{2\mu_0}$

$dU = u \, dV$, where $dV = 2\pi r l \, dr$

$$dU = \frac{1}{2\mu_0}\left(\frac{\mu_0 i}{2\pi r}\right)^2 (2\pi r l)dr = \frac{\mu_0 i^2 l}{4\pi r}dr$$

(c) $U = \int dU = \frac{\mu_0 i^2 l}{4\pi}\int_a^b \frac{dr}{r} = \frac{\mu_0 i^2 l}{4\pi}[\ln r]_a^b$

$$U = \frac{\mu_0 i^2 l}{4\pi}(\ln b - \ln a) = \frac{\mu_0 i^2 l}{4\pi}\ln\left(\frac{b}{a}\right)$$

(d) Eq. (30.9): $U = \frac{1}{2}Li^2$

Part (c): $U = \frac{\mu_0 i^2 l}{4\pi}\ln\left(\frac{b}{a}\right)$

$$\frac{1}{2}Li^2 = \frac{\mu_0 i^2 l}{4\pi}\ln\left(\frac{b}{a}\right)$$

$$L = \frac{\mu_0 l}{2\pi}\ln\left(\frac{b}{a}\right).$$

EVALUATE: The value of L we obtain from these energy considerations agrees with L calculated in part (d) of Problem 30.48 by considering flux and Eq. (30.6).

30.51. **IDENTIFY:** $U = \frac{1}{2}LI^2$. The self-inductance of a solenoid is found in Exercise 30.15 to be $L = \frac{\mu_0 AN^2}{l}$.

SET UP: The length l of the solenoid is the number of turns divided by the turns per unit length.

EXECUTE: **(a)** $L = \frac{2U}{I^2} = \frac{2(10.0\text{ J})}{(2.00\text{ A})^2} = 5.00\text{ H}.$

(b) $L = \frac{\mu_0 AN^2}{l}$. If α is the number of turns per unit length, then $N = \alpha l$ and $L = \mu_0 A\alpha^2 l$. For this coil $\alpha = 10$ coils/mm $= 10\times10^3$ coils/m. Solving for l gives

$$l = \frac{L}{\mu_0 A\alpha^2} = \frac{5.00\text{ H}}{(4\pi\times10^{-7}\text{ T}\cdot\text{m/A})\pi(0.0200\text{ m})^2(10\times10^3\text{ coils/m})^2} = 31.7\text{ m}.$$ This is not a practical length for laboratory use.

EVALUATE: The number of turns is $N = (31.7\text{ m})(10\times10^3\text{ coils/m}) = 3.17\times10^5$ turns. The length of wire in the solenoid is the circumference C of one turn times the number of turns.

$C = \pi d = \pi(4.00\times10^{-2}\text{ m}) = 0.126\text{ m}.$ The length of wire is $(0.126\text{ m})(3.17\times10^5) = 4.0\times10^4\text{ m} = 40\text{ km}.$

This length of wire will have a large resistance and I^2R electrical energy loses will be very large.

30.57. **IDENTIFY and SET UP:** Use $U_C = \frac{1}{2}CV_C^2$ (energy stored in a capacitor) to solve for C. Then use Eq. (30.22) and $\omega = 2\pi f$ to solve for the L that gives the desired current oscillation frequency.

EXECUTE: $V_C = 12.0\text{ V};\ U_C = \frac{1}{2}CV_C^2$ so $C = 2U_C/V_C^2 = 2(0.0160\text{ J})/(12.0\text{ V})^2 = 222\ \mu\text{F}$

$f = \frac{1}{2\pi\sqrt{LC}}$ so $L = \frac{1}{(2\pi f)^2 C}$

$f = 3500\text{ Hz}$ gives $L = 9.31\ \mu\text{H}$

EVALUATE: f is in Hz and ω is in rad/s; we must be careful not to confuse the two.

30.61. **IDENTIFY:** The current through an inductor doesn't change abruptly. After a long time the current isn't changing and the voltage across each inductor is zero.

SET UP: First combine the inductors.

EXECUTE: **(a)** Just after the switch is closed there is no current in the inductors. There is no current in the resistors so there is no voltage drop across either resistor. A reads zero and V reads 20.0 V.

(b) After a long time the currents are no longer changing, there is no voltage across the inductors, and the inductors can be replaced by short-circuits. The circuit becomes equivalent to the circuit shown in

Figure 30.61a. $I = (20.0 \text{ V})/(75.0 \, \Omega) = 0.267 \text{ A}$. The voltage between points a and b is zero, so the voltmeter reads zero.

(c) Combine the inductor network into its equivalent, as shown in Figure 30.61b. $R = 75.0 \, \Omega$ is the equivalent resistance. Eq. (30.14) says $i = (\mathcal{E}/R)(1 - e^{-t/\tau})$ with $\tau = L/R = (10.8 \text{ mH})/(75.0 \, \Omega) = 0.144 \text{ ms}$.

$\mathcal{E} = 20.0 \text{ V}$, $R = 75.0 \, \Omega$, $t = 0.115 \text{ ms}$ so $i = 0.147 \text{ A}$. $V_R = iR = (0.147 \text{ A})(75.0 \, \Omega) = 11.0 \text{ V}$.

$20.0 \text{ V} - V_R - V_L = 0$ and $V_L = 20.0 \text{ V} - V_R = 9.0 \text{ V}$. The ammeter reads 0.147 A and the voltmeter reads 9.0 V.

EVALUATE: The current through the battery increases from zero to a final value of 0.267 A. The voltage across the inductor network drops from 20.0 V to zero.

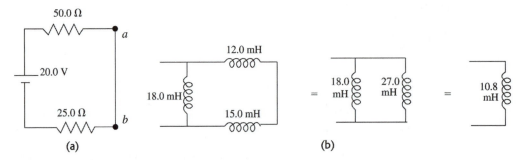

Figure 30.61

30.63. **IDENTIFY** and **SET UP:**   The current grows in the circuit as given by Eq. (30.14). In an R-L circuit the full emf initially is across the inductance and after a long time is totally across the resistance. A solenoid in a circuit is represented as a resistance in series with an inductance. Apply the loop rule to the circuit; the voltage across a resistance is given by Ohm's law.

EXECUTE: **(a)** In the R-L circuit the voltage across the resistor starts at zero and increases to the battery voltage. The voltage across the solenoid (inductor) starts at the battery voltage and decreases to zero. In the graph, the voltage drops, so the oscilloscope is across the solenoid.

(b) At $t \to \infty$ the current in the circuit approaches its final, constant value. The voltage doesn't go to zero because the solenoid has some resistance R_L. The final voltage across the solenoid is IR_L, where I is the final current in the circuit.

(c) The emf of the battery is the initial voltage across the inductor, 50 V. Just after the switch is closed, the current is zero and there is no voltage drop across any of the resistance in the circuit.

(d) As $t \to \infty$, $\mathcal{E} - IR - IR_L = 0$

$\mathcal{E} = 50 \text{ V}$ and from the graph $IR_L = 15 \text{ V}$ (the final voltage across the inductor), so

$IR_L = 35 \text{ V}$ and $I = (35 \text{ V})/R = 3.5 \text{ A}$

(e) $IR_L = 15 \text{ V}$, so $R_L = (15 \text{ V})/(3.5 \text{ A}) = 4.3 \, \Omega$

$\mathcal{E} - V_L - iR = 0$, where V_L includes the voltage across the resistance of the solenoid.

$$V_L = \mathcal{E} - iR, \, i = \frac{\mathcal{E}}{R_{\text{tot}}}(1 - e^{-t/\tau}), \text{ so } V_L = \mathcal{E}\left[1 - \frac{R}{R_{\text{tot}}}(1 - e^{-t/\tau})\right]$$

$\mathcal{E} = 50 \text{ V}$, $R = 10 \, \Omega$, $R_{\text{tot}} = 14.3 \, \Omega$, so when $t = \tau$, $V_L = 27.9 \text{ V}$. From the graph, V_L has this value when $t = 3.0 \text{ ms}$ (read approximately from the graph), so $\tau = L/R_{\text{tot}} = 3.0 \text{ ms}$. Then $L = (3.0 \text{ ms})(14.3 \, \Omega) = 43 \text{ mH}$.

EVALUATE: At $t = 0$ there is no current and the 50 V measured by the oscilloscope is the induced emf due to the inductance of the solenoid. As the current grows, there are voltage drops across the two resistances in the circuit. We derived an equation for V_L, the voltage across the solenoid. At $t = 0$ it gives $V_L = \mathcal{E}$ and at $t \to \infty$ it gives $V_L = \mathcal{E}R/R_{\text{tot}} = iR$.

30.65. **IDENTIFY** and **SET UP:**   Just after the switch is closed, the current in each branch containing an inductor is zero and the voltage across any capacitor is zero. The inductors can be treated as breaks in the circuit and the capacitors can be replaced by wires. After a long time there is no voltage across each inductor and no current in any branch containing a capacitor. The inductors can be replaced by wires and the capacitors by breaks in the circuit.

EXECUTE: **(a)** Just after the switch is closed the voltage V_5 across the capacitor is zero and there is also no current through the inductor, so $V_3 = 0$. $V_2 + V_3 = V_4 = V_5$, and since $V_5 = 0$ and $V_3 = 0$, V_4 and V_2 are also zero. $V_4 = 0$ means V_3 reads zero. V_1 then must equal 40.0 V, and this means the current read by A_1 is $(40.0\ \text{V})/(50.0\ \Omega) = 0.800\ \text{A}$. $A_2 + A_3 + A_4 = A_1$, but $A_2 = A_3 = 0$ so $A_4 = A_1 = 0.800\ \text{A}$. $A_1 = A_4 = 0.800\ \text{A}$; all other ammeters read zero. $V_1 = 40.0\ \text{V}$ and all other voltmeters read zero.

(b) After a long time the capacitor is fully charged so $A_4 = 0$, The current through the inductor isn't changing, so $V_2 = 0$. The currents can be calculated from the equivalent circuit that replaces the inductor by a short circuit, as shown in Figure 30.65a.

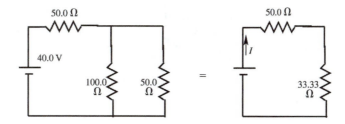

Figure 30.65a

$I = (40.0\ \text{V})/(83.33\ \Omega) = 0.480\ \text{A}$; A_1 reads 0.480 A

$V_1 = I(50.0\ \Omega) = 24.0\ \text{V}$

The voltage across each parallel branch is $40.0\ \text{V} - 24.0\ \text{V} = 16.0\ \text{V}$.

$V_2 = 0, V_3 = V_4 = V_5 = 16.0\ \text{V}$

$V_3 = 16.0\ \text{V}$ means A_2 reads 0.160 A. $V_4 = 16.0\ \text{V}$ means A_3 reads 0.320 A. A_4 reads zero. Note that $A_2 + A_3 = A_1$.

(c) $V_5 = 16.0\ \text{V}$ so $Q = CV = (12.0\ \mu\text{F})(16.0\ \text{V}) = 192\ \mu\text{C}$

(d) At $t = 0$ and $t \to \infty$, $V_2 = 0$. As the current in this branch increases from zero to 0.160 A the voltage V_2 reflects the rate of change of the current. The graph is sketched in Figure 30.65b.

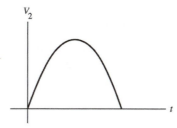

Figure 30.65b

EVALUATE: This reduction of the circuit to resistor networks only apply at $t = 0$ and $t \to \infty$. At intermediate times the analysis is complicated.

30.67. **IDENTIFY:** At $t = 0$, $i = 0$ through each inductor. At $t \to \infty$, the voltage is zero across each inductor.

SET UP: In each case redraw the circuit. At $t = 0$ replace each inductor by a break in the circuit and at $t \to \infty$ replace each inductor by a wire.

EXECUTE: **(a)** Just after the switch is closed there is no current through either inductor and they act like breaks in the circuit. The current is the same through the 40.0-Ω and 15.0-Ω resistors and is equal to $(25.0\ \text{V})/(40.0\ \Omega + 15.0\ \Omega) = 0.455\ \text{A}$. $A_1 = A_4 = 0.455\ \text{A}$; $A_2 = A_3 = 0$.

(b) After a long time the currents are constant, there is no voltage across either inductor, and each inductor can be treated as a short-circuit. The circuit is equivalent to the circuit sketched in Figure 30.67.

$I = (25.0 \text{ V})/(42.73 \text{ }\Omega) = 0.585 \text{ A}$. A_1 reads 0.585 A. The voltage across each parallel branch is

$25.0 \text{ V} - (0.585 \text{ A})(40.0 \text{ }\Omega) = 1.60 \text{ V}$. A_2 reads $(1.60 \text{ V})/(5.0 \text{ }\Omega) = 0.320 \text{ A}$. A_3 reads

$(1.60 \text{ V})/(10.0 \text{ }\Omega) = 0.160 \text{ A}$. A_4 reads $(1.60 \text{ V})/(15.0 \text{ }\Omega) = 0.107 \text{ A}$.

EVALUATE: Just after the switch is closed the current through the battery is 0.455 A. After a long time the current through the battery is 0.585 A. After a long time there are additional current paths, the equivalent resistance of the circuit is decreased and the current has increased.

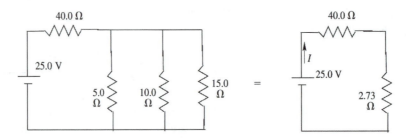

Figure 30.67

30.69.   **IDENTIFY:**   Apply the loop rule to each parallel branch. The voltage across a resistor is given by iR and the voltage across an inductor is given by $L|di/dt|$. The rate of change of current through the inductor is limited.

SET UP: With S closed the circuit is sketched in Figure 30.69a.

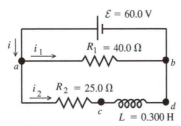

The rate of change of the current through the inductor is limited by the induced emf. Just after the switch is closed the current in the inductor has not had time to increase from zero, so $i_2 = 0$.

Figure 30.69a

EXECUTE :   **(a)** $\mathcal{E} - v_{ab} = 0$, so $v_{ab} = 60.0 \text{ V}$

(b) The voltage drops across R, as we travel through the resistor in the direction of the current, so point a is at higher potential.

(c) $i_2 = 0$ so $v_{R_2} = i_2 R_2 = 0$

$\mathcal{E} - v_{R_2} - v_L = 0$ so $v_L = \mathcal{E} = 60.0 \text{ V}$

(d) The voltage rises when we go from b to a through the emf, so it must drop when we go from a to b through the inductor. Point c must be at higher potential than point d.

(e) After the switch has been closed a long time, $\dfrac{di_2}{dt} \to 0$ so $v_L = 0$. Then $\mathcal{E} - v_{R_2} = 0$ and $i_2 R_2 = \mathcal{E}$

so $i_2 = \dfrac{\mathcal{E}}{R_2} = \dfrac{60.0 \text{ V}}{25.0 \text{ }\Omega} = 2.40 \text{ A}$.

SET UP: The rate of change of the current through the inductor is limited by the induced emf. Just after the switch is opened again the current through the inductor hasn't had time to change and is still $i_2 = 2.40 \text{ A}$. The circuit is sketched in Figure 30.69b.

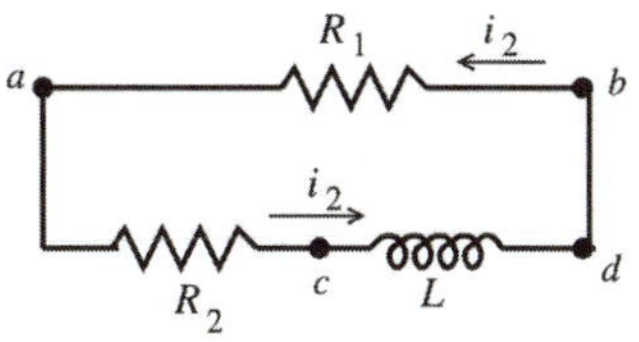

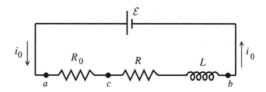

EXECUTE: The current through R_1 is $i_2 = 2.40$ A in the direction b to a.

Thus $v_{ab} = -i_2 R_1 = -(2.40$ A$)(40.0$ $\Omega)$

$v_{ab} = -96.0$ V.

Figure 30.69b

(f) Point where current enters resistor is at higher potential; point b is at higher potential.

(g) $v_L - v_{R_1} - v_{R_2} = 0$

$v_L = v_{R_1} + v_{R_2}$

$v_{R_1} = -v_{ab} = 96.0$ V; $v_{R_2} = i_2 R_2 = (2.40$ A$)(25.0$ $\Omega) = 60.0$ V

Then $v_L = v_{R_1} + v_{R_2} = 96.0$ V $+ 60.0$ V $= 156$ V.

As you travel counterclockwise around the circuit in the direction of the current, the voltage drops across each resistor, so it must rise across the inductor and point d is at higher potential than point c. The current is decreasing, so the induced emf in the inductor is directed in the direction of the current. Thus, $v_{cd} = -156$ V.

(h) Point d is at higher potential.

EVALUATE: The voltage across R_1 is constant once the switch is closed. In the branch containing R_2, just after S is closed the voltage drop is all across L and after a long time it is all across R_2. Just after S is opened the same current flows in the single loop as had been flowing through the inductor and the sum of the voltage across the resistors equals the voltage across the inductor. This voltage dies away, as the energy stored in the inductor is dissipated in the resistors.

30.71. **IDENTIFY** and **SET UP:** The circuit is sketched in Figure 30.71a. Apply the loop rule. Just after S_1 is closed, $i = 0$. After a long time i has reached its final value and $di/dt = 0$. The voltage across a resistor depends on i and the voltage across an inductor depends on di/dt.

Figure 30.71a

EXECUTE: **(a)** At time $t = 0$, $i_0 = 0$ so $v_{ac} = i_0 R_0 = 0$. By the loop rule $\mathcal{E} - v_{ac} - v_{cb} = 0$ so $v_{cb} = \mathcal{E} - v_{ac} = \mathcal{E} = 36.0$ V. ($i_0 R = 0$ so this potential difference of 36.0 V is across the inductor and is an induced emf produced by the changing current.)

(b) After a long time $\dfrac{di_0}{dt} \to 0$ so the potential $-L\dfrac{di_0}{dt}$ across the inductor becomes zero. The loop rule gives $\mathcal{E} - i_0(R_0 + R) = 0$.

$i_0 = \dfrac{\mathcal{E}}{R_0 + R} = \dfrac{36.0 \text{ V}}{50.0 \ \Omega + 150 \ \Omega} = 0.180$ A

$v_{ac} = i_0 R_0 = (0.180$ A$)(50.0$ $\Omega) = 9.0$ V

Thus $v_{cb} = i_0 R + L\dfrac{di_0}{dt} = (0.180$ A$)(150 \ \Omega) + 0 = 27.0$ V (Note that $v_{ac} + v_{cb} = \mathcal{E}$.)

(c) $\mathcal{E} - v_{ac} - v_{cb} = 0$

$\mathcal{E} - iR_0 - iR - L\dfrac{di}{dt} = 0$

$L\dfrac{di}{dt} = \mathcal{E} - i(R_0 + R)$ and $\left(\dfrac{L}{R + R_0}\right)\dfrac{di}{dt} = -i + \dfrac{\mathcal{E}}{R + R_0}$

$\dfrac{di}{-i + \mathcal{E}/(R + R_0)} = \left(\dfrac{R + R_0}{L}\right) dt$

Integrate from $t = 0$, when $i = 0$, to t, when $i = i_0$:

$\displaystyle\int_0^{i_0} \dfrac{di}{-i + \mathcal{E}/(R + R_0)} = \dfrac{R + R_0}{L}\int_0^t dt = -\ln\left[-i + \dfrac{\mathcal{E}}{R + R_0}\right]_0^{i_0} = \left(\dfrac{R + R_0}{L}\right)t$, so

$\ln\left(-i_0 + \dfrac{\mathcal{E}}{R + R_0}\right) - \ln\left(\dfrac{\mathcal{E}}{R + R_0}\right) = -\left(\dfrac{R + R_0}{L}\right)t$

$\ln\left(\dfrac{-i_0 + \mathcal{E}/(R + R_0)}{\mathcal{E}/(R + R_0)}\right) = -\left(\dfrac{R + R_0}{L}\right)t$

Taking exponentials of both sides gives $\dfrac{-i_0 + \mathcal{E}/(R + R_0)}{\mathcal{E}/(R + R_0)} = e^{-(R+R_0)t/L}$ and $i_0 = \dfrac{\mathcal{E}}{R + R_0}(1 - e^{-(R+R_0)t/L})$.

Substituting in the numerical values gives $i_0 = \dfrac{36.0\ \mathrm{V}}{50\ \Omega + 150\ \Omega}(1 - e^{-(200\ \Omega/4.00\ \mathrm{H})t}) = (0.180\ \mathrm{A})(1 - e^{-t/0.020\ \mathrm{s}})$.

At $t \to 0$, $i_0 = (0.180\ \mathrm{A})(1 - 1) = 0$ (agrees with part (a)). At $t \to \infty$, $i_0 = (0.180\ \mathrm{A})(1 - 0) = 0.180\ \mathrm{A}$ (agrees with part (b)).

$v_{ac} = i_0 R_0 = \dfrac{\mathcal{E}R_0}{R + R_0}(1 - e^{-(R+R_0)t/L}) = 9.0\ \mathrm{V}(1 - e^{-t/0.020\ \mathrm{s}})$

$v_{cb} = \mathcal{E} - v_{ac} = 36.0\ \mathrm{V} - 9.0\ \mathrm{V}(1 - e^{-t/0.020\ \mathrm{s}}) = 9.0\ \mathrm{V}(3.00 + e^{-t/0.020\ \mathrm{s}})$

At $t \to 0$, $v_{ac} = 0$, $v_{cb} = 36.0\ \mathrm{V}$ (agrees with part (a)). At $t \to \infty$, $v_{ac} = 9.0\ \mathrm{V}$, $v_{cb} = 27.0\ \mathrm{V}$ (agrees with part (b)). The graphs are given in Figure 30.71b.

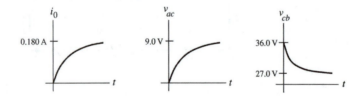

Figure 30.71b

EVALUATE: The expression for $i(t)$ we derived becomes Eq. (30.14) if the two resistors R_0 and R in series are replaced by a single equivalent resistance $R_0 + R$.

31

ALTERNATING CURRENT

31.5. **IDENTIFY:** We want the phase angle for the source voltage relative to the current, and we want the inductance if we know the current amplitude.

SET UP: $X_L = \dfrac{V}{I}$ and $X_L = 2\pi fL$.

EXECUTE: **(a)** $\phi = +90°$. The source voltage leads the current by 90°.

(b) $X_L = \dfrac{V}{I} = \dfrac{45.0 \text{ V}}{3.90 \text{ A}} = 11.54 \, \Omega$. Solving $X_L = 2\pi fL$ for f gives $f = \dfrac{X_L}{2\pi L} = \dfrac{11.54 \, \Omega}{2\pi(9.50 \times 10^{-3} \text{ H})} = 193 \text{ Hz}$.

EVALUATE: The angular frequency is about 1200 rad/s.

31.7. **IDENTIFY** and **SET UP:** Apply Eqs. (31.18) and (31.19).

EXECUTE: $V = IX_C$ so $X_C = \dfrac{V}{I} = \dfrac{170 \text{ V}}{0.850 \text{ A}} = 200 \, \Omega$

$X_C = \dfrac{1}{\omega C}$ gives $C = \dfrac{1}{2\pi fX_C} = \dfrac{1}{2\pi(60.0 \text{ Hz})(200 \, \Omega)} = 1.33 \times 10^{-5} \text{ F} = 13.3 \, \mu\text{F}$

EVALUATE: The reactance relates the voltage amplitude to the current amplitude and is similar to Ohm's law.

31.9. **IDENTIFY** and **SET UP:** Use Eqs. (31.12) and (31.18).

EXECUTE: **(a)** $X_L = \omega L = 2\pi fL = 2\pi(80.0 \text{ Hz})(3.00 \text{ H}) = 1510 \, \Omega$

(b) $X_L = 2\pi fL$ gives $L = \dfrac{X_L}{2\pi f} = \dfrac{120 \, \Omega}{2\pi(80.0 \text{ Hz})} = 0.239 \text{ H}$

(c) $X_C = \dfrac{1}{\omega C} = \dfrac{1}{2\pi fC} = \dfrac{1}{2\pi(80.0 \text{ Hz})(4.00 \times 10^{-6} \text{ F})} = 497 \, \Omega$

(d) $X_C = \dfrac{1}{2\pi fC}$ gives $C = \dfrac{1}{2\pi fX_C} = \dfrac{1}{2\pi(80.0 \text{ Hz})(120 \, \Omega)} = 1.66 \times 10^{-5} \text{ F}$

EVALUATE: X_L increases when L increases; X_C decreases when C increases.

31.15. **IDENTIFY:** Apply the equations in Section 31.3.

SET UP: $\omega = 250$ rad/s, $R = 200 \, \Omega$, $L = 0.400$ H, $C = 6.00 \, \mu\text{F}$ and $V = 30.0$ V.

EXECUTE: **(a)** $Z = \sqrt{R^2 + (\omega L - 1/\omega C)^2}$.

$Z = \sqrt{(200 \, \Omega)^2 + ((250 \text{ rad/s})(0.400 \text{ H}) - 1/((250 \text{ rad/s})(6.00 \times 10^{-6} \text{ F})))^2} = 601 \, \Omega$

(b) $I = \dfrac{V}{Z} = \dfrac{30 \text{ V}}{601 \, \Omega} = 0.0499 \text{ A}$.

(c) $\phi = \arctan\left(\dfrac{\omega L - 1/\omega C}{R}\right) = \arctan\left(\dfrac{100 \, \Omega - 667 \, \Omega}{200 \, \Omega}\right) = -70.6°$, and the voltage lags the current.

(d) $V_R = IR = (0.0499 \text{ A})(200 \,\Omega) = 9.98 \text{ V};$ $V_L = I\omega L = (0.0499 \text{ A})(250 \text{ rad/s})(0.400 \text{ H}) = 4.99 \text{ V};$

$$V_C = \frac{I}{\omega C} = \frac{(0.0499 \text{ A})}{(250 \text{ rad/s})(6.00 \times 10^{-6} \text{ F})} = 33.3 \text{ V}.$$

EVALUATE: **(e)** At any instant, $v = v_R + v_C + v_L$. But v_C and v_L are 180° out of phase, so v_C can be larger than v at a value of t, if $v_L + v_R$ is negative at that t.

31.17. **IDENTIFY and SET UP:** Use the equation that preceeds Eq. (31.20): $V^2 = V_R^2 + (V_L - V_C)^2$

EXECUTE: $V = \sqrt{(30.0 \text{ V})^2 + (50.0 \text{ V} - 90.0 \text{ V})^2} = 50.0 \text{ V}$

EVALUATE: The equation follows directly from the phasor diagrams of Fig. 31.13 (b or c) in the textbook. Note that the voltage amplitudes do not simply add to give 170.0 V for the source voltage.

31.19. **IDENTIFY:** For a pure resistance, $P_{av} = V_{rms} I_{rms} = I_{rms}^2 R$.

SET UP: 20.0 W is the average power P_{av}.

EXECUTE: **(a)** The average power is one-half the maximum power, so the maximum instantaneous power is 40.0 W.

(b) $I_{rms} = \dfrac{P_{av}}{V_{rms}} = \dfrac{20.0 \text{ W}}{120 \text{ V}} = 0.167 \text{ A}$

(c) $R = \dfrac{P_{av}}{I_{rms}^2} = \dfrac{20.0 \text{ W}}{(0.167 \text{ A})^2} = 720 \,\Omega$

EVALUATE: We can also calculate the average power as $P_{av} = \dfrac{V_{R,rms}^2}{R} = \dfrac{V_{rms}^2}{R} = \dfrac{(120 \text{ V})^2}{720 \,\Omega} = 20.0 \text{ W}.$

31.23. **IDENTIFY and SET UP:** Use the equations of Section 31.3 to calculate ϕ, Z and V_{rms}. The average power delivered by the source is given by Eq. (31.31) and the average power dissipated in the resistor is $I_{rms}^2 R$.

EXECUTE: **(a)** $X_L = \omega L = 2\pi f L = 2\pi (400 \text{ Hz})(0.120 \text{ H}) = 301.6 \,\Omega$

$$X_C = \frac{1}{\omega C} = \frac{1}{2\pi f C} = \frac{1}{2\pi(400 \text{ Hz})(7.3 \times 10^{-6} \text{ F})} = 54.51 \,\Omega$$

$\tan\phi = \dfrac{X_L - X_C}{R} = \dfrac{301.6 \,\Omega - 54.41 \,\Omega}{240 \,\Omega}$, so $\phi = +45.8°$. The power factor is $\cos\phi = +0.697$.

(b) $Z = \sqrt{R^2 + (X_L - X_C)^2} = \sqrt{(240 \,\Omega)^2 + (301.6 \,\Omega - 54.51 \,\Omega)^2} = 344 \,\Omega$

(c) $V_{rms} = I_{rms} Z = (0.450 \text{ A})(344 \,\Omega) = 155 \text{ V}$

(d) $P_{av} = I_{rms} V_{rms} \cos\phi = (0.450 \text{ A})(155 \text{ V})(0.697) = 48.6 \text{ W}$

(e) $P_{av} = I_{rms}^2 R = (0.450 \text{ A})^2 (240 \,\Omega) = 48.6 \text{ W}$

EVALUATE: The average electrical power delivered by the source equals the average electrical power consumed in the resistor.

(f) All the energy stored in the capacitor during one cycle of the current is released back to the circuit in another part of the cycle. There is no net dissipation of energy in the capacitor.

(g) The answer is the same as for the capacitor. Energy is repeatedly being stored and released in the inductor, but no net energy is dissipated there.

31.25. **IDENTIFY:** The angular frequency and the capacitance can be used to calculate the reactance X_C of the capacitor. The angular frequency and the inductance can be used to calculate the reactance X_L of the inductor. Calculate the phase angle ϕ and then the power factor is $\cos\phi$. Calculate the impedance of the circuit and then the rms current in the circuit. The average power is $P_{av} = V_{rms} I_{rms} \cos\phi$. On the average no power is consumed in the capacitor or the inductor, it is all consumed in the resistor.

SET UP: The source has rms voltage $V_{rms} = \dfrac{V}{\sqrt{2}} = \dfrac{45 \text{ V}}{\sqrt{2}} = 31.8 \text{ V}.$

EXECUTE: (a) $X_L = \omega L = (360 \text{ rad/s})(15 \times 10^{-3} \text{ H}) = 5.4 \text{ }\Omega$.

$X_C = \dfrac{1}{\omega C} = \dfrac{1}{(360 \text{ rad/s})(3.5 \times 10^{-6} \text{ F})} = 794 \text{ }\Omega$. $\tan\phi = \dfrac{X_L - X_C}{R} = \dfrac{5.4 \text{ }\Omega - 794 \text{ }\Omega}{250 \text{ }\Omega}$ and $\phi = -72.4°$.

The power factor is $\cos\phi = 0.302$.

(b) $Z = \sqrt{R^2 + (X_L - X_C)^2} = \sqrt{(250 \text{ }\Omega)^2 + (5.4 \text{ }\Omega - 794 \text{ }\Omega)^2} = 827 \text{ }\Omega$. $I_{\text{rms}} = \dfrac{V_{\text{rms}}}{Z} = \dfrac{31.8 \text{ V}}{827 \text{ }\Omega} = 0.0385 \text{ A}$.

$P_{\text{av}} = V_{\text{rms}} I_{\text{rms}} \cos\phi = (31.8 \text{ V})(0.0385 \text{ A})(0.302) = 0.370 \text{ W}$.

(c) The average power delivered to the resistor is $P_{\text{av}} = I_{\text{rms}}^2 R = (0.0385 \text{ A})^2 (250 \text{ }\Omega) = 0.370 \text{ W}$. The average power delivered to the capacitor and to the inductor is zero.

EVALUATE: On average the power delivered to the circuit equals the power consumed in the resistor. The capacitor and inductor store electrical energy during part of the current oscillation but each return the energy to the circuit during another part of the current cycle.

31.27. **IDENTIFY and SET UP:** The current is largest at the resonance frequency. At resonance, $X_L = X_C$ and $Z = R$. For part (b), calculate Z and use $I = V/Z$.

EXECUTE: (a) $f_0 = \dfrac{1}{2\pi\sqrt{LC}} = 113 \text{ Hz}$. $I = V/R = 15.0 \text{ mA}$.

(b) $X_C = 1/\omega C = 500 \text{ }\Omega$. $X_L = \omega L = 160 \text{ }\Omega$.

$Z = \sqrt{R^2 + (X_L - X_C)^2} = \sqrt{(200 \text{ }\Omega)^2 + (160 \text{ }\Omega - 500 \text{ }\Omega)^2} = 394.5 \text{ }\Omega$. $I = V/Z = 7.61 \text{ mA}$. $X_C > X_L$ so the source voltage lags the current.

EVALUATE: $\omega_0 = 2\pi f_0 = 710 \text{ rad/s}$. $\omega = 400 \text{ rad/s}$ and is less than ω_0. When $\omega < \omega_0$, $X_C > X_L$. Note that I in part (b) is less than I in part (a).

31.29. **IDENTIFY and SET UP:** At the resonance frequency, $Z = R$. Use that $V = IZ$, $V_R = IR$, $V_L = IX_L$ and $V_C = IX_C$. P_{av} is given by Eq. (31.31).

(a) EXECUTE: $V = IZ = IR = (0.500 \text{ A})(300 \text{ }\Omega) = 150 \text{ V}$

(b) $V_R = IR = 150 \text{ V}$

$X_L = \omega L = L(1/\sqrt{LC}) = \sqrt{L/C} = 2582 \text{ }\Omega$; $V_L = IX_L = 1290 \text{ V}$

$X_C = 1/(\omega C) = \sqrt{L/C} = 2582 \text{ }\Omega$; $V_C = IX_C = 1290 \text{ V}$

(c) $P_{\text{av}} = \frac{1}{2}VI\cos\phi = \frac{1}{2}I^2 R$, since $V = IR$ and $\cos\phi = 1$ at resonance.

$P_{\text{av}} = \frac{1}{2}(0.500 \text{ A})^2 (300 \text{ }\Omega) = 37.5 \text{ W}$

EVALUATE: At resonance $V_L = V_C$. Note that $V_L + V_C > V$. However, at any instant $v_L + v_C = 0$.

31.31. **IDENTIFY and SET UP:** At resonance $X_L = X_C$, $\phi = 0$ and $Z = R$. $R = 150 \text{ }\Omega$, $L = 0.750 \text{ H}$, $C = 0.0180 \text{ }\mu\text{F}$, $V = 150 \text{ V}$

EXECUTE: (a) At the resonance frequency $X_L = X_C$ and from $\tan\phi = \dfrac{X_L - X_C}{R}$ we have that $\phi = 0°$ and the power factor is $\cos\phi = 1.00$.

(b) $P_{\text{av}} = \frac{1}{2}VI\cos\phi$ (Eq. 31.31)

At the resonance frequency $Z = R$, so $I = \dfrac{V}{Z} = \dfrac{V}{R}$.

$P_{\text{av}} = \frac{1}{2}V\left(\dfrac{V}{R}\right)\cos\phi = \frac{1}{2}\dfrac{V^2}{R} = \frac{1}{2}\dfrac{(150 \text{ V})^2}{150 \text{ }\Omega} = 75.0 \text{ W}$

(c) EVALUATE: When C and f are changed but the circuit is kept on resonance, nothing changes in $P_{\text{av}} = V^2/(2R)$, so the average power is unchanged: $P_{\text{av}} = 75.0 \text{ W}$. The resonance frequency changes but since $Z = R$ at resonance the current doesn't change.

31.33.   **IDENTIFY** and **SET UP:**   The resonance angular frequency is $\omega_0 = \dfrac{1}{\sqrt{LC}}$. $X_L = \omega L$. $X_C = \dfrac{1}{\omega C}$ and

$Z = \sqrt{R^2 + (X_L - X_C)^2}$. At the resonance frequency $X_L = X_C$ and $Z = R$.

EXECUTE: **(a)** $Z = R = 115\ \Omega$

(b) $\omega_0 = \dfrac{1}{\sqrt{(4.50 \times 10^{-3}\ \text{H})(1.26 \times 10^{-6}\ \text{F})}} = 1.33 \times 10^4\ \text{rad/s}$. $\omega = 2\omega_0 = 2.66 \times 10^4\ \text{rad/s}$.

$X_L = \omega L = (2.66 \times 10^4\ \text{rad/s})(4.50 \times 10^{-3}\ \text{H}) = 120\ \Omega$. $X_C = \dfrac{1}{\omega C} = \dfrac{1}{(2.66 \times 10^4\ \text{rad/s})(1.25 \times 10^{-6}\ \text{F})} = 30\ \Omega$

$Z = \sqrt{(115\ \Omega)^2 + (120\ \Omega - 30\ \Omega)^2} = 146\ \Omega$

(c) $\omega = \omega_0/2 = 6.65 \times 10^3\ \text{rad/s}$. $X_L = 30\ \Omega$. $X_C = \dfrac{1}{\omega C} = 120\ \Omega$.

$Z = \sqrt{(115\ \Omega)^2 + (30\ \Omega - 120\ \Omega)^2} = 146\ \Omega$, the same value as in part (b).

EVALUATE: For $\omega = 2\omega_0$, $X_L > X_C$. For $\omega = \omega_0/2$, $X_L < X_C$. But $(X_L - X_C)^2$ has the same value at these two frequencies, so Z is the same.

31.35.   **IDENTIFY** and **SET UP:**   Eq. (31.35) relates the primary and secondary voltages to the number of turns in each. $I = V/R$ and the power consumed in the resistive load is $I_{rms}^2 = V_{rms}^2/R$. Let I_1, V_1 and I_2, V_2 be rms values for the primary and secondary.

EXECUTE: **(a)** $\dfrac{V_2}{V_1} = \dfrac{N_2}{N_1}$ so $\dfrac{N_1}{N_2} = \dfrac{V_1}{V_2} = \dfrac{120\ \text{V}}{12.0\ \text{V}} = 10$

(b) $I_2 = \dfrac{V_2}{R} = \dfrac{12.0\ \text{V}}{5.00\ \Omega} = 2.40\ \text{A}$

(c) $P_{av} = I_2^2 R = (2.40\ \text{A})^2 (5.00\ \Omega) = 28.8\ \text{W}$

(d) The power drawn from the line by the transformer is the 28.8 W that is delivered by the load.

$$P_{av} = \dfrac{V_1^2}{R} \text{ so } R = \dfrac{V_1^2}{P_{av}} = \dfrac{(120\ \text{V})^2}{28.8\ \text{W}} = 500\ \Omega$$

And $\left(\dfrac{N_1}{N_2}\right)^2 (5.00\ \Omega) = (10)^2 (5.00\ \Omega) = 500\ \Omega$, as was to be shown.

EVALUATE: The resistance is "transformed." A load of resistance R connected to the secondary draws the same power as a resistance $(N_1/N_2)^2 R$ connected directly to the supply line, without using the transformer.

31.37.   **IDENTIFY:**   Let I_1, V_1 and I_2, V_2 be rms values for the primary and secondary. A transformer transforms

voltages according to $\dfrac{V_2}{V_1} = \dfrac{N_2}{N_1}$. The effective resistance of a secondary circuit of resistance R is

$R_{eff} = \dfrac{R}{(N_2/N_1)^2}$. Resistance R is related to P_{av} and V_{rms} by $P_{av} = \dfrac{V_{rms}^2}{R}$. Conservation of energy requires

$P_{av,1} = P_{av,2}$ so $V_1 I_1 = V_2 I_2$.

SET UP: Let $V_1 = 240\ \text{V}$ and $V_2 = 120\ \text{V}$, so $P_{2,av} = 1600\ \text{W}$. These voltages are rms.

EXECUTE: **(a)** $V_1 = 240\ \text{V}$ and we want $V_2 = 120\ \text{V}$, so use a step-down transformer with $N_2/N_1 = \frac{1}{2}$.

(b) $P_{av} = V_1 I_1$, so $I_1 = \dfrac{P_{av}}{V_1} = \dfrac{1600\ \text{W}}{240\ \text{V}} = 6.67\ \text{A}$.

(c) The resistance R of the blower is $R = \dfrac{V_1^2}{P_{av}} = \dfrac{(120\ \text{V})^2}{1600\ \text{W}} = 9.00\ \Omega.$ The effective resistance of the blower is

$$R_{eff} = \frac{9.00\ \Omega}{(1/2)^2} = 36.0\ \Omega.$$

EVALUATE: $I_2 V_2 = (13.3\ \text{A})(120\ \text{V}) = 1600\ \text{W}.$ Energy is provided to the primary at the same rate that it is consumed in the secondary. Step-down transformers step up resistance and the current in the primary is less than the current in the secondary.

31.39. **IDENTIFY** and **SET UP:** Use Eq. (31.24) to relate L and R to ϕ. The voltage across the coil leads the current in it by 52.3°, so $\phi = +52.3°$.

EXECUTE: $\tan\phi = \dfrac{X_L - X_C}{R}.$ But there is no capacitance in the circuit so $X_C = 0.$ Thus

$\tan\phi = \dfrac{X_L}{R}$ and $X_L = R\tan\phi = (48.0\ \Omega)\tan 52.3° = 62.1\ \Omega.$

$X_L = \omega L = 2\pi f L$ so $L = \dfrac{X_L}{2\pi f} = \dfrac{62.1\ \Omega}{2\pi(80.0\ \text{Hz})} = 0.124\ \text{H}.$

EVALUATE: $\phi > 45°$ when $(X_L - X_C) > R,$ which is the case here.

31.41. **IDENTIFY:** We can use geometry to calculate the capacitance and inductance, and then use these results to calculate the resonance angular frequency.

SET UP: The capacitance of an air-filled parallel plate capacitor is $C = \dfrac{\varepsilon_0 A}{d}.$ The inductance of a long

solenoid is $L = \dfrac{\mu_0 A N^2}{l}.$ The inductor has $N = (125\ \text{coils/cm})(9.00\ \text{cm}) = 1125$ coils. The resonance

frequency is $f_0 = \dfrac{1}{2\pi\sqrt{LC}}.$ $\varepsilon_0 = 8.85\times10^{-12}\ \text{C}^2/\text{N}\cdot\text{m}^2.$ $\mu_0 = 4\pi\times10^{-7}\ \text{T}\cdot\text{m/A}.$

EXECUTE: $C = \dfrac{\varepsilon_0 A}{d} = \dfrac{(8.85\times10^{-12}\ \text{C}^2/\text{N}\cdot\text{m}^2)(4.50\times10^{-2}\ \text{m})^2}{8.00\times10^{-3}\ \text{m}} = 2.24\times10^{-12}\ \text{F}.$

$L = \dfrac{\mu_0 A N^2}{l} = \dfrac{(4\pi\times10^{-7}\ \text{T}\cdot\text{m/A})\pi(0.250\times10^{-2}\ \text{m})^2(1125)^2}{9.00\times10^{-2}\ \text{m}} = 3.47\times10^{-4}\ \text{H}.$

$\omega_0 = \dfrac{1}{\sqrt{(3.47\times10^{-4}\ \text{H})(2.24\times10^{-12}\ \text{F})}} = 3.59\times10^7\ \text{rad/s}.$

EVALUATE: The result is a rather high angular frequency.

31.45. **(a) IDENTIFY** and **SET UP:** Source voltage lags current so it must be that $X_C > X_L$ and we must add an inductor in series with the circuit. When $X_C = X_L$ the power factor has its maximum value of unity, so calculate the additional L needed to raise X_L to equal $X_C.$

(b) EXECUTE: Power factor $\cos\phi$ equals 1 so $\phi = 0$ and $X_C = X_L.$ Calculate the present value of $X_C - X_L$ to see how much more X_L is needed: $R = Z\cos\phi = (60.0\ \Omega)(0.720) = 43.2\ \Omega$

$\tan\phi = \dfrac{X_L - X_C}{R}$ so $X_L - X_C = R\tan\phi$

$\cos\phi = 0.720$ gives $\phi = -43.95°$ (ϕ is negative since the voltage lags the current)

Then $X_L - X_C = R\tan\phi = (43.2\ \Omega)\tan(-43.95°) = -41.64\ \Omega.$

Therefore need to add $41.64\ \Omega$ of $X_L.$

$X_L = \omega L = 2\pi f L$ and $L = \dfrac{X_L}{2\pi f} = \dfrac{41.64\ \Omega}{2\pi(50.0\ \text{Hz})} = 0.133\ \text{H},$ amount of inductance to add.

EVALUATE: From the information given we can't calculate the original value of L in the circuit, just how much to add. When this L is added the current in the circuit will increase.

31.47.   **IDENTIFY:**   We know the impedances and the average power consumed. From these we want to find the power factor and the rms voltage of the source.

SET UP: $P = I_{rms}^2 R.$ $\cos\phi = \dfrac{R}{Z}.$ $Z = \sqrt{R^2 + (X_L - X_C)^2}.$ $V_{rms} = I_{rms}Z.$

EXECUTE: **(a)** $I_{rms} = \sqrt{\dfrac{P}{R}} = \sqrt{\dfrac{60.0\text{ W}}{300\ \Omega}} = 0.447$ A. $Z = \sqrt{(300\ \Omega)^2 + (500\ \Omega - 300\ \Omega)^2} = 361\ \Omega.$

$\cos\phi = \dfrac{R}{Z} = \dfrac{300\ \Omega}{361\ \Omega} = 0.831.$

(b) $V_{rms} = I_{rms}Z = (0.447\text{ A})(361\ \Omega) = 161$ V.

EVALUATE: The voltage amplitude of the source is $V_{rms}\sqrt{2} = 228$ V.

31.49.   **IDENTIFY:**   The voltage and current amplitudes are the maximum values of these quantities, not necessarily the instantaneous values.

SET UP: The voltage amplitudes are $V_R = RI, V_L = X_L I,$ and $V_C = X_C I,$ where $I = V/Z$ and

$$Z = \sqrt{R^2 + \left(\omega L - \frac{1}{\omega C}\right)^2}.$$

EXECUTE: **(a)** $\omega = 2\pi f = 2\pi(1250\text{ Hz}) = 7854$ rad/s. Carrying extra figures in the calculator gives

$X_L = \omega L = (7854\text{ rad/s})(3.50\text{ mH}) = 27.5\ \Omega; XC = 1/\omega C = 1/[(7854\text{ rad/s})(10.0\ \mu\text{F})] = 12.7\ \Omega;$

$Z = \sqrt{R^2 + (X_L - X_C)^2} = \sqrt{(50.0\ \Omega)^2 + (27.5\ \Omega - 12.7\ \Omega)^2} = 57.5\ \Omega;$

$I = V/Z = (60.0\text{ V})/(52.1\ \Omega) = 1.15$ A; $V_R = RI = (50.0\ \Omega)(1.15\text{ A}) = 57.5$ V;

$V_L = X_L I = (27.5\ \Omega)(1.15\text{ A}) = 31.6$ V; $V_C = X_C I = (12.7\ \Omega)(1.15\text{ A}) = 14.6$ V.

The voltage amplitudes can add to more than 60.0 V because the voltage maxima do not all occur at the same instant of time. At any instant, the instantaneous voltages across the resistor, inductor and capacitor all add to equal the instantaneous source voltage.

(b) All of them will change because they all depend on ω. $X_L = \omega L$ will double to 55.0 Ω, and

$X_C = 1/\omega C$ will decrease by half to 6.35 Ω. Therefore $Z = \sqrt{(50.0\ \Omega)^2 + (55.0\ \Omega - 6.35\ \Omega)^2} = 69.8\ \Omega;$

$I = V/Z = (60.0\text{V})/(69.8\ \Omega) = 0.860$ A; $V_R = IR = (0.860\text{ A})(50.0\ \Omega) = 43.0$ V;

$V_L = IX_L = (0.860\text{ A})(55.0\ \Omega) = 47.3$ V; $V_C = IX_C = (0.860\text{ A})(6.35\ \Omega) = 5.46$ V.

EVALUATE: The new amplitudes in part (b) are not simple multiples of the values in part (a) because the impedance and reactances are not all the same simple multiple of the angular frequency.

31.55.   **IDENTIFY:**   We know R, X_C and ϕ so Eq. (31.24) tells us X_L. Use $P_{av} = I_{rms}^2 R$ to calculate I_{rms}. Then calculate Z and use Eq. (31.26) to calculate V_{rms} for the source.

SET UP: Source voltage lags current so $\phi = -54.0°$. $X_C = 350\ \Omega, R = 180\ \Omega, P_{av} = 140$ W

EXECUTE: **(a)** $\tan\phi = \dfrac{X_L - X_C}{R}$

$X_L = R\tan\phi + X_C = (180\ \Omega)\tan(-54.0°) + 350\ \Omega = -248\ \Omega + 350\ \Omega = 102\ \Omega$

(b) $P_{av} = V_{rms}I_{rms}\cos\phi = I_{rms}^2 R$ (Exercise 31.22). $I_{rms} = \sqrt{\dfrac{P_{av}}{R}} = \sqrt{\dfrac{140\text{ W}}{180\ \Omega}} = 0.882$ A

(c) $Z = \sqrt{R^2 + (X_L - X_C)^2} = \sqrt{(180\ \Omega)^2 + (102\ \Omega - 350\ \Omega)^2} = 306\ \Omega$

$V_{rms} = I_{rms}Z = (0.882\text{ A})(306\ \Omega) = 270$ V.

EVALUATE: We could also use Eq. (31.31): $P_{av} = V_{rms}I_{rms}\cos\phi$

$V_{rms} = \dfrac{P_{av}}{I_{rms}\cos\phi} = \dfrac{140\text{ W}}{(0.882\text{ A})\cos(-54.0°)} = 270$ V, which agrees. The source voltage lags the current

when $X_C > X_L$, and this agrees with what we found.

31.59. **IDENTIFY** and **SET UP:** Refer to the results and the phasor diagram in Problem 31.56. The source voltage is applied across each parallel branch.

EXECUTE: **(a)** $I_R = \dfrac{V}{R}; I_C = V\omega C; I_L = \dfrac{V}{\omega L}.$

(b) The graph of each current versus ω is given in Figure 31.59a.

(c) $\omega \to 0 : I_C \to 0; I_L \to \infty.$ $\omega \to \infty : I_C \to \infty; I_L \to 0.$

At low frequencies, the current is not changing much so the inductor's back-emf doesn't "resist." This allows the current to pass fairly freely. However, the current in the capacitor goes to zero because it tends to "fill up" over the slow period, making it less effective at passing charge. At high frequency, the induced emf in the inductor resists the violent changes and passes little current. The capacitor never gets a chance to fill up so passes charge freely.

(d) $\omega = \dfrac{1}{\sqrt{LC}} = \dfrac{1}{\sqrt{(2.0\text{ H})(0.50\times10^{-6}\text{ F})}} = 1000$ rad/sec and $f = 159$ Hz. The phasor diagram is sketched in Figure 31.59b.

(e) $I = \sqrt{\left(\dfrac{V}{R}\right)^2 + \left(V\omega C - \dfrac{V}{\omega L}\right)^2}.$

$I = \sqrt{\left(\dfrac{100\text{ V}}{200\ \Omega}\right)^2 + \left((100\text{ V})(1000\text{ s}^{-1})(0.50\times10^{-6}\text{ F}) - \dfrac{100\text{ V}}{(1000\text{ s}^{-1})(2.0\text{ H})}\right)^2} = 0.50$ A

(f) At resonance $I_L = I_C = V\omega C = (100\text{ V})(1000\text{ s}^{-1})(0.50\times10^{-6}\text{ F}) = 0.0500$ A and $I_R = \dfrac{V}{R} = \dfrac{100\text{ V}}{200\ \Omega} = 0.50$ A.

EVALUATE: At resonance $i_C = i_L = 0$ at all times and the current through the source equals the current through the resistor.

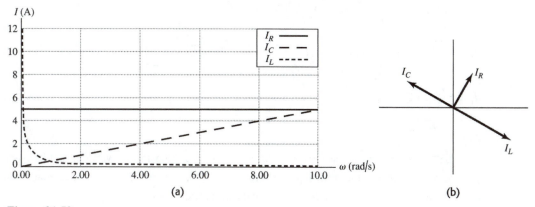

Figure 31.59

31.61. **IDENTIFY:** The resonance angular frequency is $\omega_0 = \dfrac{1}{\sqrt{LC}}$ and the resonance frequency is $f_0 = \dfrac{1}{2\pi\sqrt{LC}}.$

SET UP: ω_0 is independent of R.

EXECUTE: **(a)** ω_0 (or f_0) depends only on L and C so change these quantities.

(b) To double ω_0, decrease L and C by multiplying each of them by $\frac{1}{2}$.

EVALUATE: Increasing L and C decreases the resonance frequency; decreasing L and C increases the resonance frequency.

31.63. **IDENTIFY** and **SET UP:** Eq. (31.19) allows us to calculate I and then Eq. (31.22) gives Z. Solve Eq. (31.21) for X_L.

EXECUTE: **(a)** $V_C = IX_C$ so $I = \dfrac{V_C}{X_C} = \dfrac{360\text{ V}}{480\ \Omega} = 0.750$ A

(b) $V = IZ$ so $Z = \dfrac{V}{I} = \dfrac{120 \text{ V}}{0.750 \text{ A}} = 160 \,\Omega$

(c) $Z^2 = R^2 + (X_L - X_C)^2$

$X_L - X_C = \pm\sqrt{Z^2 - R^2}$, so

$X_L = X_C \pm \sqrt{Z^2 - R^2} = 480 \,\Omega \pm \sqrt{(160 \,\Omega)^2 - (80.0 \,\Omega)^2} = 480 \,\Omega \pm 139 \,\Omega$

$X_L = 619 \,\Omega$ or $341 \,\Omega$

(d) EVALUATE: $X_C = \dfrac{1}{\omega C}$ and $X_L = \omega L$. At resonance, $X_C = X_L$. As the frequency is lowered below the resonance frequency X_C increases and X_L decreases. Therefore, for $\omega < \omega_0$, $X_L < X_C$. So for $X_L = 341 \,\Omega$ the angular frequency is less than the resonance angular frequency. ω is greater than ω_0 when $X_L = 619 \,\Omega$. But at these two values of X_L, the magnitude of $X_L - X_C$ is the same so Z and I are the same. In one case ($X_L = 691 \,\Omega$) the source voltage leads the current and in the other ($X_L = 341 \,\Omega$) the source voltage lags the current.

31.65.   IDENTIFY and SET UP: Consider the cycle of the repeating current that lies between

$t_1 = \tau/2$ and $t_2 = 3\tau/2$. In this interval $i = \dfrac{2I_0}{\tau}(t - \tau)$. $I_{av} = \dfrac{1}{t_2 - t_1} \displaystyle\int_{t_1}^{t_2} i \, dt$ and $I_{rms}^2 = \dfrac{1}{t_2 - t_1} \displaystyle\int_{t_1}^{t_2} i^2 \, dt$.

EXECUTE: $I_{av} = \dfrac{1}{t_2 - t_1} \displaystyle\int_{t_1}^{t_2} i \, dt = \dfrac{1}{\tau} \displaystyle\int_{\tau/2}^{3\tau/2} \dfrac{2I_0}{\tau}(t - \tau) \, dt = \dfrac{2I_0}{\tau^2} \left[\dfrac{1}{2}t^2 - \tau t \right]_{\tau/2}^{3\tau/2}$

$I_{av} = \left(\dfrac{2I_0}{\tau^2} \right)\left(\dfrac{9\tau^2}{8} - \dfrac{3\tau^2}{2} - \dfrac{\tau^2}{8} + \dfrac{\tau^2}{2} \right) = (2I_0)\dfrac{1}{8}(9 - 12 - 1 + 4) = \dfrac{I_0}{4}(13 - 13) = 0.$

$I_{rms}^2 = (I^2)_{av} = \dfrac{1}{t_2 - t_1} \displaystyle\int_{t_1}^{t_2} i^2 \, dt = \dfrac{1}{\tau} \displaystyle\int_{\tau/2}^{3\tau/2} \dfrac{4I_0^2}{\tau^2}(t - \tau)^2 \, dt$

$I_{rms}^2 = \dfrac{4I_0^2}{\tau^3} \displaystyle\int_{\tau/2}^{3\tau/2} (t - \tau)^2 \, dt = \dfrac{4I_0^2}{\tau^3} \left[\dfrac{1}{3}(t - \tau)^3 \right]_{\tau/2}^{3\tau/2} = \dfrac{4I_0^2}{3\tau^3} \left[\left(\dfrac{\tau}{2} \right)^3 - \left(-\dfrac{\tau}{2} \right)^3 \right]$

$I_{rms}^2 = \dfrac{I_0^2}{6}[1 + 1] = \dfrac{1}{3}I_0^2$

$I_{rms} = \sqrt{I_{rms}^2} = \dfrac{I_0}{\sqrt{3}}.$

EVALUATE: In each cycle the current has as much negative value as positive value and its average is zero. i^2 is always positive and its average is not zero. The relation between I_{rms} and the current amplitude for this current is different from that for a sinusoidal current (Eq. 31.4).

31.71.   IDENTIFY: A transformer transforms voltages according to $\dfrac{V_2}{V_1} = \dfrac{N_2}{N_1}$. The effective resistance of a

secondary circuit of resistance R is $R_{eff} = \dfrac{R}{(N_2/N_1)^2}$.

SET UP: $N_2 = 275$ and $V_1 = 25.0$ V.

EXECUTE: (a) $V_2 = V_1(N_2/N_1) = (25.0 \text{ V})(834/275) = 75.8$ V

(b) $R_{eff} = \dfrac{R}{(N_2/N_1)^2} = \dfrac{125 \,\Omega}{(834/275)^2} = 13.6 \,\Omega$

EVALUATE: The voltage across the secondary is greater than the voltage across the primary since $N_2 > N_1$. The effective load resistance of the secondary is less than the resistance R connected across the secondary.

32

ELECTROMAGNETIC WAVES

32.3. **IDENTIFY:** $E_{max} = cB_{max}$. $\vec{E} \times \vec{B}$ is in the direction of propagation.

SET UP: $c = 3.00 \times 10^8$ m/s. $E_{max} = 4.00$ V/m.

EXECUTE: $B_{max} = E_{max}/c = 1.33 \times 10^{-8}$ T. For $\vec{E}$ in the $+x$-direction, $\vec{E} \times \vec{B}$ is in the $+z$-direction when $\vec{B}$ is in the $+y$-direction.

EVALUATE: $\vec{E}$, $\vec{B}$ and the direction of propagation are all mutually perpendicular.

32.7. **IDENTIFY:** $c = f\lambda$. $E_{max} = cB_{max}$. $k = 2\pi/\lambda$. $\omega = 2\pi f$.

SET UP: Since the wave is traveling in empty space, its wave speed is $c = 3.00 \times 10^8$ m/s.

EXECUTE: **(a)** $f = \dfrac{c}{\lambda} = \dfrac{3.00 \times 10^8 \text{ m/s}}{432 \times 10^{-9} \text{ m}} = 6.94 \times 10^{14}$ Hz

(b) $E_{max} = cB_{max} = (3.00 \times 10^8 \text{ m/s})(1.25 \times 10^{-6} \text{ T}) = 375$ V/m

(c) $k = \dfrac{2\pi}{\lambda} = \dfrac{2\pi \text{ rad}}{432 \times 10^{-9} \text{ m}} = 1.45 \times 10^7$ rad/m. $\omega = (2\pi \text{ rad})(6.94 \times 10^{14} \text{ Hz}) = 4.36 \times 10^{15}$ rad/s.

$E = E_{max} \cos(kx - \omega t) = (375 \text{ V/m}) \cos([1.45 \times 10^7 \text{ rad/m}]x - [4.36 \times 10^{15} \text{ rad/s}]t)$

$B = B_{max} \cos(kx - \omega t) = (1.25 \times 10^{-6} \text{ T}) \cos([1.45 \times 10^7 \text{ rad/m}]x - [4.36 \times 10^{15} \text{ rad/s}]t)$

EVALUATE: The $\cos(kx - \omega t)$ factor is common to both the electric and magnetic field expressions, since these two fields are in phase.

32.13. **IDENTIFY and SET UP:** $c = f\lambda$ allows calculation of λ. $k = 2\pi/\lambda$ and $\omega = 2\pi f$. Eq. (32.18) relates the electric and magnetic field amplitudes.

EXECUTE: **(a)** $c = f\lambda$ so $\lambda = \dfrac{c}{f} = \dfrac{2.998 \times 10^8 \text{ m/s}}{830 \times 10^3 \text{ Hz}} = 361$ m

(b) $k = \dfrac{2\pi}{\lambda} = \dfrac{2\pi \text{ rad}}{361 \text{ m}} = 0.0174$ rad/m

(c) $\omega = 2\pi f = (2\pi)(830 \times 10^3 \text{ Hz}) = 5.22 \times 10^6$ rad/s

(d) Eq. (32.18): $E_{max} = cB_{max} = (2.998 \times 10^8 \text{ m/s})(4.82 \times 10^{-11} \text{ T}) = 0.0144$ V/m

EVALUATE: This wave has a very long wavelength; its frequency is in the AM radio braodcast band. The electric and magnetic fields in the wave are very weak.

32.15. **IDENTIFY and SET UP:** $v = f\lambda$ relates frequency and wavelength to the speed of the wave. Use Eq. (32.22) to calculate n and K.

EXECUTE: **(a)** $\lambda = \dfrac{v}{f} = \dfrac{2.17 \times 10^8 \text{ m/s}}{5.70 \times 10^{14} \text{ Hz}} = 3.81 \times 10^{-7}$ m

(b) $\lambda = \dfrac{c}{f} = \dfrac{2.998 \times 10^8 \text{ m/s}}{5.70 \times 10^{14} \text{ Hz}} = 5.26 \times 10^{-7}$ m

(c) $n = \dfrac{c}{v} = \dfrac{2.998 \times 10^8 \text{ m/s}}{2.17 \times 10^8 \text{ m/s}} = 1.38$

(d) $n = \sqrt{KK_m} \approx \sqrt{K}$ so $K = n^2 = (1.38)^2 = 1.90$

EVALUATE: In the material $v < c$ and f is the same, so λ is less in the material than in air. $v < c$ always, so n is always greater than unity.

32.17. **IDENTIFY:** $I = P/A$. $I = \frac{1}{2}\epsilon_0 c E_{max}^2$. $E_{max} = cB_{max}$.

SET UP: The surface area of a sphere of radius r is $A = 4\pi r^2 \cdot \epsilon_0 = 8.85 \times 10^{-12} \text{ C}^2/\text{N} \cdot \text{m}^2$.

EXECUTE: **(a)** $I = \dfrac{P}{A} = \dfrac{(0.05)(75 \text{ W})}{4\pi(3.0 \times 10^{-2} \text{ m})^2} = 330 \text{ W/m}^2$.

(b) $E_{max} = \sqrt{\dfrac{2I}{\epsilon_0 c}} = \sqrt{\dfrac{2(330 \text{ W/m}^2)}{(8.85 \times 10^{-12} \text{ C}^2/\text{N} \cdot \text{m}^2)(3.00 \times 10^8 \text{ m/s})}} = 500 \text{ V/m}$.

$B_{max} = \dfrac{E_{max}}{c} = 1.7 \times 10^{-6} \text{ T} = 1.7 \ \mu\text{T}$.

EVALUATE: At the surface of the bulb the power radiated by the filament is spread over the surface of the bulb. Our calculation approximates the filament as a point source that radiates uniformly in all directions.

32.19. **IDENTIFY and SET UP:** Use Eq. (32.29) to calculate I, Eq. (32.18) to calculate B_{max}, and use $I = P_{av}/4\pi r^2$ to calculate P_{av}.

(a) EXECUTE: $I = \frac{1}{2}\epsilon_0 c E_{max}^2$; $E_{max} = 0.090 \text{ V/m}$, so $I = 1.1 \times 10^{-5} \text{ W/m}^2$

(b) $E_{max} = cB_{max}$ so $B_{max} = E_{max}/c = 3.0 \times 10^{-10} \text{ T}$

(c) $P_{av} = I(4\pi r^2) = (1.075 \times 10^{-5} \text{ W/m}^2)(4\pi)(2.5 \times 10^3 \text{ m})^2 = 840 \text{ W}$

(d) EVALUATE: The calculation in part (c) assumes that the transmitter emits uniformly in all directions.

32.21. **IDENTIFY:** $I = P_{av}/A$

SET UP: At a distance r from the star, the radiation from the star is spread over a spherical surface of area $A = 4\pi r^2$.

EXECUTE: $P_{av} = I(4\pi r^2) = (5.0 \times 10^3 \text{ W/m}^2)(4\pi)(2.0 \times 10^{10} \text{ m})^2 = 2.5 \times 10^{25} \text{ W}$

EVALUATE: The intensity decreases with distance from the star as $1/r^2$.

32.23. **IDENTIFY:** $P_{av} = IA$ and $I = E_{max}^2/2\mu_0 c$

SET UP: The surface area of a sphere is $A = 4\pi r^2$.

EXECUTE: $P_{av} = S_{av}A = \left(\dfrac{E_{max}^2}{2c\mu_0}\right)(4\pi r^2)$. $E_{max} = \sqrt{\dfrac{P_{av}c\mu_0}{2\pi r^2}} = \sqrt{\dfrac{(60.0 \text{ W})(3.00 \times 10^8 \text{ m/s})\mu_0}{2\pi(5.00 \text{ m})^2}} = 12.0 \text{ V/m}$.

$B_{max} = \dfrac{E_{max}}{c} = \dfrac{12.0 \text{ V/m}}{3.00 \times 10^8 \text{ m/s}} = 4.00 \times 10^{-8} \text{ T}$.

EVALUATE: E_{max} and B_{max} are both inversely proportional to the distance from the source.

32.25. **IDENTIFY:** Use the radiation pressure to find the intensity, and then $P_{av} = I(4\pi r^2)$.

SET UP: For a perfectly absorbing surface, $p_{rad} = \dfrac{I}{c}$.

EXECUTE: $p_{rad} = I/c$ so $I = cp_{rad} = 2.70 \times 10^3 \text{ W/m}^2$. Then $P_{av} = I(4\pi r^2) = (2.70 \times 10^3 \text{ W/m}^2)(4\pi)(5.0 \text{ m})^2 = 8.5 \times 10^5 \text{ W}$.

EVALUATE: Even though the source is very intense the radiation pressure 5.0 m from the surface is very small.

32.27. **IDENTIFY:** We know the greatest intensity that the eye can safely receive.

SET UP: $I = \dfrac{P}{A}$. $I = \frac{1}{2}\epsilon_0 c E_{max}^2$. $E_{max} = c B_{max}$.

EXECUTE: **(a)** $P = IA = (1.0 \times 10^2 \text{ W/m}^2)\pi(0.75 \times 10^{-3} \text{ m})^2 = 1.8 \times 10^{-4} \text{ W} = 0.18 \text{ mW}$.

(b) $E = \sqrt{\dfrac{2I}{\epsilon_0 c}} = \sqrt{\dfrac{2(1.0 \times 10^2 \text{ W/m}^2)}{(8.85 \times 10^{-12} \text{ C}^2/\text{N} \cdot \text{m}^2)(3.00 \times 10^8 \text{ m/s})}} = 274 \text{ V/m}$. $B_{max} = \dfrac{E_{max}}{c} = 9.13 \times 10^{-7} \text{ T}$.

(c) $P = 0.18 \text{ mW} = 0.18 \text{ mJ/s}$.

(d) $I = (1.0 \times 10^2 \text{ W/m}^2)\left(\dfrac{1 \text{ m}}{10^2 \text{ cm}}\right)^2 = 0.010 \text{ W/cm}^2$.

EVALUATE: Both the electric and magnetic fields are quite weak compared to normal laboratory fields.

32.31. **IDENTIFY:** The nodal and antinodal planes are each spaced one-half wavelength apart.

SET UP: $2\frac{1}{2}$ wavelengths fit in the oven, so $\left(2\frac{1}{2}\right)\lambda = L$, and the frequency of these waves obeys the equation $f\lambda = c$.

EXECUTE: **(a)** Since $\left(2\frac{1}{2}\right)\lambda = L$, we have $L = (5/2)(12.2 \text{ cm}) = 30.5 \text{ cm}$.

(b) Solving for the frequency gives $f = c/\lambda = (3.00 \times 10^8 \text{ m/s})/(0.122 \text{ m}) = 2.46 \times 10^9 \text{ Hz}$.

(c) $L = 35.5$ cm in this case. $\left(2\frac{1}{2}\right)\lambda = L$, so $\lambda = 2L/5 = 2(35.5 \text{ cm})/5 = 14.2 \text{ cm}$.

$$f = c/\lambda = (3.00 \times 10^8 \text{ m/s})/(0.142 \text{ m}) = 2.11 \times 10^9 \text{ Hz}$$

EVALUATE: Since microwaves have a reasonably large wavelength, microwave ovens can have a convenient size for household kitchens. Ovens using radiowaves would need to be far too large, while ovens using visible light would have to be microscopic.

32.35. **(a) IDENTIFY** and **SET UP:** The distance between adjacent nodal planes of $\vec{B}$ is $\lambda/2$. There is an antinodal plane of $\vec{B}$ midway between any two adjacent nodal planes, so the distance between a nodal plane and an adjacent antinodal plane is $\lambda/4$. Use $v = f\lambda$ to calculate λ.

EXECUTE: $\lambda = \dfrac{v}{f} = \dfrac{2.10 \times 10^8 \text{ m/s}}{1.20 \times 10^{10} \text{ Hz}} = 0.0175 \text{ m}$

$\dfrac{\lambda}{4} = \dfrac{0.0175 \text{ m}}{4} = 4.38 \times 10^{-3} \text{ m} = 4.38 \text{ mm}$

(b) IDENTIFY and **SET UP:** The nodal planes of $\vec{E}$ are at $x = 0$, $\lambda/2, \lambda, 3\lambda/2, \ldots$, so the antinodal planes of $\vec{E}$ are at $x = \lambda/4, 3\lambda/4, 5\lambda/4, \ldots$. The nodal planes of $\vec{B}$ are at $x = \lambda/4, 3\lambda/4, 5\lambda/4, \ldots$, so the antinodal planes of $\vec{B}$ are at $\lambda/2, \lambda, 3\lambda/2, \ldots$.

EXECUTE: The distance between adjacent antinodal planes of $\vec{E}$ and antinodal planes of $\vec{B}$ is therefore $\lambda/4 = 4.38$ mm.

(c) From Eqs. (32.36) and (32.37) the distance between adjacent nodal planes of $\vec{E}$ and $\vec{B}$ is $\lambda/4 = 4.38$ mm.

EVALUATE: The nodes of $\vec{E}$ coincide with the antinodes of $\vec{B}$ and conversely. The nodes of $\vec{B}$ and the nodes of $\vec{E}$ are equally spaced.

32.37. **IDENTIFY:** We know the wavelength and power of a laser beam as well as the area over which it acts and the duration of a pulse.

SET UP: The energy is $U = Pt$. For absorption the radiation pressure is $\dfrac{I}{c}$, where $I = \dfrac{P}{A}$. The wavelength in the eye is $\lambda = \dfrac{\lambda_0}{n}$. $I = \frac{1}{2}\epsilon_0 c E_{max}^2$ and $E_{max} = c B_{max}$.

EXECUTE: **(a)** $U = Pt = (250 \times 10^{-3} \text{ W})(1.50 \times 10^{-3} \text{ s}) = 3.75 \times 10^{-4} \text{ J} = 0.375 \text{ mJ}$.

(b) $I = \dfrac{P}{A} = \dfrac{250 \times 10^{-3} \text{ W}}{\pi (255 \times 10^{-6} \text{ m})^2} = 1.22 \times 10^6 \text{ W/m}^2$. The average pressure is

$\dfrac{I}{c} = \dfrac{1.22 \times 10^6 \text{ W/m}^2}{3.00 \times 10^8 \text{ m/s}} = 4.08 \times 10^{-3} \text{ Pa}$.

(c) $\lambda = \dfrac{\lambda_0}{n} = \dfrac{810 \text{ nm}}{1.34} = 604 \text{ nm}$. $f = \dfrac{v}{\lambda} = \dfrac{c}{\lambda_0} = \dfrac{3.00 \times 10^8 \text{ m/s}}{810 \times 10^{-9} \text{ m}} = 3.70 \times 10^{14} \text{ Hz}$; f is the same in the air and

in the vitreous humor.

(d) $E_{\text{max}} = \sqrt{\dfrac{2I}{\epsilon_0 c}} = \sqrt{\dfrac{2(1.22 \times 10^6 \text{ W/m}^2)}{(8.85 \times 10^{-12} \text{ C}^2/\text{N} \cdot \text{m}^2)(3.00 \times 10^8 \text{ m/s})}} = 3.03 \times 10^4 \text{ V/m}$.

$B_{\text{max}} = \dfrac{E_{\text{max}}}{c} = 1.01 \times 10^{-4} \text{ T}$.

EVALUATE: The intensity of the beam is high, as it must be to weld tissue, but the pressure it exerts on the retina is only around 10^{-8} that of atmospheric pressure. The magnetic field in the beam is about twice that of the earth's magnetic field.

32.39. **IDENTIFY:** The light exerts pressure on the paper, which produces an upward force. This force must balance the weight of the paper.

SET UP: The weight of the paper is mg. For a totally absorbing surface the radiation pressure is $\dfrac{I}{c}$ and for

a totally reflecting surface it is $\dfrac{2I}{c}$. The force is $F = PA$, and the intensity is $I = \dfrac{P}{A}$.

EXECUTE: **(a)** The radiation force must equal the weight of the paper, so $\left(\dfrac{I}{c}\right) A = mg$.

$I = \dfrac{mgc}{A} = \dfrac{(1.50 \times 10^{-3} \text{ kg})(9.80 \text{ m/s}^2)(3.00 \times 10^8 \text{ m/s})}{(0.220 \text{ m})(0.280 \text{ m})} = 7.16 \times 10^7 \text{ W/m}^2$.

(b) $I = \frac{1}{2}\epsilon_0 c E_{\text{max}}^2$. Solving for E_{max} gives

$E_{\text{max}} = \sqrt{\dfrac{2I}{\epsilon_0 c}} = \sqrt{\dfrac{2(7.16 \times 10^7 \text{ W/m}^2)}{(8.85 \times 10^{-12} \text{ C}^2/\text{N} \cdot \text{m}^2)(3.00 \times 10^8 \text{ m/s})}} = 2.32 \times 10^5 \text{ V/m}$.

$B_{\text{max}} = \dfrac{E_{\text{max}}}{c} = \dfrac{2.32 \times 10^5 \text{ V/m}}{3.00 \times 10^8 \text{ m/s}} = 7.74 \times 10^{-4} \text{ T}$.

(c) The pressure is $\dfrac{2I}{c}$, so $\left(\dfrac{2I}{c}\right) A = mg$. $I = \dfrac{mgc}{2A} = 3.58 \times 10^7 \text{ W/m}^2$.

(d) $I = \dfrac{P}{A} = \dfrac{0.500 \times 10^{-3} \text{ W}}{\pi (0.500 \times 10^{-3} \text{ m})^2} = 637 \text{ W/m}^2$.

EVALUATE: The intensity of this laser is much less than what is needed to support a sheet of paper. And to support the paper, not only must the intensity be large, it also must be over a large area.

32.41. **IDENTIFY:** The intensity of an electromagnetic wave depends on the amplitude of the electric and magnetic fields. Such a wave exerts a force because it carries energy.

SET UP: The intensity of the wave is $I = P_{\text{av}}/A = \frac{1}{2}\epsilon_0 c E_{\text{max}}^2$, and the force is $F = p_{\text{rad}} A$ where $p_{\text{rad}} = I/c$.

EXECUTE: **(a)** $I = P_{\text{av}}/A = (25{,}000 \text{ W})/[4\pi(575 \text{ m})^2] = 0.00602 \text{ W/m}^2$

(b) $I = \frac{1}{2}\epsilon_0 c E_{\text{max}}^2$, so $E_{\text{max}} = \sqrt{\dfrac{2I}{\epsilon_0 c}} = \sqrt{\dfrac{2(0.00602 \text{ W/m}^2)}{(8.85 \times 10^{-12} \text{ C}^2/\text{N} \cdot \text{m}^2)(3.00 \times 10^8 \text{ m/s})}} = 2.13 \text{ N/C}$.

$B_{\text{max}} = E_{\text{max}}/c = (2.13 \text{ N/C})/(3.00 \times 10^8 \text{ m/s}) = 7.10 \times 10^{-9} \text{ T}$

(c) $F = p_{\text{rad}} A = (I/c)A = (0.00602 \text{ W/m}^2)(0.150 \text{ m})(0.400 \text{ m})/(3.00 \times 10^8 \text{ m/s}) = 1.20 \times 10^{-12} \text{ N}$

EVALUATE: The fields are very weak compared to ordinary laboratory fields, and the force is hardly worth worrying about!

32.47.   **IDENTIFY:**   The same intensity light falls on both reflectors, but the force on the reflecting surface will be twice as great as the force on the absorbing surface. Therefore there will be a net torque about the rotation axis.

SET UP: For a totally absorbing surface, $F = p_{rad}A = (I/c)A$, while for a totally reflecting surface the force will be twice as great. The intensity of the wave is $I = \frac{1}{2}\epsilon_0 cE_{max}^2$. Once we have the torque, we can use the rotational form of Newton's second law, $\tau_{net} = I\alpha$, to find the angular acceleration.

EXECUTE: The force on the absorbing reflector is $F_{Abs} = p_{rad}A = (I/c)A = \dfrac{\frac{1}{2}\epsilon_0 cE_{max}^2 A}{c} = \frac{1}{2}\epsilon_0 AE_{max}^2$.

For a totally reflecting surface, the force will be twice as great, which is $\epsilon_0 cE_{max}^2$. The net torque is therefore $\tau_{net} = F_{Refl}(L/2) - F_{Abs}(L/2) = \epsilon_0 AE_{max}^2 L/4$.

Newton's second law for rotation gives $\tau_{net} = I\alpha$. $\epsilon_0 AE_{max}^2 L/4 = 2m(L/2)^2 \alpha$.

Solving for α gives

$$\alpha = \epsilon_0 AE_{max}^2/(2mL) = \frac{(8.85 \times 10^{-12} \text{ C}^2/\text{N} \cdot \text{m}^2)(0.0150 \text{ m})^2(1.25 \text{ N/C})^2}{(2)(0.00400 \text{ kg})(1.00 \text{ m})} = 3.89 \times 10^{-13} \text{ rad/s}^2.$$

EVALUATE: This is an extremely small angular acceleration. To achieve a larger value, we would have to greatly increase the intensity of the light wave or decrease the mass of the reflectors.

32.49.   **IDENTIFY and SET UP:**   In the wire the electric field is related to the current density by Eq. (25.7). Use Ampere's law to calculate $\vec{B}$. The Poynting vector is given by Eq. (32.28) and the equation that follows it relates the energy flow through a surface to $\vec{S}$.

EXECUTE: **(a)** The direction of $\vec{E}$ is parallel to the axis of the cylinder, in the direction of the current. From Eq. (25.7), $E = \rho J = \rho I/\pi a^2$. ($E$ is uniform across the cross section of the conductor.)

(b) A cross-sectional view of the conductor is given in Figure 32.49a; take the current to be coming out of the page.

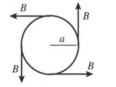

Apply Ampere's law to a circle of radius a.

$\oint \vec{B} \cdot d\vec{l} = B(2\pi a)$

$I_{encl} = I$

Figure 32.49a

$\oint \vec{B} \cdot d\vec{l} = \mu_0 I_{encl}$ gives $B(2\pi a) = \mu_0 I$ and $B = \dfrac{\mu_0 I}{2\pi a}$

The direction of $\vec{B}$ is counterclockwise around the circle.

(c) The directions of $\vec{E}$ and $\vec{B}$ are shown in Figure 32.49b.

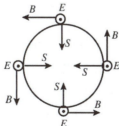

The direction of $\vec{S} = \dfrac{1}{\mu_0} \vec{E} \times \vec{B}$

is radially inward.

$S = \dfrac{1}{\mu_0} EB = \dfrac{1}{\mu_0} \left(\dfrac{\rho I}{\pi a^2} \right) \left(\dfrac{\mu_0 I}{2\pi a} \right)$

$S = \dfrac{\rho I^2}{2\pi^2 a^3}$

Figure 32.49b

(d) EVALUATE: Since S is constant over the surface of the conductor, the rate of energy flow P is given by S times the surface of a length l of the conductor: $P = SA = S(2\pi al) = \dfrac{\rho I^2}{2\pi^2 a^3}(2\pi al) = \dfrac{\rho l I^2}{\pi a^2}$. But $R = \dfrac{\rho l}{\pi a^2}$, so the result from the Poynting vector is $P = RI^2$. This agrees with $P_R = I^2 R$, the rate at which electrical energy is being dissipated by the resistance of the wire. Since $\vec{S}$ is radially inward at the surface of the wire and has magnitude equal to the rate at which electrical energy is being dissipated in the wire, this energy can be thought of as entering through the cylindrical sides of the conductor.

32.51.   IDENTIFY and SET UP: Find the force on you due to the momentum carried off by the light. Express this force in terms of the radiated power of the flashlight. Use this force to calculate your acceleration and use a constant acceleration equation to find the time.

(a) EXECUTE: $p_{rad} = I/c$ and $F = p_{rad}A$ gives $F = IA/c = P_{av}/c$

$a_x = F/m = P_{av}/(mc) = (200 \text{ W})/[(150 \text{ kg})(3.00 \times 10^8 \text{ m/s})] = 4.44 \times 10^{-9} \text{ m/s}^2$

Then $x - x_0 = v_{0x}t + \frac{1}{2}a_x t^2$ gives

$t = \sqrt{2(x - x_0)/a_x} = \sqrt{2(16.0 \text{ m})/(4.44 \times 10^{-9} \text{ m/s}^2)} = 8.49 \times 10^4 \text{ s} = 23.6 \text{ h}$

EVALUATE: The radiation force is very small. In the calculation we have ignored any other forces on you.

(b) You could throw the flashlight in the direction away from the ship. By conservation of linear momentum you would move toward the ship with the same magnitude of momentum as you gave the flashlight.

32.53.   IDENTIFY: The orbiting satellite obeys Newton's second law of motion. The intensity of the electromagnetic waves it transmits obeys the inverse-square distance law, and the intensity of the waves depends on the amplitude of the electric and magnetic fields.

SET UP: Newton's second law applied to the satellite gives $mv^2/r = GmM/r^2$, where M is the mass of the earth and m is the mass of the satellite. The intensity I of the wave is $I = S_{av} = \frac{1}{2}\epsilon_0 c E_{max}^2$, and by definition, $I = P_{av}/A$.

EXECUTE: (a) The period of the orbit is 12 hr. Applying Newton's second law to the satellite gives $mv^2/r = GmM/r^2$, which gives $\dfrac{m(2\pi r/T)^2}{r} = \dfrac{GmM}{r^2}$. Solving for r, we get

$r = \left(\dfrac{GMT^2}{4\pi^2}\right)^{1/3} = \left[\dfrac{(6.67 \times 10^{-11} \text{ N} \cdot \text{m}^2/\text{kg}^2)(5.97 \times 10^{24} \text{ kg})(12 \times 3600 \text{ s})^2}{4\pi^2}\right]^{1/3} = 2.66 \times 10^7 \text{ m}$

The height above the surface is $h = 2.66 \times 10^7 \text{ m} - 6.38 \times 10^6 \text{ m} = 2.02 \times 10^7 \text{ m}$. The satellite only radiates its energy to the lower hemisphere, so the area is 1/2 that of a sphere. Thus, from the definition of intensity, the intensity at the ground is

$I = P_{av}/A = P_{av}/(2\pi h^2) = (25.0 \text{ W})/[2\pi(2.02 \times 10^7 \text{ m})^2] = 9.75 \times 10^{-15} \text{ W/m}^2$

(b) $I = S_{av} = \frac{1}{2}\epsilon_0 c E_{max}^2$, so $E_{max} = \sqrt{\dfrac{2I}{\epsilon_0 c}} = \sqrt{\dfrac{2(9.75 \times 10^{-15} \text{ W/m}^2)}{(8.85 \times 10^{-12} \text{ C}^2/\text{N} \cdot \text{m}^2)(3.00 \times 10^8 \text{ m/s})}} = 2.71 \times 10^{-6} \text{ N/C}$

$B_{max} = E_{max}/c = (2.71 \times 10^{-6} \text{ N/C})/(3.00 \times 10^8 \text{ m/s}) = 9.03 \times 10^{-15} \text{ T}$

$t = d/c = (2.02 \times 10^7 \text{ m})/(3.00 \times 10^8 \text{ m/s}) = 0.0673 \text{ s}$

(c) $p_{rad} = I/c = (9.75 \times 10^{-15} \text{ W/m}^2)/(3.00 \times 10^8 \text{ m/s}) = 3.25 \times 10^{-23} \text{ Pa}$

(d) $\lambda = c/f = (3.00 \times 10^8 \text{ m/s})/(1575.42 \times 10^6 \text{ Hz}) = 0.190 \text{ m}$

EVALUATE: The fields and pressures due to these waves are very small compared to typical laboratory quantities.

THE NATURE AND PROPAGATION OF LIGHT

33.7. **IDENTIFY:** Apply Eqs. (33.2) and (33.4) to calculate θ_r and θ_b. The angles in these equations are measured with respect to the normal, not the surface.

(a) SET UP: The incident, reflected and refracted rays are shown in Figure 33.7.

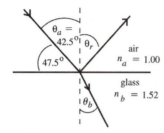

EXECUTE: $\theta_r = \theta_a = 42.5°$ The reflected ray makes an angle of $90.0° - \theta_r = 47.5°$ with the surface of the glass.

Figure 33.7

(b) $n_a \sin\theta_a = n_b \sin\theta_b$, where the angles are measured from the normal to the interface.

$$\sin\theta_b = \frac{n_a \sin\theta_a}{n_b} = \frac{(1.00)(\sin 42.5°)}{1.66} = 0.4070$$

$\theta_b = 24.0°$

The refracted ray makes an angle of $90.0° - \theta_b = 66.0°$ with the surface of the glass.

EVALUATE: The light is bent toward the normal when the light enters the material of larger refractive index.

33.9. **IDENTIFY and SET UP:** Use Snell's law to find the index of refraction of the plastic and then use Eq. (33.1) to calculate the speed v of light in the plastic.

EXECUTE: $n_a \sin\theta_a = n_b \sin\theta_b$

$$n_b = n_a\left(\frac{\sin\theta_a}{\sin\theta_b}\right) = 1.00\left(\frac{\sin 62.7°}{\sin 48.1°}\right) = 1.194$$

$n = \dfrac{c}{v}$ so $v = \dfrac{c}{n} = (3.00\times10^8 \text{ m/s})/1.194 = 2.51\times10^8$ m/s

EVALUATE: Light is slower in plastic than in air. When the light goes from air into the plastic it is bent toward the normal.

33.11. **IDENTIFY:** The figure shows the angle of incidence and angle of refraction for light going from the water into material X. Snell's law applies at the air-water and water-X boundaries.

SET UP: Snell's law says $n_a \sin\theta_a = n_b \sin\theta_b$. Apply Snell's law to the refraction from material X into the water and then from the water into the air.

EXECUTE: **(a)** Material X to water: $n_a = n_X$, $n_b = n_w = 1.333$. $\theta_a = 25°$ and $\theta_b = 48°$.

$$n_a = n_b\left(\frac{\sin\theta_b}{\sin\theta_a}\right) = (1.333)\left(\frac{\sin 48°}{\sin 25°}\right) = 2.34.$$

(b) Water to air: As Figure 33.11 shows, $\theta_a = 48°$. $n_a = 1.333$ and $n_b = 1.00$.

$$\sin\theta_b = \left(\frac{n_a}{n_b}\right)\sin\theta_a = (1.333)\sin 48° = 82°.$$

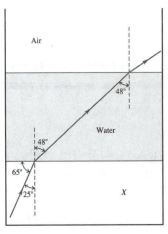

Figure 33.11

EVALUATE: $n > 1$ for material X, as it must be.

33.15. **IDENTIFY:** Apply $n_a \sin\theta_a = n_b \sin\theta_b$.

SET UP: $n_a = 1.70$, $\theta_a = 62.0°$. $n_b = 1.58$.

EXECUTE: $\sin\theta_b = \left(\frac{n_a}{n_b}\right)\sin\theta_a = \left(\frac{1.70}{1.58}\right)\sin 62.0° = 0.950$ and $\theta_b = 71.8°$.

EVALUATE: The ray refracts into a material of smaller n, so it is bent away from the normal.

33.17. **IDENTIFY:** The critical angle for total internal reflection is θ_a that gives $\theta_b = 90°$ in Snell's law.

SET UP: In Figure 33.17 the angle of incidence θ_a is related to angle θ by $\theta_a + \theta = 90°$.

EXECUTE: **(a)** Calculate θ_a that gives $\theta_b = 90°$. $n_a = 1.60$, $n_b = 1.00$ so $n_a \sin\theta_a = n_b \sin\theta_b$ gives

$(1.60)\sin\theta_a = (1.00)\sin 90°$. $\sin\theta_a = \frac{1.00}{1.60}$ and $\theta_a = 38.7°$. $\theta = 90° - \theta_a = 51.3°$.

(b) $n_a = 1.60$, $n_b = 1.333$. $(1.60)\sin\theta_a = (1.333)\sin 90°$. $\sin\theta_a = \frac{1.333}{1.60}$ and $\theta_a = 56.4°$.

$\theta = 90° - \theta_a = 33.6°$.

EVALUATE: The critical angle increases when the ratio $\frac{n_a}{n_b}$ decreases.

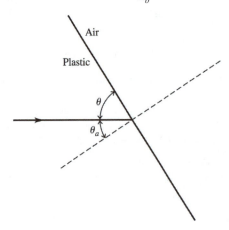

Figure 33.17

33.19. **IDENTIFY:** Use the critical angle to find the index of refraction of the liquid.

SET UP: Total internal reflection requires that the light be incident on the material with the larger n, in this case the liquid. Apply $n_a \sin \theta_a = n_b \sin \theta_b$ with a = liquid and b = air, so $n_a = n_{\text{liq}}$ and $n_b = 1.0$.

EXECUTE: $\theta_a = \theta_{\text{crit}}$ when $\theta_b = 90°$, so $n_{\text{liq}} \sin \theta_{\text{crit}} = (1.0) \sin 90°$

$$n_{\text{liq}} = \frac{1}{\sin \theta_{\text{crit}}} = \frac{1}{\sin 42.5°} = 1.48.$$

(a) $n_a \sin \theta_a = n_b \sin \theta_b$ (a = liquid, b = air)

$$\sin \theta_b = \frac{n_a \sin \theta_a}{n_b} = \frac{(1.48) \sin 35.0°}{1.0} = 0.8489 \text{ and } \theta_b = 58.1°$$

(b) Now $n_a \sin \theta_a = n_b \sin \theta_b$ with a = air, b = liquid

$$\sin \theta_b = \frac{n_a \sin \theta_a}{n_b} = \frac{(1.0) \sin 35.0°}{1.48} = 0.3876 \text{ and } \theta_b = 22.8°$$

EVALUATE: For light traveling liquid $\to$ air the light is bent away from the normal. For light traveling air $\to$ liquid the light is bent toward the normal.

33.23. **IDENTIFY:** Total internal reflection must be occurring at the glass-water boundary. Snell's law applies there.

SET UP: $n_a \sin \theta_a = n_b \sin \theta_b$. $\lambda = \lambda_0 / n$.

EXECUTE: Apply Snell's law to find n_{gl}: $n_{\text{gl}} \sin 62.0° = n_w \sin 90.0°$ and $n_{\text{gl}} = 1.510$. Then

$$\lambda_w n_w = \lambda_{\text{gl}} n_{\text{gl}} \text{ and } \lambda_w = \lambda_{\text{gl}} \left(\frac{n_{\text{gl}}}{n_w} \right) = (408 \text{ nm}) \left(\frac{1.510}{1.333} \right) = 462 \text{ nm}.$$

EVALUATE: The wavelength is greater in the water than it is in the glass, as it must be, since $n_w < n_{\text{gl}}$.

33.25. **IDENTIFY:** The index of refraction depends on the wavelength of light, so the light from the red and violet ends of the spectrum will be bent through different angles as it passes into the glass. Snell's law applies at the surface.

SET UP: $n_a \sin \theta_a = n_b \sin \theta_b$. From the graph in Figure 33.18 in the textbook, for $\lambda = 400$ nm (the violet end of the visible spectrum), $n = 1.67$ and for $\lambda = 700$ nm (the red end of the visible spectrum), $n = 1.62$. The path of a ray with a single wavelength is sketched in Figure 33.25.

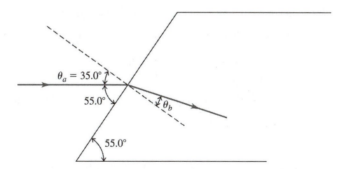

Figure 33.25

EXECUTE: For $\lambda = 400$ nm, $\sin \theta_b = \dfrac{n_a}{n_b} \sin \theta_a = \dfrac{1.00}{1.67} \sin 35.0°$, so $\theta_b = 20.1°$. For $\lambda = 700$ nm,

$\sin \theta_b = \dfrac{1.00}{1.62} \sin 35.0°$, so $\theta_b = 20.7°$. $\Delta \theta$ is about $0.6°$.

EVALUATE: This angle is small, but the separation of the beams could be fairly large if the light travels through a fairly large slab.

33.29. **IDENTIFY:** When unpolarized light passes through a polarizer the intensity is reduced by a factor of $\frac{1}{2}$ and the transmitted light is polarized along the axis of the polarizer. When polarized light of intensity I_{max} is incident on a polarizer, the transmitted intensity is $I = I_{\text{max}} \cos^2 \phi$, where ϕ is the angle between the polarization direction of the incident light and the axis of the filter.

SET UP: For the second polarizer $\phi = 60°$. For the third polarizer, $\phi = 90° - 60° = 30°$.

EXECUTE: (a) At point A the intensity is $I_0/2$ and the light is polarized along the vertical direction. At point B the intensity is $(I_0/2)(\cos 60°)^2 = 0.125 I_0$, and the light is polarized along the axis of the second polarizer. At point C the intensity is $(0.125 I_0)(\cos 30°)^2 = 0.0938 I_0$.

(b) Now for the last filter $\phi = 90°$ and $I = 0$.

EVALUATE: Adding the middle filter increases the transmitted intensity.

33.31.   **IDENTIFY** and **SET UP:**   Reflected beam completely linearly polarized implies that the angle of incidence equals the polarizing angle, so $\theta_p = 54.5°$. Use Eq. (33.8) to calculate the refractive index of the glass. Then use Snell's law to calculate the angle of refraction.

EXECUTE: (a) $\tan \theta_p = \dfrac{n_b}{n_a}$ gives $n_{glass} = n_{air} \tan \theta_p = (1.00) \tan 54.5° = 1.40$.

(b) $n_a \sin \theta_a = n_b \sin \theta_b$

$\sin \theta_b = \dfrac{n_a \sin \theta_a}{n_b} = \dfrac{(1.00)\sin 54.5°}{1.40} = 0.5815$ and $\theta_b = 35.5°$

EVALUATE:

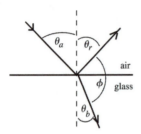

Note: $\phi = 180.0° - \theta_r - \theta_b$ and $\theta_r = \theta_a$.
Thus $\phi = 180.0° - 54.5° - 35.5° = 90.0°$;
the reflected ray and the refracted ray are perpendicular to each other. This agrees with Figure 33.28 in the text book.

Figure 33.31

33.35.   **IDENTIFY:**   When unpolarized light of intensity I_0 is incident on a polarizing filter, the transmitted light has intensity $\frac{1}{2} I_0$ and is polarized along the filter axis. When polarized light of intensity I_0 is incident on a polarizing filter the transmitted light has intensity $I_0 \cos^2 \phi$.

SET UP: For the second filter, $\phi = 62.0° - 25.0° = 37.0°$.

EXECUTE: After the first filter the intensity is $\frac{1}{2} I_0 = 10.0 \text{ W/m}^2$ and the light is polarized along the axis of the first filter. The intensity after the second filter is $I = I_0 \cos^2 \phi$, where $I_0 = 10.0 \text{ W/m}^2$ and $\phi = 37.0°$. This gives $I = 6.38 \text{ W/m}^2$.

EVALUATE: The transmitted intensity depends on the angle between the axes of the two filters.

33.37.   **IDENTIFY** and **SET UP:**   Apply Eq. (33.7) to polarizers #2 and #3. The light incident on the first polarizer is unpolarized, so the transmitted light has half the intensity of the incident light, and the transmitted light is polarized.

(a) EXECUTE: The axes of the three filters are shown in Figure 33.37a.

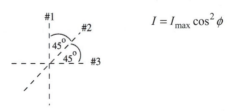

$I = I_{max} \cos^2 \phi$

Figure 33.37a

After the first filter the intensity is $I_1 = \frac{1}{2}I_0$ and the light is linearly polarized along the axis of the first polarizer. After the second filter the intensity is $I_2 = I_1 \cos^2 \phi = (\frac{1}{2}I_0)(\cos 45.0°)^2 = 0.250 I_0$ and the light is linearly polarized along the axis of the second polarizer. After the third filter the intensity is $I_3 = I_2 \cos^2 \phi = 0.250 I_0 (\cos 45.0°)^2 = 0.125 I_0$ and the light is linearly polarized along the axis of the third polarizer.

(b) The axes of the remaining two filters are shown in Figure 33.37b.

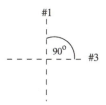

After the first filter the intensity is $I_1 = \frac{1}{2}I_0$ and the light is linearly polarized along the axis of the first polarizer.

Figure 33.37b

After the next filter the intensity is $I_3 = I_1 \cos^2 \phi = \left(\frac{1}{2}I_0\right)(\cos 90.0°)^2 = 0$. No light is passed.

EVALUATE: Light is transmitted through all three filters, but no light is transmitted if the middle polarizer is removed.

33.41. **IDENTIFY:** Snell's law applies to the sound waves in the heart. (See Exercise 33.24.)

SET UP: $n_a \sin \theta_a = n_b \sin \theta_b$. If θ_a is the critical angle then $\theta_b = 90°$. For air, $n_{air} = 1.00$. For heart muscle, $n_{mus} = \dfrac{344 \text{ m/s}}{1480 \text{ m/s}} = 0.2324$.

EXECUTE: **(a)** $n_a \sin \theta_a = n_b \sin \theta_b$ gives $(1.00)\sin(9.73°) = (0.2324)\sin \theta_b$. $\sin \theta_b = \dfrac{\sin(9.73°)}{0.2324}$ so $\theta_b = 46.7°$.

(b) $(1.00)\sin \theta_{crit} = (0.2324)\sin 90°$ gives $\theta_{crit} = 13.4°$.

EVALUATE: To interpret a sonogram, it should be important to know the true direction of travel of the sound waves within muscle. This would require knowledge of the refractive index of the muscle.

33.43. **IDENTIFY:** The angle of incidence at A is to be the critical angle. Apply Snell's law at the air to glass refraction at the top of the block.

SET UP: The ray is sketched in Figure 33.43.

EXECUTE: For glass $\to$ air at point A, Snell's law gives $(1.38)\sin \theta_{crit} = (1.00)\sin 90°$ and $\theta_{crit} = 46.4°$. $\theta_b = 90° - \theta_{crit} = 43.6°$. Snell's law applied to the refraction from air to glass at the top of the block gives $(1.00)\sin \theta_a = (1.38)\sin(43.6°)$ and $\theta_a = 72.1°$.

EVALUATE: If θ_a is larger than $72.1°$ then the angle of incidence at point A is less than the initial critical angle and total internal reflection doesn't occur.

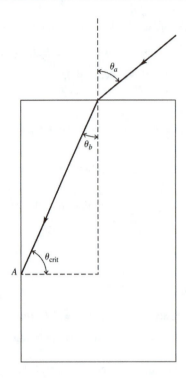

Figure 33.43

33.45.   **IDENTIFY:**   For total internal reflection, the angle of incidence must be at least as large as the critical angle.
SET UP: The angle of incidence for the glass-oil interface must be the critical angle, so $\theta_b = 90°$.

$n_a \sin\theta_a = n_b \sin\theta_b$.

EXECUTE: $n_a \sin\theta_a = n_b \sin\theta_b$ gives $(1.52)\sin 57.2° = n_{oil} \sin 90°$. $n_{oil} = (1.52)\sin 57.2° = 1.28$.

EVALUATE: $n_{oil} > 1$, which it must be, and 1.28 is a reasonable value for an oil.

33.47.   **IDENTIFY:**   Find the critical angle for glass $\rightarrow$ air. Light incident at this critical angle is reflected back to the edge of the halo.
SET UP: The ray incident at the critical angle is sketched in Figure 33.47.

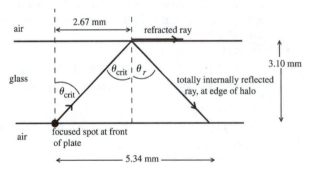

Figure 33.47

EXECUTE: From the distances given in the sketch, $\tan\theta_{crit} = \dfrac{2.67 \text{ mm}}{3.10 \text{ mm}} = 0.8613; \theta_{crit} = 40.7°$.

Apply Snell's law to the total internal reflection to find the refractive index of the glass:

$n_a \sin\theta_a = n_b \sin\theta_b$ $n_{glass} \sin\theta_{crit} = 1.00 \sin 90°$

$n_{glass} = \dfrac{1}{\sin\theta_{crit}} = \dfrac{1}{\sin 40.7°} = 1.53$

EVALUATE: Light incident on the back surface is also totally reflected if it is incident at angles greater than θ_{crit}. If it is incident at less than θ_{crit} it refracts into the air and does not reflect back to the emulsion.

33.49. **IDENTIFY:** Use Snell's law to determine the effect of the liquid on the direction of travel of the light as it enters the liquid.

SET UP: Use geometry to find the angles of incidence and refraction. Before the liquid is poured in, the ray along your line of sight has the path shown in Figure 33.49a.

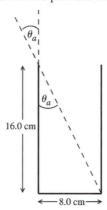

$$\tan\theta_a = \frac{8.0 \text{ cm}}{16.0 \text{ cm}} = 0.500$$

$$\theta_a = 26.57°$$

Figure 33.49a

After the liquid is poured in, θ_a is the same and the refracted ray passes through the center of the bottom of the glass, as shown in Figure 33.49b.

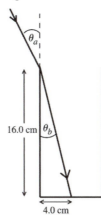

$$\tan\theta_b = \frac{4.0 \text{ cm}}{16.0 \text{ cm}} = 0.250$$

$$\theta_b = 14.04°$$

Figure 33.49b

EXECUTE: Use Snell's law to find n_b, the refractive index of the liquid:

$$n_a \sin\theta_a = n_b \sin\theta_b$$

$$n_b = \frac{n_a \sin\theta_a}{\sin\theta_b} = \frac{(1.00)(\sin 26.57°)}{\sin 14.04°} = 1.84$$

EVALUATE: When the light goes from air to liquid (larger refractive index) it is bent toward the normal.

33.51. **IDENTIFY:** Apply Snell's law to the water → ice and ice → air interfaces.

(a) SET UP: Consider the ray shown in Figure 33.51.

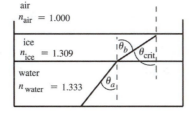

We want to find the incident angle θ_a at the water-ice interface that causes the incident angle at the ice-air interface to be the critical angle.

Figure 33.51

EXECUTE: ice-air interface: $n_{ice} \sin \theta_{crit} = 1.0 \sin 90°$

$n_{ice} \sin \theta_{crit} = 1.0$ so $\sin \theta_{crit} = \dfrac{1}{n_{ice}}$

But from the diagram we see that $\theta_b = \theta_{crit}$, so $\sin \theta_b = \dfrac{1}{n_{ice}}$.

water-ice interface: $n_w \sin \theta_a = n_{ice} \sin \theta_b$

But $\sin \theta_b = \dfrac{1}{n_{ice}}$ so $n_w \sin \theta_a = 1.0$. $\sin \theta_a = \dfrac{1}{n_w} = \dfrac{1}{1.333} = 0.7502$ and $\theta_a = 48.6°$.

(b) EVALUATE: The angle calculated in part (a) is the critical angle for a water-air interface; the answer would be the same if the ice layer wasn't there!

33.53. **IDENTIFY:** Apply Snell's law to the refraction of each ray as it emerges from the glass. The angle of incidence equals the angle $A = 25.0°$.

SET UP: The paths of the two rays are sketched in Figure 33.53.

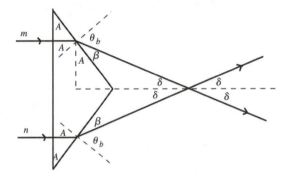

Figure 33.53

EXECUTE: $n_a \sin \theta_a = n_b \sin \theta_b$

$n_{glass} \sin 25.0° = 1.00 \sin \theta_b$

$\sin \theta_b = n_{glass} \sin 25.0°$

$\sin \theta_b = 1.66 \sin 25.0° = 0.7015$

$\theta_b = 44.55°$

$\beta = 90.0° - \theta_b = 45.45°$

Then $\delta = 90.0° - A - \beta = 90.0° - 25.0° - 45.45° = 19.55°$. The angle between the two rays is $2\delta = 39.1°$.

EVALUATE: The light is incident normally on the front face of the prism so the light is not bent as it enters the prism.

33.63. **IDENTIFY:** The reflected light is totally polarized when light strikes a surface at Brewster's angle.

SET UP: At the plastic wall, Brewster's angle obeys the equation $\tan \theta_p = n_b / n_a$, and Snell's law,

$n_a \sin \theta_a = n_b \sin \theta_b$, applies at the air-water surface.

EXECUTE: To be totally polarized, the reflected sunlight must have struck the wall at Brewster's angle. $\tan \theta_p = n_b / n_a = (1.61)/(1.00)$ and $\theta_p = 58.15°$.

This is the angle of incidence at the wall. A little geometry tells us that the angle of incidence at the water surface is $90.00° - 58.15° = 31.85°$. Applying Snell's law at the water surface gives

$$(1.00) \sin 31.85° = 1.333 \sin \theta \text{ and } \theta = 23.3°$$

EVALUATE: We have two different principles involved here: Reflection at Brewster's angle at the wall and Snell's law at the water surface.

GEOMETRIC OPTICS

34.5. **IDENTIFY** and **SET UP:** Use Eq. (34.6) to calculate s' and use Eq. (34.7) to calculate y'. The image is real if s' is positive and is erect if $m > 0$. Concave means R and f are positive, $R = +22.0$ cm; $f = R/2 = +11.0$ cm.

EXECUTE: **(a)**

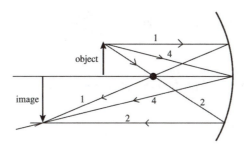

Figure 34.5

Three principal rays, numbered as in Section 34.2, are shown in Figure 34.5. The principal ray diagram shows that the image is real, inverted, and enlarged.

(b) $\dfrac{1}{s} + \dfrac{1}{s'} = \dfrac{1}{f}$

$\dfrac{1}{s'} = \dfrac{1}{f} - \dfrac{1}{s} = \dfrac{s-f}{sf}$ so $s' = \dfrac{sf}{s-f} = \dfrac{(16.5 \text{ cm})(11.0 \text{ cm})}{16.5 \text{ cm} - 11.0 \text{ cm}} = +33.0$ cm

$s' > 0$ so real image, 33.0 cm to left of mirror vertex

$m = -\dfrac{s'}{s} = -\dfrac{33.0 \text{ cm}}{16.5 \text{ cm}} = -2.00$ ($m < 0$ means inverted image) $|y'| = |m||y| = 2.00(0.600 \text{ cm}) = 1.20$ cm

EVALUATE: The image is 33.0 cm to the left of the mirror vertex. It is real, inverted, and is 1.20 cm tall (enlarged). The calculation agrees with the image characterization from the principal ray diagram. A concave mirror used alone always forms a real, inverted image if $s > f$ and the image is enlarged if $f < s < 2f$.

34.7. **IDENTIFY:** $\dfrac{1}{s} + \dfrac{1}{s'} = \dfrac{1}{f}$. $m = -\dfrac{s'}{s}$. $|m| = \dfrac{|y'|}{y}$. Find m and calculate y'.

SET UP: $f = +1.75$ m.

EXECUTE: $s \gg f$ so $s' = f = 1.75$ m.

$m = -\dfrac{s'}{s} = -\dfrac{1.75 \text{ m}}{5.58 \times 10^{10} \text{ m}} = -3.14 \times 10^{-11}$.

$|y'| = |m||y| = (3.14 \times 10^{-11})(6.794 \times 10^6 \text{ m}) = 2.13 \times 10^{-4}$ m $= 0.213$ mm.

EVALUATE: The image is real and is 1.75 m in front of the mirror.

34.9.   **IDENTIFY:**   The shell behaves as a spherical mirror.

   SET UP: The equation relating the object and image distances to the focal length of a spherical mirror is

   $\dfrac{1}{s}+\dfrac{1}{s'}=\dfrac{1}{f}$, and its magnification is given by $m=-\dfrac{s'}{s}$.

   EXECUTE: $\dfrac{1}{s}+\dfrac{1}{s'}=\dfrac{1}{f} \Rightarrow \dfrac{1}{s}=\dfrac{2}{-18.0\text{ cm}}-\dfrac{1}{-6.00\text{ cm}} \Rightarrow s=18.0\text{ cm}$ from the vertex.

   $m=-\dfrac{s'}{s}=-\dfrac{-6.00\text{ cm}}{18.0\text{ cm}}=\dfrac{1}{3} \Rightarrow y'=\dfrac{1}{3}(1.5\text{ cm})=0.50\text{ cm}.$ The image is 0.50 cm tall, erect and virtual.

   EVALUATE: Since the magnification is less than one, the image is smaller than the object.

34.13.   **IDENTIFY:** $\dfrac{1}{s}+\dfrac{1}{s'}=\dfrac{1}{f}$ and $m=\dfrac{y'}{y}=-\dfrac{s'}{s}.$

   SET UP: $m=+2.00$ and $s=1.25$ cm. An erect image must be virtual.

   EXECUTE: (a) $s'=\dfrac{sf}{s-f}$ and $m=-\dfrac{f}{s-f}.$ For a concave mirror, m can be larger than 1.00. For a convex

   mirror, $|f|=-f$ so $m=+\dfrac{|f|}{s+|f|}$ and m is always less than 1.00. The mirror must be concave $(f>0).$

   (b) $\dfrac{1}{f}=\dfrac{s'+s}{ss'}.$ $f=\dfrac{ss'}{s+s'}.$ $m=-\dfrac{s'}{s}=+2.00$ and $s'=-2.00s.$ $f=\dfrac{s(-2.00s)}{s-2.00s}=+2.00s=+2.50$ cm.

   $R=2f=+5.00$ cm.

   (c) The principal-ray diagram is drawn in Figure 34.13.

   EVALUATE: The principal-ray diagram agrees with the description from the equations.

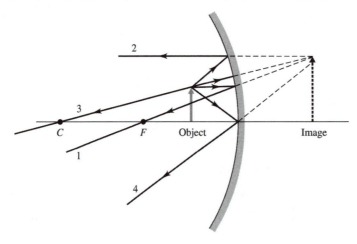

   Figure 34.13

34.17.   **IDENTIFY:**   Think of the surface of the water as a section of a sphere having an infinite radius of curvature.

   SET UP: $\dfrac{n_a}{s}+\dfrac{n_b}{s'}=0.$ $n_a=1.00.$ $n_b=1.333.$

   EXECUTE: The image is $5.20\text{ m}-0.80\text{ m}=4.40\text{ m}$ above the surface of the water, so $s'=-4.40$ m.

   $s=-\dfrac{n_a}{n_b}s'=-\left(\dfrac{1.00}{1.333}\right)(-4.40\text{ m})=+3.30\text{ m}.$

   EVALUATE: The diving board is closer to the water than it looks to the swimmer.

34.19.   **IDENTIFY:** $\dfrac{n_a}{s}+\dfrac{n_b}{s'}=\dfrac{n_b-n_a}{R}.$ $m=-\dfrac{n_a s'}{n_b s}.$ Light comes from the fish to the person's eye.

   SET UP: $R=-14.0$ cm. $s=+14.0$ cm. $n_a=1.333$ (water). $n_b=1.00$ (air). Figure 34.19 shows the object and the refracting surface.

EXECUTE: **(a)** $\dfrac{1.333}{14.0 \text{ cm}} + \dfrac{1.00}{s'} = \dfrac{1.00 - 1.333}{-14.0 \text{ cm}}$. $s' = -14.0$ cm. $m = -\dfrac{(1.333)(-14.0 \text{ cm})}{(1.00)(14.0 \text{ cm})} = +1.33$.

The fish's image is 14.0 cm to the left of the bowl surface so is at the center of the bowl and the magnification is 1.33.

(b) The focal point is at the image location when $s \to \infty$. $\dfrac{n_b}{s'} = \dfrac{n_b - n_a}{R}$. $n_a = 1.00$. $n_b = 1.333$.

$R = +14.0$ cm. $\dfrac{1.333}{s'} = \dfrac{1.333 - 1.00}{14.0 \text{ cm}}$. $s' = +56.0$ cm. s' is greater than the diameter of the bowl, so the surface facing the sunlight does not focus the sunlight to a point inside the bowl. The focal point is outside the bowl and there is no danger to the fish.

EVALUATE: In part (b) the rays refract when they exit the bowl back into the air so the image we calculated is not the final image.

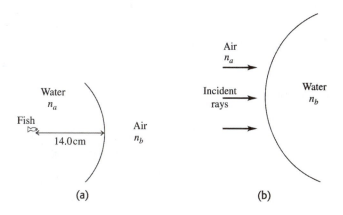

Figure 34.19

34.21. **IDENTIFY:** The hemispherical glass surface forms an image by refraction. The location of this image depends on the curvature of the surface and the indices of refraction of the glass and oil.

SET UP: The image and object distances are related to the indices of refraction and the radius of curvature by the equation $\dfrac{n_a}{s} + \dfrac{n_b}{s'} = \dfrac{n_b - n_a}{R}$.

EXECUTE: $\dfrac{n_a}{s} + \dfrac{n_b}{s'} = \dfrac{n_b - n_a}{R} \Rightarrow \dfrac{1.45}{s} + \dfrac{1.60}{1.20 \text{ m}} = \dfrac{0.15}{0.0300 \text{ m}} \Rightarrow s = 39.5$ cm

EVALUATE: The presence of the oil changes the location of the image.

34.27. **IDENTIFY:** Use the lensmaker's equation to calculate f.

SET UP: The lensmaker's equation is $\dfrac{1}{s} + \dfrac{1}{s'} = (n-1)\left(\dfrac{1}{R_1} - \dfrac{1}{R_2}\right)$, and the magnification of the lens is

$m = -\dfrac{s'}{s}$.

EXECUTE: **(a)** $\dfrac{1}{s} + \dfrac{1}{s'} = (n-1)\left(\dfrac{1}{R_1} - \dfrac{1}{R_2}\right) \Rightarrow \dfrac{1}{24.0 \text{ cm}} + \dfrac{1}{s'} = (1.52 - 1)\left(\dfrac{1}{-7.00 \text{ cm}} - \dfrac{1}{-4.00 \text{ cm}}\right)$

$\Rightarrow s' = 71.2$ cm, to the right of the lens.

(b) $m = -\dfrac{s'}{s} = -\dfrac{71.2 \text{ cm}}{24.0 \text{ cm}} = -2.97$

EVALUATE: Since the magnification is negative, the image is inverted.

34.29. **IDENTIFY:** The thin-lens equation applies in this case.

SET UP: The thin-lens equation is $\dfrac{1}{s} + \dfrac{1}{s'} = \dfrac{1}{f}$, and the magnification is $m = -\dfrac{s'}{s} = \dfrac{y'}{y}$.

EXECUTE: $m = \dfrac{y'}{y} = \dfrac{34.0 \text{ mm}}{8.00 \text{ mm}} = 4.25 = -\dfrac{s'}{s} = -\dfrac{-12.0 \text{ cm}}{s} \Rightarrow s = 2.82 \text{ cm}.$ The thin-lens equation gives

$\dfrac{1}{s} + \dfrac{1}{s'} = \dfrac{1}{f} \Rightarrow f = 3.69 \text{ cm}.$

EVALUATE: Since the focal length is positive, this is a converging lens. The image distance is negative because the object is inside the focal point of the lens.

34.31.   **IDENTIFY:**   Apply $\dfrac{1}{f} = (n-1)\left(\dfrac{1}{R_1} - \dfrac{1}{R_2}\right).$

SET UP: For a distant object the image is at the focal point of the lens. Therefore, $f = 1.87$ cm. For the double-convex lens, $R_1 = +R$ and $R_2 = -R$, where $R = 2.50$ cm.

EXECUTE: $\dfrac{1}{f} = (n-1)\left(\dfrac{1}{R} - \dfrac{1}{-R}\right) = \dfrac{2(n-1)}{R}.$ $n = \dfrac{R}{2f} + 1 = \dfrac{2.50 \text{ cm}}{2(1.87 \text{ cm})} + 1 = 1.67.$

EVALUATE: $f > 0$ and the lens is converging. A double-convex lens is always converging.

34.33.   **IDENTIFY:**   First use the lensmaker's formula to find the radius of curvature of the cornea.

SET UP: $\dfrac{1}{f} = (n-1)\left(\dfrac{1}{R_1} - \dfrac{1}{R_2}\right).$ $R_1 = +5.0$ mm. $\dfrac{1}{s} + \dfrac{1}{s'} = \dfrac{1}{f}.$ $m = \dfrac{y'}{y} = -\dfrac{s'}{s}.$

EXECUTE: **(a)** $\dfrac{1}{f(n-1)} = \dfrac{1}{R_1} - \dfrac{1}{R_2}.$ $\dfrac{1}{R_2} = \dfrac{1}{R_1} - \dfrac{1}{f(n-1)} = \dfrac{1}{+5.0 \text{ mm}} - \dfrac{1}{(18.0 \text{ mm})(0.38)}$ so $R_2 = 18.6$ mm.

(b) $\dfrac{1}{s'} = \dfrac{1}{f} - \dfrac{1}{s} = \dfrac{s-f}{sf}.$ $s' = \dfrac{sf}{s-f} = \dfrac{(25 \text{ cm})(1.8 \text{ cm})}{25 \text{ cm} - 1.8 \text{ cm}} = 1.9 \text{ cm} = 19 \text{ mm}.$

(c) $m = -\dfrac{s'}{s} = -\dfrac{1.9 \text{ cm}}{25 \text{ cm}} = -0.076.$ $y' = my = (-0.076)(8.0 \text{ mm}) = -0.61 \text{ mm}.$ $s' > 0$ so the image is real. $m < 0$ so the image is inverted.

EVALUATE: The cornea alone would focus an object at a distance of 19 mm, which is not at the retina. We must consider the effects of the lens of the eye and the fact that the eye is filled with liquid having an index of refraction.

34.39.   **IDENTIFY:**   The first lens forms an image that is then the object for the second lens.

SET UP: Apply $\dfrac{1}{s} + \dfrac{1}{s'} = \dfrac{1}{f}$ to each lens. $m_1 = \dfrac{y_1'}{y_1}$ and $m_2 = \dfrac{y_2'}{y_2}.$

EXECUTE: **(a)** *Lens 1:* $\dfrac{1}{s} + \dfrac{1}{s'} = \dfrac{1}{f}$ gives $s_1' = \dfrac{s_1 f_1}{s_1 - f_1} = \dfrac{(50.0 \text{ cm})(40.0 \text{ cm})}{50.0 \text{ cm} - 40.0 \text{ cm}} = +200 \text{ cm}.$

$m_1 = -\dfrac{s_1'}{s_1} = -\dfrac{200 \text{ cm}}{50 \text{ cm}} = -4.00.$ $y_1' = m_1 y_1 = (-4.00)(1.20 \text{ cm}) = -4.80 \text{ cm}.$ The image I_1 is 200 cm to the right of lens 1, is 4.80 cm tall and is inverted.

(b) *Lens 2:* $y_2 = -4.80$ cm. The image I_1 is $300 \text{ cm} - 200 \text{ cm} = 100 \text{ cm}$ to the left of lens 2, so

$s_2 = +100$ cm. $s_2' = \dfrac{s_2 f_2}{s_2 - f_2} = \dfrac{(100 \text{ cm})(60.0 \text{ cm})}{100 \text{ cm} - 60.0 \text{ cm}} = +150 \text{ cm}.$ $m_2 = -\dfrac{s_2'}{s_2} = -\dfrac{150 \text{ cm}}{100 \text{ cm}} = -1.50.$

$y_2' = m_2 y_2 = (-1.50)(-4.80 \text{ cm}) = +7.20 \text{ cm}.$ The image is 150 cm to the right of the second lens, is 7.20 cm tall, and is erect with respect to the original object.

EVALUATE: The overall magnification of the lens combination is $m_{\text{tot}} = m_1 m_2.$

34.41.   **IDENTIFY:**   The first lens forms an image that is then the object for the second lens. We follow the same general procedure as in Problem 34.39.

SET UP: $m_{\text{tot}} = m_1 m_2.$ $\dfrac{1}{s} + \dfrac{1}{s'} = \dfrac{1}{f}$ gives $s' = \dfrac{sf}{s-f}.$

EXECUTE: **(a)** *Lens 1:* $f_1 = -12.0$ cm, $s_1 = 20.0$ cm. $s' = \dfrac{(20.0\text{ cm})(-12.0\text{ cm})}{20.0\text{ cm} + 12.0\text{ cm}} = -7.5$ cm.

$m_1 = -\dfrac{s_1'}{s_1} = -\dfrac{-7.5\text{ cm}}{20.0\text{ cm}} = +0.375$.

Lens 2: The image of lens 1 is 7.5 cm to the left of lens 1 so is $7.5\text{ cm} + 9.00\text{ cm} = 16.5$ cm to the left of lens 2.

$s_2 = +16.5$ cm. $f_2 = +12.0$ cm. $s_2' = \dfrac{(16.5\text{ cm})(12.0\text{ cm})}{16.5\text{ cm} - 12.0\text{ cm}} = 44.0$ cm. $m_2 = -\dfrac{s_2'}{s_2} = -\dfrac{44.0\text{ cm}}{16.5\text{ cm}} = -2.67$. The

final image is 44.0 cm to the right of lens 2 so is 53.0 cm to the right of the first lens.

(b) $s_2' > 0$ so the final image is real.

(c) $m_{\text{tot}} = m_1 m_2 = (+0.375)(-2.67) = -1.00$. The image is 2.50 mm tall and is inverted.

EVALUATE: The light travels through the lenses in the direction from left to right. A real image for the second lens is to the right of that lens and a virtual image is to the left of the second lens.

34.47. **IDENTIFY** and **SET UP:** Find the lateral magnification that results in this desired image size. Use Eq. (34.17) to relate m and s' and Eq. (34.16) to relate s and s' to f.

EXECUTE: **(a)** We need $m = -\dfrac{24 \times 10^{-3}\text{ m}}{160\text{ m}} = -1.5 \times 10^{-4}$. Alternatively, $m = -\dfrac{36 \times 10^{-3}\text{ m}}{240\text{ m}} = -1.5 \times 10^{-4}$.

$s \gg f$ so $s' \approx f$

Then $m = -\dfrac{s'}{s} = -\dfrac{f}{s} = -1.5 \times 10^{-4}$ and $f = (1.5 \times 10^{-4})(600\text{ m}) = 0.090\text{ m} = 90$ mm.

A smaller f means a smaller s' and a smaller m, so with $f = 85$ mm the object's image nearly fills the picture area.

(b) We need $m = -\dfrac{36 \times 10^{-3}\text{ m}}{9.6\text{ m}} = -3.75 \times 10^{-3}$. Then, as in part (a), $\dfrac{f}{s} = 3.75 \times 10^{-3}$ and

$f = (40.0\text{ m})(3.75 \times 10^{-3}) = 0.15\text{ m} = 150$ mm. Therefore use the 135-mm lens.

EVALUATE: When $s \gg f$ and $s' \approx f$, $y' = -f(y/s)$. For the mobile home y/s is smaller so a larger f is needed. Note that m is very small; the image is much smaller than the object.

34.49. **IDENTIFY:** The f-number of a lens is the ratio of its focal length to its diameter. To maintain the same exposure, the amount of light passing through the lens during the exposure must remain the same.
SET UP: The f-number is f/D.

EXECUTE: **(a)** $f\text{-number} = \dfrac{f}{D} \Rightarrow f\text{-number} = \dfrac{180.0\text{ mm}}{16.36\text{ mm}} \Rightarrow f\text{-number} = f/11$. (The f-number is an integer.)

(b) $f/11$ to $f/2.8$ is four steps of 2 in intensity, so one needs $1/16^{\text{th}}$ the exposure. The exposure should be $1/480\text{ s} = 2.1 \times 10^{-3}\text{ s} = 2.1$ ms.

EVALUATE: When opening the lens from $f/11$ to $f/2.8$, the area increases by a factor of 16, so 16 times as much light is allowed in. Therefore the exposure time must be decreased by a factor of 1/16 to maintain the same exposure on the film or light receptors of a digital camera.

34.53. **(a)** **IDENTIFY:** The purpose of the corrective lens is to take an object 25 cm from the eye and form a virtual image at the eye's near point. Use Eq. (34.16) to solve for the image distance when the object distance is 25 cm.

SET UP: $\dfrac{1}{f} = +2.75$ diopters means $f = +\dfrac{1}{2.75}\text{ m} = +0.3636$ m (converging lens)

$f = 36.36$ cm; $s = 25$ cm; $s' = ?$

EXECUTE: $\dfrac{1}{s} + \dfrac{1}{s'} = \dfrac{1}{f}$ so

$s' = \dfrac{sf}{s - f} = \dfrac{(25\text{ cm})(36.36\text{ cm})}{25\text{ cm} - 36.36\text{ cm}} = -80.0$ cm

The eye's near point is 80.0 cm from the eye.

(b) **IDENTIFY:** The purpose of the corrective lens is to take an object at infinity and form a virtual image of it at the eye's far point. Use Eq. (34.16) to solve for the image distance when the object is at infinity.

SET UP: $\dfrac{1}{f} = -1.30$ diopters means $f = -\dfrac{1}{1.30}$ m $= -0.7692$ m (diverging lens)

$f = -76.92$ cm; $s = \infty$; $s' = ?$

EXECUTE: $\dfrac{1}{s} + \dfrac{1}{s'} = \dfrac{1}{f}$ and $s = \infty$ says $\dfrac{1}{s'} = \dfrac{1}{f}$ and $s' = f = -76.9$ cm. The eye's far point is 76.9 cm

from the eye.

EVALUATE: In each case a virtual image is formed by the lens. The eye views this virtual image instead of the object. The object is at a distance where the eye can't focus on it, but the virtual image is at a distance where the eye can focus.

34.55. **IDENTIFY** and **SET UP:** For an object 25.0 cm from the eye, the corrective lens forms a virtual image at the near point of the eye. The distances from the corrective lens are $s = 23.0$ cm and $s' = -43.0$ cm.

$\dfrac{1}{s} + \dfrac{1}{s'} = \dfrac{1}{f}$. P(in diopters)$ = 1/f$ (in m).

EXECUTE: Solving $\dfrac{1}{s} + \dfrac{1}{s'} = \dfrac{1}{f}$ for f gives $f = \dfrac{ss'}{s + s'} = \dfrac{(23.0 \text{ cm})(-43.0 \text{ cm})}{23.0 \text{ cm} - 43.0 \text{ cm}} = +49.4$ cm. The power is

$\dfrac{1}{0.494 \text{ m}} = 2.02$ diopters.

EVALUATE: In Problem 34.54 the contact lenses have power 1.78 diopters. The power of the lenses is different for ordinary glasses versus contact lenses.

34.57. **IDENTIFY** and **SET UP:** For an object very far from the eye, the corrective lens forms a virtual image at the far point of the eye. The distances from the lens are $s \to \infty$ and $s' = -73.0$ cm.

$\dfrac{1}{s} + \dfrac{1}{s'} = \dfrac{1}{f}$. P(in diopters)$ = 1/f$ (in m).

EXECUTE: In $\dfrac{1}{s} + \dfrac{1}{s'} = \dfrac{1}{f}$, $s \to \infty$, so $f = s' = -73.0$ cm. The power is $\dfrac{1}{-0.730 \text{ m}} = -1.37$ diopters.

EVALUATE: A diverging lens is needed to form a virtual image of a distant object. A converging lens could not do this since distant objects cannot be inside its focal point.

34.59. **IDENTIFY:** Use Eqs. (34.16) and (34.17) to calculate s and y'.

(a) SET UP: $f = 8.00$ cm; $s' = -25.0$ cm; $s = ?$

$\dfrac{1}{s} + \dfrac{1}{s'} = \dfrac{1}{f}$, so $\dfrac{1}{s} = \dfrac{1}{f} - \dfrac{1}{s'} = \dfrac{s' - f}{s'f}$

EXECUTE: $s = \dfrac{s'f}{s' - f} = \dfrac{(-25.0 \text{ cm})(+8.00 \text{ cm})}{-25.0 \text{ cm} - 8.00 \text{ cm}} = +6.06$ cm

(b) $m = -\dfrac{s'}{s} = -\dfrac{-25.0 \text{ cm}}{6.06 \text{ cm}} = +4.125$

$|m| = \dfrac{|y'|}{|y|}$ so $|y'| = |m||y| = (4.125)(1.00 \text{ mm}) = 4.12$ mm

EVALUATE: The lens allows the object to be much closer to the eye than the near point. The lens allows the eye to view an image at the near point rather than the object.

34.63. **(a) IDENTIFY** and **SET UP:**

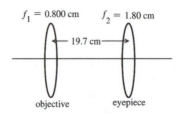

Figure 34.63

Final image is at ∞ so the object for the eyepiece is at its focal point. But the object for the eyepiece is the image of the objective so the image formed by the objective is $19.7 \text{ cm} - 1.80 \text{ cm} = 17.9 \text{ cm}$ to the right of the lens. Apply Eq. (34.16) to the image formation by the objective, solve for the object distance s.

$f = 0.800 \text{ cm}; \ s' = 17.9 \text{ cm}; \ s = ?$

$$\frac{1}{s} + \frac{1}{s'} = \frac{1}{f}, \text{ so } \frac{1}{s} = \frac{1}{f} - \frac{1}{s'} = \frac{s' - f}{s'f}$$

EXECUTE: $s = \dfrac{s'f}{s' - f} = \dfrac{(17.9 \text{ cm})(+0.800 \text{ cm})}{17.9 \text{ cm} - 0.800 \text{ cm}} = +8.37 \text{ mm}$

(b) SET UP: Use Eq. (34.17).

EXECUTE: $m_1 = -\dfrac{s'}{s} = -\dfrac{17.9 \text{ cm}}{0.837 \text{ cm}} = -21.4$

The linear magnification of the objective is 21.4.

(c) SET UP: Use Eq. (34.24): $M = m_1 M_2$

EXECUTE: $M_2 = \dfrac{25 \text{ cm}}{f_2} = \dfrac{25 \text{ cm}}{1.80 \text{ cm}} = 13.9$

$M = m_1 M_2 = (-21.4)(13.9) = -297$

EVALUATE: M is not accurately given by $(25 \text{ cm})s_1'/f_1 f_2 = 311$, because the object is not quite at the focal point of the objective ($s_1 = 0.837 \text{ cm}$ and $f_1 = 0.800 \text{ cm}$).

34.65.   **(a) IDENTIFY and SET UP:** Use Eq. (34.25), with $f_1 = 95.0 \text{ cm}$ (objective) and $f_2 = 15.0 \text{ cm}$ (eyepiece).

EXECUTE: $M = -\dfrac{f_1}{f_2} = -\dfrac{95.0 \text{ cm}}{15.0 \text{ cm}} = -6.33$

(b) IDENTIFY: Use Eq. (34.17) to calculate y'.

SET UP: $s = 3.00 \times 10^3 \text{ m}$

$s' = f_1 = 95.0 \text{ cm}$ (since s is very large, $s' \approx f$)

EXECUTE: $m = -\dfrac{s'}{s} = -\dfrac{0.950 \text{ m}}{3.00 \times 10^3 \text{ m}} = -3.167 \times 10^{-4}$

$|y'| = |m||y| = (3.167 \times 10^{-4})(60.0 \text{ m}) = 0.0190 \text{ m} = 1.90 \text{ cm}$

(c) IDENTIFY: Use Eq. (34.21) and the angular magnification M obtained in part (a) to calculate θ'. The angular size θ of the image formed by the objective (object for the eyepiece) is its height divided by its distance from the objective.

EXECUTE: The angular size of the object for the eyepiece is $\theta = \dfrac{0.0190 \text{ m}}{0.950 \text{ m}} = 0.0200 \text{ rad}$.

(Note that this is also the angular size of the object for the objective: $\theta = \dfrac{60.0 \text{ m}}{3.00 \times 10^3 \text{ m}} = 0.0200 \text{ rad}$. For a thin lens the object and image have the same angular size and the image of the objective is the object for the eyepiece.) $M = \dfrac{\theta'}{\theta}$ (Eq. 34.21) so the angular size of the image is $\theta' = M\theta = -(6.33)(0.0200 \text{ rad}) = -0.127 \text{ rad}$. (The minus sign shows that the final image is inverted.)

EVALUATE: The lateral magnification of the objective is small; the image it forms is much smaller than the object. But the total angular magnification is larger than 1.00; the angular size of the final image viewed by the eye is 6.33 times larger than the angular size of the original object, as viewed by the unaided eye.

34.67.   **IDENTIFY:** $f = R/2$ and $M = -\dfrac{f_1}{f_2}$.

SET UP: For object and image both at infinity, $f_1 + f_2$ equals the distance d between the eyepiece and the mirror vertex. $f_2 = 1.10 \text{ cm}$. $R_1 = 1.30 \text{ m}$.

EXECUTE: **(a)** $f_1 = \dfrac{R_1}{2} = 0.650 \text{ m} \Rightarrow d = f_1 + f_2 = 0.661 \text{ m}$.

(b) $|M| = \dfrac{f_1}{f_2} = \dfrac{0.650 \text{ m}}{0.011 \text{ m}} = 59.1.$

EVALUATE: For a telescope, $f_1 \gg f_2$.

34.73. **IDENTIFY:** We are given the image distance, the image height, and the object height. Use Eq. (34.7) to calculate the object distance s. Then use Eq. (34.4) to calculate R.

SET UP: The image is to be formed on screen so it is a real image; $s' > 0$. The mirror-to-screen distance is 8.00 m, so $s' = +800$ cm. $m = -\dfrac{s'}{s} < 0$ since both s and s' are positive.

EXECUTE: **(a)** $|m| = \dfrac{|y'|}{|y|} = \dfrac{24.0 \text{ cm}}{0.600 \text{ cm}} = 40.0,$ so $m = -40.0.$ Then $m = -\dfrac{s'}{s}$ gives

$s = -\dfrac{s'}{m} = -\dfrac{800 \text{ cm}}{-40.0} = +20.0 \text{ cm}.$

(b) $\dfrac{1}{s} + \dfrac{1}{s'} = \dfrac{2}{R},$ so $\dfrac{2}{R} = \dfrac{s+s'}{ss'}.$ $R = 2\left(\dfrac{ss'}{s+s'}\right) = 2\left(\dfrac{(20.0 \text{ cm})(800 \text{ cm})}{20.0 \text{ cm} + 800 \text{ cm}}\right) = 39.0 \text{ cm}.$

EVALUATE: R is calculated to be positive, which is correct for a concave mirror. Also, in part (a) s is calculated to be positive, as it should be for a real object.

34.77. **IDENTIFY:** Since the truck is moving toward the mirror, its image will also be moving toward the mirror.

SET UP: The equation relating the object and image distances to the focal length of a spherical mirror is $\dfrac{1}{s} + \dfrac{1}{s'} = \dfrac{1}{f},$ where $f = R/2.$

EXECUTE: Since the mirror is convex, $f = R/2 = (-1.50 \text{ m})/2 = -0.75 \text{ m}.$ Applying the equation for a spherical mirror gives $\dfrac{1}{s} + \dfrac{1}{s'} = \dfrac{1}{f} \Rightarrow s' = \dfrac{fs}{s-f}.$ Using the chain rule from calculus and the fact that

$v = ds/dt,$ we have $v' = \dfrac{ds'}{dt} = \dfrac{ds'}{ds}\dfrac{ds}{dt} = v\dfrac{f^2}{(s-f)^2}.$ Solving for v gives

$v = v'\left(\dfrac{s-f}{f}\right)^2 = (1.9 \text{ m/s})\left[\dfrac{2.0 \text{ m} - (-0.75 \text{ m})}{-0.75 \text{ m}}\right]^2 = 25.5 \text{ m/s}.$ This is the velocity of the truck relative to the

mirror, so the truck is approaching the mirror at 25.5 m/s. You are traveling at 25 m/s, so the truck must be traveling at 25 m/s + 25.5 m/s = 51 m/s relative to the highway.

EVALUATE: Even though the truck and car are moving at constant speed, the image of the truck is *not* moving at constant speed because its location depends on the distance from the mirror to the truck.

34.81. **IDENTIFY:** Apply Eq. (34.11) to the image formed by refraction at the front surface of the sphere.

SET UP: Let n_g be the index of refraction of the glass. The image formation is shown in Figure 34.81.

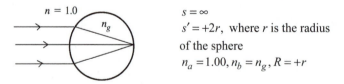

$s = \infty$
$s' = +2r,$ where r is the radius
of the sphere
$n_a = 1.00, n_b = n_g, R = +r$

Figure 34.81

$\dfrac{n_a}{s} + \dfrac{n_b}{s'} = \dfrac{n_b - n_a}{R}$

EXECUTE: $\dfrac{1}{\infty} + \dfrac{n_g}{2r} = \dfrac{n_g - 1.00}{r}$

$\dfrac{n_g}{2r} = \dfrac{n_g}{r} - \dfrac{1}{r}; \dfrac{n_g}{2r} = \dfrac{1}{r}$ and $n_g = 2.00$

EVALUATE: The required refractive index of the glass does not depend on the radius of the sphere.

34.83. **IDENTIFY:** Apply Eqs. (34.11) and (34.12) to the refraction as the light enters the rod and as it leaves the rod. The image formed by the first surface serves as the object for the second surface. The total magnification is $m_{tot} = m_1 m_2$, where m_1 and m_2 are the magnifications for each surface.

SET UP: The object and rod are shown in Figure 34.83.

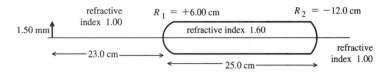

Figure 34.83

(a) image formed by refraction at first surface (left end of rod):

$s = +23.0$ cm; $n_a = 1.00$; $n_b = 1.60$; $R = +6.00$ cm

$$\frac{n_a}{s} + \frac{n_b}{s'} = \frac{n_b - n_a}{R}$$

EXECUTE: $\dfrac{1}{23.0 \text{ cm}} + \dfrac{1.60}{s'} = \dfrac{1.60 - 1.00}{6.00 \text{ cm}}$

$\dfrac{1.60}{s'} = \dfrac{1}{10.0 \text{ cm}} - \dfrac{1}{23.0 \text{ cm}} = \dfrac{23 - 10}{230 \text{ cm}} = \dfrac{13}{230 \text{ cm}}$

$s' = 1.60\left(\dfrac{230 \text{ cm}}{13}\right) = +28.3$ cm; image is 28.3 cm to right of first vertex.

This image serves as the object for the refraction at the second surface (right-hand end of rod). It is 28.3 cm − 25.0 cm = 3.3 cm to the right of the second vertex. For the second surface $s = -3.3$ cm (virtual object).

(b) **EVALUATE:** Object is on side of outgoing light, so is a virtual object.

(c) **SET UP:** Image formed by refraction at second surface (right end of rod):

$s = -3.3$ cm; $n_a = 1.60$; $n_b = 1.00$; $R = -12.0$ cm

$$\frac{n_a}{s} + \frac{n_b}{s'} = \frac{n_b - n_a}{R}$$

EXECUTE: $\dfrac{1.60}{-3.3 \text{ cm}} + \dfrac{1.00}{s'} = \dfrac{1.00 - 1.60}{-12.0 \text{ cm}}$

$s' = +1.9$ cm; $s' > 0$ so image is 1.9 cm to right of vertex at right-hand end of rod.

(d) $s' > 0$ so final image is real.

Magnification for first surface:

$$m_1 = -\frac{n_a s'}{n_b s} = -\frac{(1.00)(+28.3 \text{ cm})}{(1.60)(+23.0 \text{ cm})} = -0.769$$

Magnification for second surface:

$$m_2 = -\frac{n_a s'}{n_b s} = -\frac{(1.60)(+1.9 \text{ cm})}{(1.00)(-3.3 \text{ cm})} = +0.92$$

The overall magnification is $m_{tot} = m_1 m_2 = (-0.769)(+0.92) = -0.71$ $m_{tot} < 0$ so final image is inverted with respect to the original object.

(e) $y' = m_{tot} y = (-0.71)(1.50 \text{ mm}) = -1.06$ mm

The final image has a height of 1.06 mm.

EVALUATE: The two refracting surfaces are not close together and Eq. (34.18) does not apply.

34.85. **IDENTIFY:** $\dfrac{1}{s} + \dfrac{1}{s'} = \dfrac{1}{f}$. The type of lens determines the sign of f. $m = \dfrac{y'}{y} = -\dfrac{s'}{s}$. The sign of s' depends on whether the image is real or virtual. $s = 16.0$ cm.

SET UP: $s' = -22.0$ cm; s' is negative because the image is on the same side of the lens as the object.

EXECUTE: **(a)** $\dfrac{1}{f} = \dfrac{s+s'}{ss'}$ and $f = \dfrac{ss'}{s+s'} = \dfrac{(16.0 \text{ cm})(-22.0 \text{ cm})}{16.0 \text{ cm} - 22.0 \text{ cm}} = +58.7$ cm. f is positive so the lens is converging.

(b) $m = -\dfrac{s'}{s} = -\dfrac{-22.0 \text{ cm}}{16.0 \text{ cm}} = 1.38$. $y' = my = (1.38)(3.25 \text{ mm}) = 4.48$ mm. $s' < 0$ and the image is virtual.

EVALUATE: A converging lens forms a virtual image when the object is closer to the lens than the focal point.

34.87. **IDENTIFY:** The image formed by refraction at the surface of the eye is located by $\dfrac{n_a}{s} + \dfrac{n_b}{s'} = \dfrac{n_b - n_a}{R}$.

SET UP: $n_a = 1.00$, $n_b = 1.35$. $R > 0$. For a distant object, $s \approx \infty$ and $\dfrac{1}{s} \approx 0$.

EXECUTE: **(a)** $s \approx \infty$ and $s' = 2.5$ cm: $\dfrac{1.35}{2.5 \text{ cm}} = \dfrac{1.35 - 1.00}{R}$ and $R = 0.648$ cm $= 6.48$ mm.

(b) $R = 0.648$ cm and $s = 25$ cm: $\dfrac{1.00}{25 \text{ cm}} + \dfrac{1.35}{s'} = \dfrac{1.35 - 1.00}{0.648}$. $\dfrac{1.35}{s'} = 0.500$ and $s' = 2.70$ cm $= 27.0$ mm. The image is formed behind the retina.

(c) Calculate s' for $s \approx \infty$ and $R = 0.50$ cm: $\dfrac{1.35}{s'} = \dfrac{1.35 - 1.00}{0.50 \text{ cm}}$. $s' = 1.93$ cm $= 19.3$ mm. The image is formed in front of the retina.

EVALUATE: The cornea alone cannot achieve focus of both close and distant objects.

34.89. **IDENTIFY:** Apply $\dfrac{n_a}{s} + \dfrac{n_b}{s'} = \dfrac{n_b - n_a}{R}$ to each surface. The image of the first surface is the object for the second surface. The relation between s_1' and s_2 involves the length d of the rod.

SET UP: For the first surface, $n_a = 1.00$, $n_b = 1.55$ and $R = +6.00$ cm. For the second surface, $n_a = 1.55$, $n_b = 1.00$ and $R = -6.00$ cm.

EXECUTE: We have images formed from both ends. From the first surface:

$$\dfrac{n_a}{s} + \dfrac{n_b}{s'} = \dfrac{n_b - n_a}{R} \Rightarrow \dfrac{1}{25.0 \text{ cm}} + \dfrac{1.55}{s'} = \dfrac{0.55}{6.00 \text{ cm}} \Rightarrow s' = 30.0 \text{ cm}.$$

This image becomes the object for the second end:

$$\dfrac{n_a}{s} + \dfrac{n_b}{s'} = \dfrac{n_b - n_a}{R} \Rightarrow \dfrac{1.55}{d - 30.0 \text{ cm}} + \dfrac{1}{65.0 \text{ cm}} = \dfrac{-0.55}{-6.00 \text{ cm}}.$$

$d - 30.0 \text{ cm} = 20.3 \text{ cm} \Rightarrow d = 50.3 \text{ cm}.$

EVALUATE: The final image is real. The first image is 20.3 cm to the left of the second surface and serves as a real object.

34.91. **IDENTIFY and SET UP:** Apply Eq. (34.16) for each lens position. The lens to screen distance in each case is the image distance. There are two unknowns, the original object distance x and the focal length f of the lens. But each lens position gives an equation, so there are two equations for these two unknowns. The object, lens and screen before and after the lens is moved are shown in Figure 34.91.

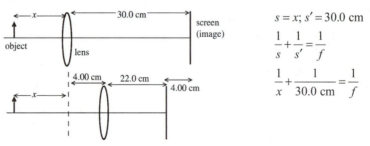

Figure 34.91

$s = x + 4.00$ cm; $s' = 22.0$ cm

$\dfrac{1}{s} + \dfrac{1}{s'} = \dfrac{1}{f}$ gives $\dfrac{1}{x + 4.00 \text{ cm}} + \dfrac{1}{22.0 \text{ cm}} = \dfrac{1}{f}$

EXECUTE: Equate these two expressions for $1/f$:

$\dfrac{1}{x} + \dfrac{1}{30.0 \text{ cm}} = \dfrac{1}{x + 4.00 \text{ cm}} + \dfrac{1}{22.0 \text{ cm}}$

$\dfrac{1}{x} - \dfrac{1}{x + 4.00 \text{ cm}} = \dfrac{1}{22.0 \text{ cm}} - \dfrac{1}{30.0 \text{ cm}}$

$\dfrac{x + 4.00 \text{ cm} - x}{x(x + 4.00 \text{ cm})} = \dfrac{30.0 - 22.0}{660 \text{ cm}}$ and $\dfrac{4.00 \text{ cm}}{x(x + 4.00 \text{ cm})} = \dfrac{8}{660 \text{ cm}}$

$x^2 + (4.00 \text{ cm})x - 330 \text{ cm}^2 = 0$ and $x = \dfrac{1}{2}(-4.00 \pm \sqrt{16.0 + 4(330)}) \text{ cm}$

x must be positive so $x = \dfrac{1}{2}(-4.00 + 36.55) \text{ cm} = 16.28$ cm

Then $\dfrac{1}{x} + \dfrac{1}{30.0 \text{ cm}} = \dfrac{1}{f}$ and $\dfrac{1}{f} = \dfrac{1}{16.28 \text{ cm}} + \dfrac{1}{30.0 \text{ cm}}$

$f = +10.55$ cm, which rounds to 10.6 cm. $f > 0$; the lens is converging.

EVALUATE: We can check that $s = 16.28$ cm and $f = 10.55$ cm gives $s' = 30.0$ cm and that $s = (16.28 + 4.0)$ cm $= 20.28$ cm and $f = 10.55$ cm gives $s' = 22.0$ cm.

34.95. **IDENTIFY:** Apply $\dfrac{n_a}{s} + \dfrac{n_b}{s'} = \dfrac{n_b - n_a}{R}$ to each case.

SET UP: $s = 20.0$ cm. $R > 0$. Use $s' = +9.12$ cm to find R. For this calculation, $n_a = 1.00$ and $n_b = 1.55$. Then repeat the calculation with $n_a = 1.33$.

EXECUTE: $\dfrac{n_a}{s} + \dfrac{n_b}{s'} = \dfrac{n_b - n_a}{R}$ gives $\dfrac{1.00}{20.0 \text{ cm}} + \dfrac{1.55}{9.12 \text{ cm}} = \dfrac{1.55 - 1.00}{R}$. $R = 2.50$ cm.

Then $\dfrac{1.33}{20.0 \text{ cm}} + \dfrac{1.55}{s'} = \dfrac{1.55 - 1.33}{2.50 \text{ cm}}$ gives $s' = 72.1$ cm. The image is 72.1 cm to the right of the surface vertex.

EVALUATE: With the rod in air the image is real and with the rod in water the image is also real.

34.97. **IDENTIFY:** Apply Eq. (34.11) with $R \to \infty$ to the refraction at each surface. For refraction at the first surface the point P serves as a virtual object. The image formed by the first refraction serves as the object for the second refraction.

SET UP: The glass plate and the two points are shown in Figure 34.97.

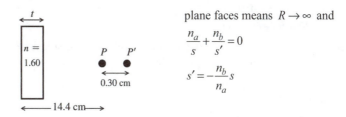

plane faces means $R \to \infty$ and

$\dfrac{n_a}{s} + \dfrac{n_b}{s'} = 0$

$s' = -\dfrac{n_b}{n_a}s$

Figure 34.97

EXECUTE: refraction at the first (left-hand) surface of the piece of glass:
The rays converging toward point P constitute a virtual object for this surface, so $s = -14.4$ cm.

$n_a = 1.00$, $n_b = 1.60$.

$$s' = -\frac{1.60}{1.00}(-14.4 \text{ cm}) = +23.0 \text{ cm}$$

This image is 23.0 cm to the right of the first surface so is a distance $23.0 \text{ cm} - t$ to the right of the second surface. This image serves as a virtual object for the second surface.

<u>refraction at the second (right-hand) surface of the piece of glass:</u>
The image is at P' so $s' = 14.4 \text{ cm} + 0.30 \text{ cm} - t = 14.7 \text{ cm} - t$. $s = -(23.0 \text{ cm} - t)$; $n_a = 1.60$; $n_b = 1.00$

$s' = -\dfrac{n_b}{n_a}s$ gives $14.7 \text{ cm} - t = -\left(\dfrac{1.00}{1.60}\right)(-[23.0 \text{ cm} - t])$. $14.7 \text{ cm} - t = +14.4 \text{ cm} - 0.625t$.

$0.375t = 0.30 \text{ cm}$ and $t = 0.80 \text{ cm}$

EVALUATE: The overall effect of the piece of glass is to diverge the rays and move their convergence point to the right. For a real object, refraction at a plane surface always produces a virtual image, but with a virtual object the image can be real.

34.99.   **IDENTIFY:** Apply $\dfrac{1}{s} + \dfrac{1}{s'} = \dfrac{1}{f}$.

SET UP: The image formed by the converging lens is 30.0 cm from the converging lens, and becomes a virtual object for the diverging lens at a position 15.0 cm to the right of the diverging lens. The final image is projected $15 \text{ cm} + 19.2 \text{ cm} = 34.2 \text{ cm}$ from the diverging lens.

EXECUTE: $\dfrac{1}{s} + \dfrac{1}{s'} = \dfrac{1}{f} \Rightarrow \dfrac{1}{-15.0 \text{ cm}} + \dfrac{1}{34.2 \text{ cm}} = \dfrac{1}{f} \Rightarrow f = -26.7 \text{ cm}$.

EVALUATE: Our calculation yields a negative value of f, which should be the case for a diverging lens.

34.105.   **IDENTIFY:** Apply Eq. (34.16) to calculate the image distance for each lens. The image formed by the first lens serves as the object for the second lens, and the image formed by the second lens serves as the object for the third lens.

SET UP: The positions of the object and lenses are shown in Figure 34.105.

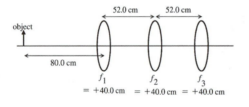

Figure 34.105

EXECUTE: <u>lens #1</u>
$s = +80.0 \text{ cm}$; $f = +40.0 \text{ cm}$

$$s' = \frac{sf}{s-f} = \frac{(+80.0 \text{ cm})(+40.0 \text{ cm})}{+80.0 \text{ cm} - 40.0 \text{ cm}} = +80.0 \text{ cm}$$

The image formed by the first lens is 80.0 cm to the right of the first lens, so it is
$80.0 \text{ cm} - 52.0 \text{ cm} = 28.0 \text{ cm}$ to the right of the second lens.

<u>lens #2</u>
$s = -28.0 \text{ cm}$; $f = +40.0 \text{ cm}$

$$s' = \frac{sf}{s-f} = \frac{(-28.0 \text{ cm})(+40.0 \text{ cm})}{-28.0 \text{ cm} - 40.0 \text{ cm}} = +16.47 \text{ cm}$$

The image formed by the second lens is 16.47 cm to the right of the second lens, so it is
$52.0 \text{ cm} - 16.47 \text{ cm} = 35.53 \text{ cm}$ to the left of the third lens.

lens #3

$s = +35.53$ cm; $f = +40.0$ cm

$$s' = \frac{sf}{s-f} = \frac{(+35.53 \text{ cm})(+40.0 \text{ cm})}{+35.53 \text{ cm} - 40.0 \text{ cm}} = -318 \text{ cm}$$

The final image is 318 cm to the left of the third lens, so it is $318 \text{ cm} - 52 \text{ cm} - 52 \text{ cm} - 80 \text{ cm} = 134 \text{ cm}$ to the left of the object.

EVALUATE: We used the separation between the lenses and the sign conventions for s and s' to determine the object distances for the second and third lenses. The final image is virtual since the final s' is negative.

34.107. **IDENTIFY** and **SET UP:** The generalization of Eq. (34.22) is $M = \dfrac{\text{near point}}{f}$, so $f = \dfrac{\text{near point}}{M}$.

EXECUTE: (a) age 10, near point = 7 cm

$$f = \frac{7 \text{ cm}}{2.0} = 3.5 \text{ cm}$$

(b) age 30, near point = 14 cm

$$f = \frac{14 \text{ cm}}{2.0} = 7.0 \text{ cm}$$

(c) age 60, near point = 200 cm

$$f = \frac{200 \text{ cm}}{2.0} = 100 \text{ cm}$$

(d) $f = 3.5$ cm (from part (a)) and near point = 200 cm (for 60-year-old)

$$M = \frac{200 \text{ cm}}{3.5 \text{ cm}} = 57$$

(e) **EVALUATE:** No. The reason $f = 3.5$ cm gives a larger M for a 60-year-old than for a 10-year-old is that the eye of the older person can't focus on as close an object as the younger person can. The unaided eye of the 60-year-old must view a much smaller angular size, and that is why the same f gives a much larger M. The angular size of the image depends only on f and is the same for the two ages.

34.109. **IDENTIFY:** Apply $\dfrac{1}{s} + \dfrac{1}{s'} = \dfrac{1}{f}$. The near point is at infinity, so that is where the image must be formed for any objects that are close.

SET UP: The power in diopters equals $\dfrac{1}{f}$, with f in meters.

EXECUTE: $\dfrac{1}{f} = \dfrac{1}{s} + \dfrac{1}{s'} = \dfrac{1}{24 \text{ cm}} + \dfrac{1}{-\infty} = \dfrac{1}{0.24 \text{ m}} = 4.17$ diopters.

EVALUATE: To focus on closer objects, the power must be increased.

34.111. **IDENTIFY** and **SET UP:** The person's eye cannot focus on anything closer than 85.0 cm. The problem asks us to find the location of an object such that his old lenses produce a virtual image 85.0 cm from his eye.

$\dfrac{1}{s} + \dfrac{1}{s'} = \dfrac{1}{f}$. P (in diopters) $= 1/f$ (in m).

EXECUTE: (a) $\dfrac{1}{f} = 2.25$ diopters so $f = 44.4$ cm. The image is 85.0 cm from his eye so is 83.0 cm from the eyeglass lens. Solving $\dfrac{1}{s} + \dfrac{1}{s'} = \dfrac{1}{f}$ for s gives $s = \dfrac{s'f}{s'-f} = \dfrac{(-83.0 \text{ cm})(44.4 \text{ cm})}{-83.0 \text{ cm} - 44.4 \text{ cm}} = +28.9$ cm. The object is 28.9 cm from the eyeglasses so is 30.9 cm from his eyes.

(b) Now $s' = -85.0$ cm. $s = \dfrac{s'f}{s'-f} = \dfrac{(-85.0 \text{ cm})(44.4 \text{ cm})}{-85.0 \text{ cm} - 44.4 \text{ cm}} = +29.2$ cm.

EVALUATE: The old glasses allow him to focus on objects as close as about 30 cm from his eyes. This is much better than a closest distance of 85 cm with no glasses, but his current glasses probably allow him to focus as close as 25 cm.

34.115.   **IDENTIFY** and **SET UP:**   The image formed by the objective is the object for the eyepiece. The total lateral magnification is $m_{\text{tot}} = m_1 m_2$. $f_1 = 8.00$ mm (objective); $f_2 = 7.50$ cm (eyepiece)

(a) The locations of the object, lenses and screen are shown in Figure 34.115.

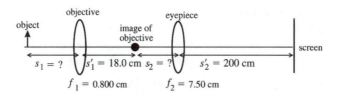

Figure 34.115

EXECUTE: Find the object distance s_1 for the objective:

$s_1' = +18.0$ cm, $f_1 = 0.800$ cm, $s_1 = ?$

$$\frac{1}{s_1} + \frac{1}{s_1'} = \frac{1}{f_1}, \text{ so } \frac{1}{s_1} = \frac{1}{f_1} - \frac{1}{s_1'} = \frac{s_1' - f_1}{s_1' f_1}$$

$$s_1 = \frac{s_1' f_1}{s_1' - f_1} = \frac{(18.0 \text{ cm})(0.800 \text{ cm})}{18.0 \text{ cm} - 0.800 \text{ cm}} = 0.8372 \text{ cm}$$

Find the object distance s_2 for the eyepiece:

$s_2' = +200$ cm, $f_2 = 7.50$ cm, $s_2 = ?$

$$\frac{1}{s_2} + \frac{1}{s_2'} = \frac{1}{f_2}$$

$$s_2 = \frac{s_2' f_2}{s_2' - f_2} = \frac{(200 \text{ cm})(7.50 \text{ cm})}{200 \text{ cm} - 7.50 \text{ cm}} = 7.792 \text{ cm}$$

Now we calculate the magnification for each lens:

$$m_1 = -\frac{s_1'}{s_1} = -\frac{18.0 \text{ cm}}{0.8372 \text{ cm}} = -21.50$$

$$m_2 = -\frac{s_2'}{s_2} = -\frac{200 \text{ cm}}{7.792 \text{ cm}} = -25.67$$

$m_{\text{tot}} = m_1 m_2 = (-21.50)(-25.67) = 552.$

(b) From the sketch we can see that the distance between the two lenses is

$s_1' + s_2 = 18.0 \text{ cm} + 7.792 \text{ cm} = 25.8 \text{ cm}.$

EVALUATE: The microscope is not being used in the conventional way; it merely serves as a two-lens system. In particular, the final image formed by the eyepiece in the problem is real, not virtual as is the case normally for a microscope. Eq. (34.24) does not apply here, and in any event gives the angular not the lateral magnification.

INTERFERENCE

35.5. **IDENTIFY:** Use $c = f\lambda$ to calculate the wavelength of the transmitted waves. Compare the difference in the distance from A to P and from B to P. For constructive interference this path difference is an integer multiple of the wavelength.

SET UP: Consider Figure 35.5.

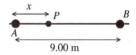

The distance of point P from each coherent source is $r_A = x$ and $r_B = 9.00 \text{ m} - x$.

Figure 35.5

EXECUTE: The path difference is $r_B - r_A = 9.00 \text{ m} - 2x$.

$r_B - r_A = m\lambda, m = 0, \pm 1, \pm 2, \ldots$

$\lambda = \dfrac{c}{f} = \dfrac{2.998 \times 10^8 \text{ m/s}}{120 \times 10^6 \text{ Hz}} = 2.50 \text{ m}$

Thus $9.00 \text{ m} - 2x = m(2.50 \text{ m})$ and $x = \dfrac{9.00 \text{ m} - m(2.50 \text{ m})}{2} = 4.50 \text{ m} - (1.25 \text{ m})m$. x must lie in the range

0 to 9.00 m since P is said to be between the two antennas.

$m = 0$ gives $x = 4.50 \text{ m}$

$m = +1$ gives $x = 4.50 \text{ m} - 1.25 \text{ m} = 3.25 \text{ m}$

$m = +2$ gives $x = 4.50 \text{ m} - 2.50 \text{ m} = 2.00 \text{ m}$

$m = +3$ gives $x = 4.50 \text{ m} - 3.75 \text{ m} = 0.75 \text{ m}$

$m = -1$ gives $x = 4.50 \text{ m} + 1.25 \text{ m} = 5.75 \text{ m}$

$m = -2$ gives $x = 4.50 \text{ m} + 2.50 \text{ m} = 7.00 \text{ m}$

$m = -3$ gives $x = 4.50 \text{ m} + 3.75 \text{ m} = 8.25 \text{ m}$

All other values of m give values of x out of the allowed range. Constructive interference will occur for $x = 0.75 \text{ m}, 2.00 \text{ m}, 3.25 \text{ m}, 4.50 \text{ m}, 5.75 \text{ m}, 7.00 \text{ m}$ and 8.25 m.

EVALUATE: Constructive interference occurs at the midpoint between the two sources since that point is the same distance from each source. The other points of constructive interference are symmetrically placed relative to this point.

35.11. **IDENTIFY** and **SET UP:** The dark lines correspond to destructive interference and hence are located by Eq. (35.5):

$$d \sin\theta = \left(m + \frac{1}{2}\right)\lambda \text{ so } \sin\theta = \frac{\left(m + \frac{1}{2}\right)\lambda}{d}, m = 0, \pm 1, \pm 2, \ldots$$

Solve for θ that locates the second and third dark lines. Use $y = R\tan\theta$ to find the distance of each of the dark lines from the center of the screen.

EXECUTE: 1st dark line is for $m = 0$

2nd dark line is for $m = 1$ and $\sin\theta_1 = \dfrac{3\lambda}{2d} = \dfrac{3(500\times10^{-9}\text{ m})}{2(0.450\times10^{-3}\text{ m})} = 1.667\times10^{-3}$ and $\theta_1 = 1.667\times10^{-3}$ rad

3rd dark line is for $m = 2$ and $\sin\theta_2 = \dfrac{5\lambda}{2d} = \dfrac{5(500\times10^{-9}\text{ m})}{2(0.450\times10^{-3}\text{ m})} = 2.778\times10^{-3}$ and $\theta_2 = 2.778\times10^{-3}$ rad

(Note that θ_1 and θ_2 are small so that the approximation $\theta \approx \sin\theta \approx \tan\theta$ is valid.) The distance of each dark line from the center of the central bright band is given by $y_m = R\tan\theta$, where $R = 0.850$ m is the distance to the screen.
$\tan\theta \approx \theta$ so $y_m = R\theta_m$

$y_1 = R\theta_1 = (0.750\text{ m})(1.667\times10^{-3}\text{ rad}) = 1.25\times10^{-3}$ m

$y_2 = R\theta_2 = (0.750\text{ m})(2.778\times10^{-3}\text{ rad}) = 2.08\times10^{-3}$ m

$\Delta y = y_2 - y_1 = 2.08\times10^{-3}\text{ m} - 1.25\times10^{-3}\text{ m} = 0.83$ mm

EVALUATE: Since θ_1 and θ_2 are very small we could have used Eq. (35.6), generalized to destructive

interference: $y_m = R\left(m + \dfrac{1}{2}\right)\lambda/d$.

35.13.   **IDENTIFY:**   Bright fringes are located at angles θ given by $d\sin\theta = m\lambda$.
SET UP: The largest value $\sin\theta$ can have is 1.00.

EXECUTE: **(a)** $m = \dfrac{d\sin\theta}{\lambda}$. For $\sin\theta = 1$, $m = \dfrac{d}{\lambda} = \dfrac{0.0116\times10^{-3}\text{ m}}{5.85\times10^{-7}\text{ m}} = 19.8$. Therefore, the largest m for

fringes on the screen is $m = 19$. There are $2(19) + 1 = 39$ bright fringes, the central one and 19 above and 19 below it.

(b) The most distant fringe has $m = \pm19$. $\sin\theta = m\dfrac{\lambda}{d} = \pm19\left(\dfrac{5.85\times10^{-7}\text{ m}}{0.0116\times10^{-3}\text{ m}}\right) = \pm0.958$ and $\theta = \pm73.3°$.

EVALUATE: For small θ the spacing Δy between adjacent fringes is constant but this is no longer the case for larger angles.

35.15.   **IDENTIFY and SET UP:**   The dark lines are located by $d\sin\theta = \left(m + \dfrac{1}{2}\right)\lambda$. The distance of each line from

the center of the screen is given by $y = R\tan\theta$.
EXECUTE: First dark line is for $m = 0$ and $d\sin\theta_1 = \lambda/2$.

$\sin\theta_1 = \dfrac{\lambda}{2d} = \dfrac{550\times10^{-9}\text{ m}}{2(1.80\times10^{-6}\text{ m})} = 0.1528$ and $\theta_1 = 8.789°$. Second dark line is for $m = 1$ and $d\sin\theta_2 = 3\lambda/2$.

$\sin\theta_2 = \dfrac{3\lambda}{2d} = 3\left(\dfrac{550\times10^{-9}\text{ m}}{2(1.80\times10^{-6}\text{ m})}\right) = 0.4583$ and $\theta_2 = 27.28°$.

$y_1 = R\tan\theta_1 = (0.350\text{ m})\tan 8.789° = 0.0541$ m

$y_2 = R\tan\theta_2 = (0.350\text{ m})\tan 27.28° = 0.1805$ m

The distance between the lines is $\Delta y = y_2 - y_1 = 0.1805\text{ m} - 0.0541\text{ m} = 0.126\text{ m} = 12.6$ cm.

EVALUATE: $\sin\theta_1 = 0.1528$ and $\tan\theta_1 = 0.1546$. $\sin\theta_2 = 0.4583$ and $\tan\theta_2 = 0.5157$. As the angle increases, $\sin\theta \approx \tan\theta$ becomes a poorer approximation.

35.17.   **IDENTIFY and SET UP:**   Use the information given about the bright fringe to find the distance d between the two slits. Then use Eq. (35.5) and $y = R\tan\theta$ to calculate λ for which there is a first-order dark fringe at this same place on the screen.

EXECUTE: $y_1 = \dfrac{R\lambda_1}{d}$, so $d = \dfrac{R\lambda_1}{y_1} = \dfrac{(3.00 \text{ m})(600 \times 10^{-9} \text{ m})}{4.84 \times 10^{-3} \text{ m}} = 3.72 \times 10^{-4}$ m. (R is much greater than d, so

Eq. 35.6 is valid.) The dark fringes are located by $d\sin\theta = \left(m + \dfrac{1}{2}\right)\lambda$, $m = 0, \pm 1, \pm 2, \ldots$ The first-order dark

fringe is located by $\sin\theta = \lambda_2/2d$, where λ_2 is the wavelength we are seeking.

$$y = R\tan\theta \approx R\sin\theta = \dfrac{\lambda_2 R}{2d}$$

We want λ_2 such that $y = y_1$. This gives $\dfrac{R\lambda_1}{d} = \dfrac{R\lambda_2}{2d}$ and $\lambda_2 = 2\lambda_1 = 1200$ nm.

EVALUATE: For $\lambda = 600$ nm the path difference from the two slits to this point on the screen is 600 nm. For this same path difference (point on the screen) the path difference is $\lambda/2$ when $\lambda = 1200$ nm.

35.19. **IDENTIFY:** Eq. (35.10): $I = I_0 \cos^2(\phi/2)$. Eq. (35.11): $\phi = (2\pi/\lambda)(r_2 - r_1)$.

SET UP: ϕ is the phase difference and $(r_2 - r_1)$ is the path difference.

EXECUTE: **(a)** $I = I_0(\cos 30.0°)^2 = 0.750 I_0$

(b) $60.0° = (\pi/3)$ rad. $(r_2 - r_1) = (\phi/2\pi)\lambda = [(\pi/3)/2\pi]\lambda = \lambda/6 = 80$ nm.

EVALUATE: $\phi = 360°/6$ and $(r_2 - r_1) = \lambda/6$.

35.21. **IDENTIFY and SET UP:** The phase difference ϕ is given by $\phi = (2\pi d/\lambda)\sin\theta$ (Eq. 35.13.)

EXECUTE: $\phi = [2\pi(0.340 \times 10^{-3} \text{ m})/(500 \times 10^{-9} \text{ m}) \sin 23.0° = 1670$ rad

EVALUATE: The mth bright fringe occurs when $\phi = 2\pi m$, so there are a large number of bright fringes within $23.0°$ from the centerline. Note that Eq. (35.13) gives ϕ in radians.

35.23. **IDENTIFY:** The intensity decreases as we move away from the central maximum.

SET UP: The intensity is given by $I = I_0 \cos^2\left(\dfrac{\pi d y}{\lambda R}\right)$.

EXECUTE: First find the wavelength: $\lambda = c/f = (3.00 \times 10^8 \text{ m/s})/(12.5 \text{ MHz}) = 24.00$ m

At the farthest the receiver can be placed, $I = I_0/4$, which gives

$$\dfrac{I_0}{4} = I_0 \cos^2\left(\dfrac{\pi d y}{\lambda R}\right) \Rightarrow \cos^2\left(\dfrac{\pi d y}{\lambda R}\right) = \dfrac{1}{4} \Rightarrow \cos\left(\dfrac{\pi d y}{\lambda R}\right) = \pm\dfrac{1}{2}$$

The solutions are $\pi d y/\lambda R = \pi/3$ and $2\pi/3$. Using $\pi/3$, we get

$$y = \lambda R/3d = (24.00 \text{ m})(500 \text{ m})/[3(56.0 \text{ m})] = 71.4 \text{ m}$$

It must remain within 71.4 m of point C.

EVALUATE: Using $\pi d y/\lambda R = 2\pi/3$ gives $y = 142.8$ m. But to reach this point, the receiver would have to go beyond 71.4 m from C, where the signal would be too weak, so this second point is not possible.

35.27. **IDENTIFY:** The fringes are produced by interference between light reflected from the top and bottom surfaces of the air wedge. The refractive index of glass is greater than that of air, so the waves reflected from the top surface of the air wedge have no reflection phase shift, and the waves reflected from the bottom surface of the air wedge do have a half-cycle reflection phase shift. The condition for constructive interference (bright fringes) is therefore $2t = (m + \frac{1}{2})\lambda$.

SET UP: The geometry of the air wedge is sketched in Figure 35.27. At a distance x from the point of contact of the two plates, the thickness of the air wedge is t.

EXECUTE: $\tan\theta = \dfrac{t}{x}$ so $t = x\tan\theta$. $t_m = (m + \frac{1}{2})\dfrac{\lambda}{2}$. $x_m = (m + \frac{1}{2})\dfrac{\lambda}{2\tan\theta}$ and $x_{m+1} = (m + \frac{3}{2})\dfrac{\lambda}{2\tan\theta}$. The

distance along the plate between adjacent fringes is $\Delta x = x_{m+1} - x_m = \dfrac{\lambda}{2\tan\theta}$. 15.0 fringes/cm $= \dfrac{1.00}{\Delta x}$ and

$$\Delta x = \frac{1.00}{15.0 \text{ fringes/cm}} = 0.0667 \text{ cm.} \quad \tan\theta = \frac{\lambda}{2\Delta x} = \frac{546\times10^{-9} \text{ m}}{2(0.0667\times10^{-2} \text{ m})} = 4.09\times10^{-4}. \text{ The angle of the}$$

wedge is 4.09×10^{-4} rad $= 0.0234°$.

EVALUATE: The fringes are equally spaced; Δx is independent of m.

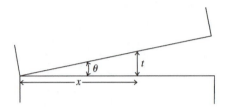

Figure 35.27

35.29.   **IDENTIFY:** The light reflected from the top of the TiO_2 film interferes with the light reflected from the top of the glass surface. These waves are out of phase due to the path difference in the film and the phase differences caused by reflection.

SET UP: There is a π phase change at the TiO_2 surface but none at the glass surface, so for destructive interference the path difference must be $m\lambda$ in the film.

EXECUTE: (a) Calling T the thickness of the film gives $2T = m\lambda_0/n$, which yields $T = m\lambda_0/(2n)$. Substituting the numbers gives

$$T = m\ (520.0 \text{ nm})/[2(2.62)] = 99.237 nm$$

T must be greater than 1036 nm, so $m = 11$, which gives $T = 1091.6$ nm, since we want to know the minimum thickness to add.

$$\Delta T = 1091.6 \text{ nm} - 1036 \text{ nm} = 55.6 \text{ nm.}$$

(b) (i) Path difference $= 2T = 2(1092 \text{ nm}) = 2184 \text{ nm} = 2180 \text{ nm.}$

(ii) The wavelength in the film is $\lambda = \lambda_0/n = (520.0 \text{ nm})/2.62 = 198.5 \text{ nm.}$

Path difference $= (2180 \text{ nm})/[(198.5 \text{ nm})/\text{wavelength}] = 11.0$ wavelengths

EVALUATE: Because the path difference in the film is 11.0 wavelengths, the light reflected off the top of the film will be 180° out of phase with the light that traveled through the film and was reflected off the glass due to the phase change at reflection off the top of the film.

35.33.   **IDENTIFY:** Require destructive interference between light reflected from the two points on the disc.

SET UP: Both reflections occur for waves in the plastic substrate reflecting from the reflective coating, so they both have the same phase shift upon reflection and the condition for destructive interference (cancellation) is $2t = (m + \frac{1}{2})\lambda$, where t is the depth of the pit. $\lambda = \frac{\lambda_0}{n}$. The minimum pit depth is for $m = 0$.

EXECUTE: $2t = \frac{\lambda}{2}$. $t = \frac{\lambda}{4} = \frac{\lambda_0}{4n} = \frac{790 \text{ nm}}{4(1.8)} = 110 \text{ nm} = 0.11 \mu\text{m.}$

EVALUATE: The path difference occurs in the plastic substrate and we must compare the wavelength in the substrate to the path difference.

35.37.   **IDENTIFY:** Consider the interference between light reflected from the top and bottom surfaces of the air film between the lens and the glass plate.

SET UP: For maximum intensity, with a net half-cycle phase shift due to reflections,

$$2t = \left(m + \frac{1}{2}\right)\lambda. \quad t = R - \sqrt{R^2 - r^2}.$$

EXECUTE: $\dfrac{(2m+1)\lambda}{4} = R - \sqrt{R^2 - r^2} \Rightarrow \sqrt{R^2 - r^2} = R - \dfrac{(2m+1)\lambda}{4}$

$$\Rightarrow R^2 - r^2 = R^2 + \left[\frac{(2m+1)\lambda}{4}\right]^2 - \frac{(2m+1)\lambda R}{2} \Rightarrow r = \sqrt{\frac{(2m+1)\lambda R}{2} - \left[\frac{(2m+1)\lambda}{4}\right]^2}$$

$$\Rightarrow r \approx \sqrt{\frac{(2m+1)\lambda R}{2}}, \text{ for } R \gg \lambda.$$

The second bright ring is when $m = 1$:

$$r \approx \sqrt{\frac{[2(1)+1](5.80 \times 10^{-7}\text{ m})(0.684\text{ m})}{2}} = 7.71 \times 10^{-4}\text{ m} = 0.771\text{ mm}. \text{ So the diameter of the second bright}$$

ring is 1.54 mm.

EVALUATE: The diameter of the mth ring is proportional to $\sqrt{2m+1}$, so the rings get closer together as m increases. This agrees with Figure 35.16b in the textbook.

35.39. **IDENTIFY** and **SET UP:** Consider the interference of the rays reflected from each side of the film. At the front of the film light in air reflects off the film ($n = 1.432$) and there is a 180° phase shift. At the back of the film light in the film ($n = 1.432$) reflects off the glass ($n = 1.62$) and there is a 180° phase shift. Therefore, the reflections introduce no net phase shift. The path difference is $2t$, where t is the thickness of the film. The wavelength in the film is $\lambda = \frac{\lambda_{\text{air}}}{n}$.

EXECUTE: **(a)** Since there is no net phase difference produced by the reflections, the condition for destructive interference is $2t = (m + \frac{1}{2})\lambda$. $t = (m + \frac{1}{2})\frac{\lambda}{2}$ and the minimum thickness is $t = \frac{\lambda}{4} = \frac{\lambda_{\text{air}}}{4n} = \frac{550\text{ nm}}{4(1.432)} = 96.0\text{ nm}.$

(b) For destructive interference, $2t = (m + \frac{1}{2})\frac{\lambda_{\text{air}}}{n}$ and $\lambda_{\text{air}} = \frac{2tn}{m + \frac{1}{2}} = \frac{275\text{ nm}}{m + \frac{1}{2}}$. $m = 0$: $\lambda_{\text{air}} = 550\text{ nm}.$

$m = 1$: $\lambda_{\text{air}} = 183\text{ nm}.$ All other λ_{air} values are shorter. For constructive interference, $2t = m\frac{\lambda_{\text{air}}}{n}$ and $\lambda_{\text{air}} = \frac{2tn}{m} = \frac{275\text{ nm}}{m}$. For $m = 1$, $\lambda_{\text{air}} = 275\text{ nm}$ and all other λ_{air} values are shorter.

EVALUATE: The only visible wavelength in air for which there is destructive interference is 550 nm. There are no visible wavelengths in air for which there is constructive interference.

35.41. **IDENTIFY:** The insertion of the metal foil produces a wedge of air, which is an air film of varying thickness. This film causes a path difference between light reflected off the top and bottom of this film.
SET UP: The two sheets of glass are sketched in Figure 35.41. The thickness of the air wedge at a distance x from the line of contact is $t = x \tan\theta$. Consider rays 1 and 2 that are reflected from the top and bottom surfaces, respectively, of the air film. Ray 1 has no phase change when it reflects and ray 2 has a 180° phase change when it reflects, so the reflections introduce a net 180° phase difference. The path difference is $2t$ and the wavelength in the film is $\lambda = \lambda_{\text{air}}$.

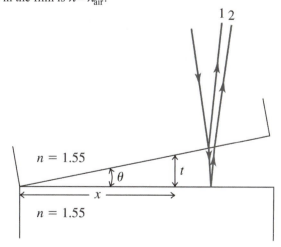

Figure 35.41

EXECUTE: (a) Since there is a 180° phase difference from the reflections, the condition for constructive interference is $2t = (m + \frac{1}{2})\lambda$. The positions of first enhancement correspond to $m = 0$ and $2t = \dfrac{\lambda}{2}$.

$x \tan\theta = \dfrac{\lambda}{4}$. θ is a constant, so $\dfrac{x_1}{\lambda_1} = \dfrac{x_2}{\lambda_2}$. $x_1 = 1.15$ mm, $\lambda_1 = 400.0$ nm. $x_2 = x_1\left(\dfrac{\lambda_2}{\lambda_1}\right)$. For

$\lambda_2 = 550$ nm (green), $x_2 = (1.15 \text{ mm})\left(\dfrac{550 \text{ nm}}{400 \text{ nm}}\right) = 1.58$ mm. For $\lambda_2 = 600$ nm (orange),

$x_2 = (1.15 \text{ mm})\left(\dfrac{600 \text{ nm}}{400 \text{ nm}}\right) = 1.72$ mm.

(b) The positions of next enhancement correspond to $m = 1$ and $2t = \dfrac{3\lambda}{2}$. $x \tan\theta = \dfrac{3\lambda}{4}$. The values of x are 3 times what they are in part (a). Violet: 3.45 mm; green: 4.74 mm; orange: 5.16 mm.

(c) $\tan\theta = \dfrac{\lambda}{4x} = \dfrac{400.0 \times 10^{-9} \text{ m}}{4(1.15 \times 10^{-3} \text{ m})} = 8.70 \times 10^{-5}$. $\tan\theta = \dfrac{t_{\text{foil}}}{11.0 \text{ cm}}$, so $t_{\text{foil}} = 9.57 \times 10^{-4}$ cm $= 9.57 \ \mu$m.

EVALUATE: The thickness of the foil must be very small to cause these observable interference effects. If it is too thick, the film is no longer a "thin film."

35.43.   IDENTIFY: The liquid alters the wavelength of the light and that affects the locations of the interference minima.

SET UP: The interference minima are located by $d \sin\theta = (m + \frac{1}{2})\lambda$. For a liquid with refractive index n,

$\lambda_{\text{liq}} = \dfrac{\lambda_{\text{air}}}{n}$.

EXECUTE: $\dfrac{\sin\theta}{\lambda} = \dfrac{(m + \frac{1}{2})}{d} = $ constant, so $\dfrac{\sin\theta_{\text{air}}}{\lambda_{\text{air}}} = \dfrac{\sin\theta_{\text{liq}}}{\lambda_{\text{liq}}}$. $\dfrac{\sin\theta_{\text{air}}}{\lambda_{\text{air}}} = \dfrac{\sin\theta_{\text{liq}}}{\lambda_{\text{air}}/n}$ and

$n = \dfrac{\sin\theta_{\text{air}}}{\sin\theta_{\text{liq}}} = \dfrac{\sin 35.20°}{\sin 19.46°} = 1.730$.

EVALUATE: In the liquid the wavelength is shorter and $\sin\theta = (m + \frac{1}{2})\dfrac{\lambda}{d}$ gives a smaller θ than in air, for the same m.

35.45.   IDENTIFY: *Both* frequencies will interfere constructively when the path difference from both of them is an integral number of wavelengths.

SET UP: Constructive interference occurs when $\sin\theta = m\lambda/d$.

EXECUTE: First find the two wavelengths.

$$\lambda_1 = v/f_1 = (344 \text{ m/s})/(900 \text{ Hz}) = 0.3822 \text{ m}$$

$$\lambda_2 = v/f_2 = (344 \text{ m/s})/(1200 \text{ Hz}) = 0.2867 \text{ m}$$

To interfere constructively at the same angle, the angles must be the same, and hence the sines of the angles must be equal. Each sine is of the form $\sin\theta = m\lambda/d$, so we can equate the sines to get

$$m_1\lambda_1/d = m_2\lambda_2/d$$
$$m_1(0.3822 \text{ m}) = m_2(0.2867 \text{ m})$$
$$m_2 = 4/3 \ m_1$$

Since both m_1 and m_2 must be integers, the allowed pairs of values of m_1 and m_2 are

$$m_1 = m_2 = 0$$
$$m_1 = 3, \ m_2 = 4$$
$$m_1 = 6, \ m_2 = 8$$
$$m_1 = 9, \ m_2 = 12$$
$$\text{etc.}$$

For $m_1 = m_2 = 0$, we have $\theta = 0$,

For $m_1 = 3$, $m_2 = 4$, we have $\sin\theta_1 = (3)(0.3822 \text{ m})/(2.50 \text{ m})$, giving $\theta_1 = 27.3°$.

For $m_1 = 6$, $m_2 = 8$, we have $\sin\theta_1 = (6)(0.3822 \text{ m})/(2.50 \text{ m})$, giving $\theta_1 = 66.5°$.

For $m_1 = 9$, $m_2 = 12$, we have $\sin\theta_1 = (9)(0.3822 \text{ m})/(2.50 \text{ m}) = 1.38 > 1$, so no angle is possible.

EVALUATE: At certain other angles, one frequency will interfere constructively, but the other will not.

35.47. **IDENTIFY:** The two scratches are parallel slits, so the light that passes through them produces an interference pattern. However, the light is traveling through a medium (plastic) that is different from air.
SET UP: The central bright fringe is bordered by a dark fringe on each side of it. At these dark fringes, $d \sin\theta = \frac{1}{2} \lambda/n$, where n is the refractive index of the plastic.
EXECUTE: First use geometry to find the angles at which the two dark fringes occur. At the first dark fringe $\tan\theta = [(5.82 \text{ mm})/2]/(3250 \text{ mm})$, giving $\theta = \pm 0.0513°$.

For destructive interference, we have $d \sin\theta = \frac{1}{2} \lambda/n$ and

$$n = \lambda/(2d \sin\theta) = (632.8 \text{ nm})/[2(0.000225 \text{ m})(\sin 0.0513°)] = 1.57$$

EVALUATE: The wavelength of the light in the plastic is reduced compared to what it would be in air.

35.49. **IDENTIFY:** For destructive interference the net phase difference must be 180°, which is one-half a period, or $\lambda/2$. Part of this phase difference is due to the fact that the speakers are $\frac{1}{4}$ of a period out of phase, and the rest is due to the path difference between the sound from the two speakers.
SET UP: The phase of A is 90° or, $\lambda/4$, ahead of B. At points above the centerline, points are closer to A than to B and the signal from A gains phase relative to B because of the path difference. Destructive interference will occur when $d \sin\theta = (m + \frac{1}{4})\lambda$, $m = 0, 1, 2, \ldots$. At points at an angle θ below the centerline, the signal from B gains phase relative to A because of the phase difference. Destructive interference will occur when $d \sin\theta = (m + \frac{3}{4})\lambda$, $m = 0, 1, 2, \ldots$. $\lambda = \dfrac{v}{f}$.

EXECUTE: $\lambda = \dfrac{340 \text{ m/s}}{444 \text{ Hz}} = 0.766 \text{ m}$.

Points above the centerline: $\sin\theta = (m + \frac{1}{4})\dfrac{\lambda}{d} = (m + \frac{1}{4})\left(\dfrac{0.766 \text{ m}}{3.50 \text{ m}}\right) = 0.219(m + \frac{1}{4})$. $m = 0$: $\theta = 3.14°$;

$m = 1$: $\theta = 15.9°$; $m = 2$: $\theta = 29.5°$; $m = 3$: $\theta = 45.4°$; $m = 4$: $\theta = 68.6°$.

Points below the centerline: $\sin\theta = (m + \frac{3}{4})\dfrac{\lambda}{d} = (m + \frac{3}{4})\left(\dfrac{0.766 \text{ m}}{3.50 \text{ m}}\right) = 0.219(m + \frac{3}{4})$. $m = 0$: $\theta = 9.45°$;

$m = 1$: $\theta = 22.5°$; $m = 2$: $\theta = 37.0°$; $m = 3$: $\theta = 55.2°$.

EVALUATE: It is *not* always true that the path difference for destructive interference must be $(m + \frac{1}{2})\lambda$,

but it *is* always true that the phase difference must be 180° (or odd multiples of 180°).

35.51. **IDENTIFY and SET UP:** Consider interference between rays reflected from the upper and lower surfaces of the film to relate the thickness of the film to the wavelengths for which there is destructive interference. The thermal expansion of the film changes the thickness of the film when the temperature changes.
EXECUTE: For this film on this glass, there is a net $\lambda/2$ phase change due to reflection and the condition for destructive interference is $2t = m(\lambda/n)$, where $n = 1.750$.

Smallest nonzero thickness is given by $t = \lambda/2n$.

At 20.0°C, $t_0 = (582.4 \text{ nm})/[(2)(1.750)] = 166.4 \text{ nm}$.

At 170°C, $t = (588.5 \text{ nm})/[(2)(1.750)] = 168.1 \text{ nm}$.

$t = t_0(1 + \alpha\Delta T)$ so

$\alpha = (t - t_0)/(t_0\Delta T) = (1.7 \text{ nm})/[(166.4 \text{ nm})(150\text{C}°)] = 6.8 \times 10^{-5} (\text{C}°)^{-1}$

EVALUATE: When the film is heated its thickness increases, and it takes a larger wavelength in the film to equal 2t. The value we calculated for α is the same order of magnitude as those given in Table 17.1.

35.57.   **IDENTIFY:** The slits will produce an interference pattern, but in the liquid, the wavelength of the light will be less than it was in air.

SET UP: The first bright fringe occurs when $d \sin\theta = \lambda/n$.

EXECUTE: In air: $d\sin 18.0° = \lambda$. In the liquid: $d\sin 12.6° = \lambda/n$. Dividing the equations gives

$$n = (\sin 18.0°)/(\sin 12.6°) = 1.42$$

EVALUATE: It was not necessary to know the spacing of the slits, since it was the same in both air and the liquid.

36

DIFFRACTION

36.5. **IDENTIFY:** The minima are located by $\sin\theta = \dfrac{m\lambda}{a}$.

SET UP: $a = 12.0$ cm. $x = 8.00$ m.

EXECUTE: The angle to the first minimum is $\theta = \arcsin\left(\dfrac{\lambda}{a}\right) = \arcsin\left(\dfrac{9.00\text{ cm}}{12.00\text{ cm}}\right) = 48.6°$.

So the distance from the central maximum to the first minimum is just
$y_1 = x\tan\theta = (8.00\text{ m})\tan(48.6°) = \pm(9.07\text{ m})$.

EVALUATE: $2\lambda/a$ is greater than 1, so only the $m = 1$ minimum is seen.

36.7. **IDENTIFY:** We can model the hole in the concrete barrier as a single slit that will produce a single-slit diffraction pattern of the water waves on the shore.

SET UP: For single-slit diffraction, the angles at which destructive interference occurs are given by $\sin\theta_m = m\lambda/a$, where $m = 1, 2, 3, \dots$.

EXECUTE: **(a)** The frequency of the water waves is $f = 75.0\text{ min}^{-1} = 1.25\text{ s}^{-1} = 1.25$ Hz, so their wavelength is $\lambda = v/f = (15.0\text{ cm/s})/(1.25\text{ Hz}) = 12.0$ cm.

At the first point for which destructive interference occurs, we have
$\tan\theta = (0.613\text{ m})/(3.20\text{ m}) \Rightarrow \theta = 10.84°$. $a\sin\theta = \lambda$ and
$$a = \lambda/\sin\theta = (12.0\text{ cm})/(\sin 10.84°) = 63.8\text{ cm}.$$

(b) First find the angles at which destructive interference occurs.
$$\sin\theta_2 = 2\lambda/a = 2(12.0\text{ cm})/(63.8\text{ cm}) \rightarrow \theta_2 = \pm 22.1°$$
$$\sin\theta_3 = 3\lambda/a = 3(12.0\text{ cm})/(63.8\text{ cm}) \rightarrow \theta_3 = \pm 34.3°$$
$$\sin\theta_4 = 4\lambda/a = 4(12.0\text{ cm})/(63.8\text{ cm}) \rightarrow \theta_4 = \pm 48.8°$$
$$\sin\theta_5 = 5\lambda/a = 5(12.0\text{ cm})/(63.8\text{ cm}) \rightarrow \theta_5 = \pm 70.1°$$

EVALUATE: These are large angles, so we cannot use the approximation that $\theta_m \approx m\lambda/a$.

36.9. **IDENTIFY and SET UP:** $v = f\lambda$ gives λ. The person hears no sound at angles corresponding to diffraction minima. The diffraction minima are located by $\sin\theta = m\lambda/a$, $m = \pm 1, \pm 2, \dots$ Solve for θ.

EXECUTE: $\lambda = v/f = (344\text{ m/s})/(1250\text{ Hz}) = 0.2752$ m; $a = 1.00$ m. $m = \pm 1$, $\theta = \pm 16.0°$; $m = \pm 2$, $\theta = \pm 33.4°$; $m = \pm 3$, $\theta = \pm 55.6°$; no solution for larger m

EVALUATE: $\lambda/a = 0.28$ so for the large wavelength sound waves diffraction by the doorway is a large effect. Diffraction would not be observable for visible light because its wavelength is much smaller and $\lambda/a \ll 1$.

36.11. **IDENTIFY and SET UP:** $\sin\theta = \lambda/a$ locates the first minimum. $y = x\tan\theta$.

EXECUTE: $\tan\theta = y/x = (36.5\text{ cm})/(40.0\text{ cm})$ and $\theta = 42.38°$.

$a = \lambda/\sin\theta = (620\times10^{-9}\text{ m})/(\sin 42.38°) = 0.920\ \mu\text{m}$

EVALUATE: $\theta = 0.74$ rad and $\sin\theta = 0.67$, so the approximation $\sin\theta \approx \theta$ would not be accurate.

36.15.   **(a) IDENTIFY:**   Use Eq. (36.2) with $m = 1$ to locate the angular position of the first minimum and then use $y = x \tan \theta$ to find its distance from the center of the screen.

SET UP: The diffraction pattern is sketched in Figure 36.15.

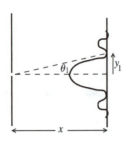

$$\sin \theta_1 = \frac{\lambda}{a} = \frac{540 \times 10^{-9} \text{ m}}{0.240 \times 10^{-3} \text{ m}} = 2.25 \times 10^{-3}$$

$$\theta_1 = 2.25 \times 10^{-3} \text{ rad}$$

Figure 36.15

$y_1 = x \tan \theta_1 = (3.00 \text{ m}) \tan(2.25 \times 10^{-3} \text{ rad}) = 6.75 \times 10^{-3} \text{ m} = 6.75 \text{ mm}$

(b) IDENTIFY and SET UP: Use Eqs. (36.5) and (36.6) to calculate the intensity at this point.

EXECUTE: Midway between the center of the central maximum and the first minimum implies

$y = \dfrac{1}{2}(6.75 \text{ mm}) = 3.375 \times 10^{-3} \text{ m}.$

$\tan \theta = \dfrac{y}{x} = \dfrac{3.375 \times 10^{-3} \text{ m}}{3.00 \text{ m}} = 1.125 \times 10^{-3}; \theta = 1.125 \times 10^{-3} \text{ rad}$

The phase angle β at this point on the screen is

$$\beta = \left(\frac{2\pi}{\lambda}\right) a \sin \theta = \frac{2\pi}{540 \times 10^{-9} \text{ m}} (0.240 \times 10^{-3} \text{ m}) \sin(1.125 \times 10^{-3} \text{ rad}) = \pi.$$

Then $I = I_0 \left(\dfrac{\sin \beta/2}{\beta/2}\right)^2 = (6.00 \times 10^{-6} \text{ W/m}^2) \left(\dfrac{\sin \pi/2}{\pi/2}\right)^2.$

$I = \left(\dfrac{4}{\pi^2}\right)(6.00 \times 10^{-6} \text{ W/m}^2) = 2.43 \times 10^{-6} \text{ W/m}^2.$

EVALUATE: The intensity at this point midway between the center of the central maximum and the first minimum is less than half the maximum intensity. Compare this result to the corresponding one for the two-slit pattern, Exercise 35.22.

36.19.   **IDENTIFY:**   The space between the skyscrapers behaves like a single slit and diffracts the radio waves.

SET UP: Cancellation of the waves occurs when $a \sin \theta = m\lambda$, $m = 1, 2, 3, \ldots$, and the intensity of the

waves is given by $I_0 \left(\dfrac{\sin \beta/2}{\beta/2}\right)^2$, where $\beta/2 = \dfrac{\pi a \sin \theta}{\lambda}.$

EXECUTE: **(a)** First find the wavelength of the waves:

$\lambda = c/f = (3.00 \times 10^8 \text{ m/s})/(88.9 \text{ MHz}) = 3.375 \text{ m}$

For no signal, $a \sin \theta = m\lambda$.

$m = 1$: $\sin \theta_1 = (1)(3.375 \text{ m})/(15.0 \text{ m}) \Rightarrow \theta_1 = \pm 13.0°$

$m = 2$: $\sin \theta_2 = (2)(3.375 \text{ m})/(15.0 \text{ m}) \Rightarrow \theta_2 = \pm 26.7°$

$m = 3$: $\sin \theta_3 = (3)(3.375 \text{ m})/(15.0 \text{ m}) \Rightarrow \theta_3 = \pm 42.4°$

$m = 4$: $\sin \theta_4 = (4)(3.375 \text{ m})/(15.0 \text{ m}) \Rightarrow \theta_4 = \pm 64.1°$

(b) $I_0 \left(\dfrac{\sin \beta/2}{\beta/2}\right)^2$, where $\beta/2 = \dfrac{\pi a \sin \theta}{\lambda} = \dfrac{\pi(15.0 \text{ m})\sin(5.00°)}{3.375 \text{ m}} = 1.217 \text{ rad}$

$$I = (3.50 \text{ W/m}^2)\left[\frac{\sin(1.217 \text{ rad})}{1.217 \text{ rad}}\right]^2 = 2.08 \text{ W/m}^2$$

EVALUATE: The wavelength of the radio waves is very long compared to that of visible light, but it is still considerably shorter than the distance between the buildings.

36.23.   **(a) IDENTIFY and SET UP:** If the slits are very narrow then the central maximum of the diffraction pattern for each slit completely fills the screen and the intensity distribution is given solely by the two-slit interference. The maxima are given by $d \sin \theta = m\lambda$ so $\sin \theta = m\lambda/d$. Solve for θ.

EXECUTE: 1st order maximum: $m = 1$, so $\sin \theta = \dfrac{\lambda}{d} = \dfrac{580 \times 10^{-9} \text{ m}}{0.530 \times 10^{-3} \text{ m}} = 1.094 \times 10^{-3}$; $\theta = 0.0627°$

2nd order maximum: $m = 2$, so $\sin \theta = \dfrac{2\lambda}{d} = 2.188 \times 10^{-3}$; $\theta = 0.125°$

(b) IDENTIFY and SET UP: The intensity is given by Eq. (36.12): $I = I_0 \cos^2(\phi/2) \left(\dfrac{\sin \beta/2}{\beta/2} \right)^2$. Calculate ϕ and β at each θ from part (a).

EXECUTE: $\phi = \left(\dfrac{2\pi d}{\lambda} \right) \sin \theta = \left(\dfrac{2\pi d}{\lambda} \right) \left(\dfrac{m\lambda}{d} \right) = 2\pi m$, so $\cos^2(\phi/2) = \cos^2(m\pi) = 1$

(Since the angular positions in part (a) correspond to interference maxima.)

$\beta = \left(\dfrac{2\pi a}{\lambda} \right) \sin \theta = \left(\dfrac{2\pi a}{\lambda} \right) \left(\dfrac{m\lambda}{d} \right) = 2\pi m(a/d) = m2\pi \left(\dfrac{0.320 \text{ mm}}{0.530 \text{ mm}} \right) = m(3.794 \text{ rad})$

1st order maximum: $m = 1$, so $I = I_0(1) \left(\dfrac{\sin(3.794/2)\text{rad}}{(3.794/2)\text{rad}} \right)^2 = 0.249 I_0$

2nd order maximum: $m = 2$, so $I = I_0(1) \left(\dfrac{\sin 3.794 \text{ rad}}{3.794 \text{ rad}} \right)^2 = 0.0256 I_0$

EVALUATE: The first diffraction minimum is at an angle θ given by $\sin \theta = \lambda/a$ so $\theta = 0.104°$. The first order fringe is within the central maximum and the second order fringe is inside the first diffraction maximum on one side of the central maximum. The intensity here at this second fringe is much less than I_0.

36.27.   **IDENTIFY:** The diffraction minima are located by $\sin \theta = \dfrac{m_d \lambda}{a}$ and the two-slit interference maxima are located by $\sin \theta = \dfrac{m_i \lambda}{d}$. The third bright band is missing because the first order single-slit minimum occurs at the same angle as the third order double-slit maximum.

SET UP: The pattern is sketched in Figure 36.27. $\tan \theta = \dfrac{3 \text{ cm}}{90 \text{ cm}}$, so $\theta = 1.91°$.

EXECUTE: Single-slit dark spot: $a \sin \theta = \lambda$ and $a = \dfrac{\lambda}{\sin \theta} = \dfrac{500 \text{ nm}}{\sin 1.91°} = 1.50 \times 10^4 \text{ nm} = 15.0 \text{ } \mu\text{m}$ (width)

Double-slit bright fringe: $d \sin \theta = 3\lambda$ and $d = \dfrac{3\lambda}{\sin \theta} = \dfrac{3(500 \text{ nm})}{\sin 1.91°} = 4.50 \times 10^4 \text{ nm} = 45.0 \text{ } \mu\text{m}$ (separation).

EVALUATE: Note that $d/a = 3.0$.

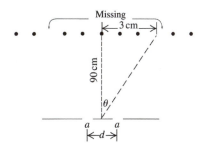

Figure 36.27

36.33. **IDENTIFY:** Knowing the wavelength of the light and the location of the first interference maxima, we can calculate the line density of the grating.

SET UP: The line density in lines/cm is $1/d$, with d in cm. The bright spots are located by $d\sin\theta = m\lambda$, $m = 0, \pm 1, \pm 2, \dots$.

EXECUTE: **(a)** $d = \dfrac{m\lambda}{\sin\theta} = \dfrac{(1)(632.8\times10^{-9}\text{ m})}{\sin 17.8°} = 2.07\times10^{-6}\text{ m} = 2.07\times10^{-4}\text{ cm}.$ $\dfrac{1}{d} = 4830$ lines/cm.

(b) $\sin\theta = \dfrac{m\lambda}{d} = m\left(\dfrac{632.8\times10^{-9}\text{ m}}{2.07\times10^{-6}\text{ m}}\right) = m(0.3057).$ For $m = \pm 2,$ $\theta = \pm 37.7°.$ For $m = \pm 3,$ $\theta = \pm 66.5°.$

EVALUATE: The angles are large, so they are not equally spaced; $37.7° \neq 2(17.8°)$ and $66.5° \neq 3(17.8°)$

36.37. **IDENTIFY:** The resolving power depends on the line density and the width of the grating.

SET UP: The resolving power is given by $R = Nm == \lambda/\Delta\lambda.$

EXECUTE: **(a)** $R = Nm = (5000\text{ lines/cm})(3.50\text{ cm})(1) = 17{,}500$

(b) The resolving power needed to resolve the sodium doublet is

$$R = \lambda/\Delta\lambda = (589\text{ nm})/(589.59\text{ nm} - 589.00\text{ nm}) = 998$$

so this grating can easily resolve the doublet.

(c) (i) $R = \lambda/\Delta\lambda.$ Since $R = 17{,}500$ when $m = 1,$ $R = 2\times17{,}500 = 35{,}000$ for $m = 2.$ Therefore

$$\Delta\lambda = \lambda/R = (587.8\text{ nm})/35{,}000 = 0.0168\text{ nm}$$

$$\lambda_{\min} = \lambda + \Delta\lambda = 587.8002\text{ nm} + 0.0168\text{ nm} = 587.8170\text{ nm}$$

(ii) $\lambda_{\max} = \lambda - \Delta\lambda = 587.8002\text{ nm} - 0.0168\text{ nm} = 587.7834\text{ nm}$

EVALUATE: (iii) Therefore the range of resolvable wavelengths is $587.7834\text{ nm} < \lambda < 587.8170\text{ nm}.$

36.41. **IDENTIFY:** The crystal behaves like a diffraction grating.

SET UP: The maxima are at angles θ given by $2d\sin\theta = m\lambda,$ where $d = 0.440$ nm.

EXECUTE: $m = 1.$ $\lambda = \dfrac{2d\sin\theta}{1} = 2(0.440\text{ nm})\sin 39.4° = 0.559\text{ nm}.$

EVALUATE: The result is a reasonable x ray wavelength.

36.43. **IDENTIFY:** Apply $\sin\theta = 1.22\dfrac{\lambda}{D}.$

SET UP: $\theta = \dfrac{W}{h},$ where $W = 28$ km and $h = 1200$ km. θ is small, so $\sin\theta \approx \theta.$

EXECUTE: $D = \dfrac{1.22\lambda}{\sin\theta} = 1.22\lambda\dfrac{h}{W} = 1.22(0.036\text{ m})\dfrac{1.2\times10^6\text{ m}}{2.8\times10^4\text{ m}} = 1.88\text{ m}$

EVALUATE: D must be significantly larger than the wavelength, so a much larger diameter is needed for microwaves than for visible wavelengths.

36.45. **IDENTIFY and SET UP:** The angular size of the first dark ring is given by $\sin\theta_1 = 1.22\lambda/D$ (Eq. 36.17). Calculate $\theta_1,$ and then the diameter of the ring on the screen is $2(4.5\text{ m})\tan\theta_1.$

EXECUTE: $\sin\theta_1 = 1.22\left(\dfrac{620\times10^{-9}\text{ m}}{7.4\times10^{-6}\text{ m}}\right) = 0.1022;$ $\theta_1 = 0.1024$ rad

The radius of the Airy disk (central bright spot) is $r = (4.5\text{ m})\tan\theta_1 = 0.462\text{ m}.$ The diameter is $2r = 0.92\text{ m} = 92\text{ cm}.$

EVALUATE: $\lambda/D = 0.084.$ For this small D the central diffraction maximum is broad.

36.47. **IDENTIFY and SET UP:** Resolved by Rayleigh's criterion means angular separation θ of the objects equals $1.22\lambda/D.$ The angular separation θ of the objects is their linear separation divided by their distance from the telescope.

EXECUTE: $\theta = \dfrac{250\times10^3\text{ m}}{5.93\times10^{11}\text{ m}},$ where 5.93×10^{11} m is the distance from earth to Jupiter. Thus

$\theta = 4.216\times10^{-7}.$

Then $\theta = 1.22\dfrac{\lambda}{D}$ and $D = \dfrac{1.22\lambda}{\theta} = \dfrac{1.22(500\times10^{-9}\ \text{m})}{4.216\times10^{-7}} = 1.45\ \text{m}$

EVALUATE: This is a very large telescope mirror. The greater the angular resolution the greater the diameter the lens or mirror must be.

36.49. **IDENTIFY** and **SET UP:** Let y be the separation between the two points being resolved and let s be their distance from the telescope. Then the limit of resolution corresponds to $1.22\dfrac{\lambda}{D} = \dfrac{y}{s}$.

EXECUTE: **(a)** Let the two points being resolved be the opposite edges of the crater, so y is the diameter of the crater. For the moon, $s = 3.8\times10^{8}\ \text{m}$. $y = 1.22\lambda s / D$.

Hubble: $D = 2.4\ \text{m}$ and $\lambda = 400\ \text{nm}$ gives the maximum resolution, so $y = 77\ \text{m}$

Arecibo: $D = 305\ \text{m}$ and $\lambda = 0.75\ \text{m}$; $y = 1.1\times10^{6}\ \text{m}$

(b) $s = \dfrac{yD}{1.22\lambda}$. Let $y \approx 0.30\ \text{m}$ (the size of a license plate).

$s = (0.30\ \text{m})(2.4\ \text{m})/[(1.22)(400\times10^{-9}\ \text{m})] = 1500\ \text{km}$.

EVALUATE: D/λ is much larger for the optical telescope and it has a much larger resolution even though the diameter of the radio telescope is much larger.

36.51. **IDENTIFY:** We can apply the equation for single-slit diffraction to the hair, with the thickness of the hair replacing the thickness of the slit.

SET UP: The dark fringes are located by $\sin\theta = m\dfrac{\lambda}{a}$. The first dark fringes are for $m = \pm1$. $y = R\tan\theta$ is the distance from the center of the screen. From the center to one minimum is 2.61 cm.

EXECUTE: $\tan\theta = \dfrac{y}{R} = \dfrac{2.61\ \text{cm}}{125\ \text{cm}} = 0.02088$ so $\theta = 1.20°$. $a = \dfrac{\lambda}{\sin\theta} = \dfrac{632.8\times10^{-9}\ \text{m}}{\sin 1.20°} = 30.2\ \mu\text{m}$.

EVALUATE: Although the thickness of human hairs can vary considerably, $30\ \mu\text{m}$ is a reasonable thickness.

36.53. **IDENTIFY:** In the single-slit diffraction pattern, the intensity is a maximum at the center and zero at the dark spots. At other points, it depends on the angle at which one is observing the light.

SET UP: Dark fringes occur when $\sin\theta_m = m\lambda/a$, where $m = 1, 2, 3, \ldots$, and the intensity is given by

$I_0\left(\dfrac{\sin\beta/2}{\beta/2}\right)^2$, where $\beta/2 = \dfrac{\pi a\sin\theta}{\lambda}$.

EXECUTE: **(a)** At the maximum possible angle, $\theta = 90°$, so

$m_{\text{max}} = (a\sin 90°)/\lambda = (0.0250\ \text{mm})/(632.8\ \text{nm}) = 39.5$

Since m must be an integer and $\sin\theta$ must be ≤ 1, $m_{\text{max}} = 39$. The total number of dark fringes is 39 on each side of the central maximum for a total of 78.

(b) The farthest dark fringe is for $m = 39$, giving

$\sin\theta_{39} = (39)(632.8\ \text{nm})/(0.0250\ \text{mm}) \Rightarrow \theta_{39} = \pm80.8°$

(c) The next closer dark fringe occurs at $\sin\theta_{38} = (38)(632.8\ \text{nm})/(0.0250\ \text{mm}) \Rightarrow \theta_{38} = 74.1°$.

The angle midway these two extreme fringes is $(80.8° + 74.1°)/2 = 77.45°$, and the intensity at this angle is

$I = I_0\left(\dfrac{\sin\beta/2}{\beta/2}\right)^2$, where $\beta/2 = \dfrac{\pi a\sin\theta}{\lambda} = \dfrac{\pi(0.0250\ \text{mm})\sin(77.45°)}{632.8\ \text{nm}} = 121.15\ \text{rad}$, which gives

$I = (8.50\ \text{W/m}^2)\left[\dfrac{\sin(121.15\ \text{rad})}{121.15\ \text{rad}}\right]^2 = 5.55\times10^{-4}\ \text{W/m}^2$.

EVALUATE: At the angle in part (c), the intensity is so low that the light would be barely perceptible.

36.55. **IDENTIFY** and **SET UP:** $\sin\theta = \lambda/a$ locates the first dark band. In the liquid the wavelength changes and this changes the angular position of the first diffraction minimum.

EXECUTE: $\sin\theta_{air} = \dfrac{\lambda_{air}}{a}$; $\sin\theta_{liquid} = \dfrac{\lambda_{liquid}}{a}$. $\lambda_{liquid} = \lambda_{air}\left(\dfrac{\sin\theta_{liquid}}{\sin\theta_{air}}\right) = \lambda_{air}\dfrac{\sin 21.6°}{\sin 38.2°} = 0.5953\lambda_{air}.$

$\lambda_{liquid} = \lambda_{air}/n$ (Eq. 33.5), so $n = \lambda_{air}/\lambda_{liquid} = \dfrac{\lambda_{air}}{0.5953\lambda_{air}} = 1.68.$

EVALUATE: Light travels faster in air and n must be >1.00. The smaller λ in the liquid reduces θ that located the first dark band.

36.57. **(a) IDENTIFY** and **SET UP:** The angular position of the first minimum is given by $a\sin\theta = m\lambda$ (Eq. 36.2), with $m = 1$. The distance of the minimum from the center of the pattern is given by $y = x\tan\theta$.

$\sin\theta = \dfrac{\lambda}{a} = \dfrac{540\times10^{-9}\text{ m}}{0.360\times10^{-3}\text{ m}} = 1.50\times10^{-3}$; $\theta = 1.50\times10^{-3}$ rad

$y_1 = x\tan\theta = (1.20\text{ m})\tan(1.50\times10^{-3}\text{ rad}) = 1.80\times10^{-3}$ m $= 1.80$ mm.

(Note that θ is small enough for $\theta \approx \sin\theta \approx \tan\theta$, and Eq. (36.3) applies.)

(b) IDENTIFY and **SET UP:** Find the phase angle β where $I = I_0/2$. Then use Eq. (36.6) to solve for θ and $y = x\tan\theta$ to find the distance.

EXECUTE: Eq. (36.5) gives that $I = \dfrac{1}{2}I_0$ when $\beta = 2.78$ rad.

$\beta = \left(\dfrac{2\pi}{\lambda}\right)a\sin\theta$ (Eq. (36.6)), so $\sin\theta = \dfrac{\beta\lambda}{2\pi a}.$

$y = x\tan\theta \approx x\sin\theta = \dfrac{\beta\lambda x}{2\pi a} = \dfrac{(2.78\text{ rad})(540\times10^{-9}\text{ m})(1.20\text{ m})}{2\pi(0.360\times10^{-3}\text{ m})} = 7.96\times10^{-4}$ m $= 0.796$ mm

EVALUATE: The point where $I = I_0/2$ is not midway between the center of the central maximum and the first minimum; see Exercise 36.15.

36.63. **IDENTIFY** and **SET UP:** The condition for an intensity maximum is $d\sin\theta = m\lambda$, $m = 0, \pm1, \pm2,...$ Third order means $m = 3$. The longest observable wavelength is the one that gives $\theta = 90°$ and hence $\theta = 1$.

EXECUTE: 9200 lines/cm so 9.2×10^5 lines/m and $d = \dfrac{1}{9.2\times10^5}$ m $= 1.087\times10^{-6}$ m.

$\lambda = \dfrac{d\sin\theta}{m} = \dfrac{(1.087\times10^{-6}\text{ m})(1)}{3} = 3.6\times10^{-7}$ m $= 360$ nm.

EVALUATE: The longest wavelength that can be obtained decreases as the order increases.

36.65. **IDENTIFY:** The maxima are given by $d\sin\theta = m\lambda$. We need $\sin\theta = \dfrac{m\lambda}{d} \leq 1$ in order for all the visible wavelengths to be seen.

SET UP: For 650 slits/mm $\Rightarrow d = \dfrac{1}{6.50\times10^5\text{ m}^{-1}} = 1.53\times10^{-6}$ m.

EXECUTE: $\lambda_1 = 4.00\times10^{-7}$m: $m = 1$: $\dfrac{\lambda_1}{d} = 0.26$; $m = 2$: $\dfrac{2\lambda_1}{d} = 0.52$; $m = 3$: $\dfrac{3\lambda_1}{d} = 0.78$.

$\lambda_2 = 7.00\times10^{-7}$m: $m = 1$: $\dfrac{\lambda_2}{d} = 0.46$; $m = 2$: $\dfrac{2\lambda_2}{d} = 0.92$; $m = 3$: $\dfrac{3\lambda_2}{d} = 1.37$. So, the third order does not contain the violet end of the spectrum, and therefore only the first- and second-order diffraction patterns contain all colors of the spectrum.

EVALUATE: θ for each maximum is larger for longer wavelengths.

36.69. **IDENTIFY:** The diameter D of the aperture limits the resolution due to diffraction, by Rayleigh's criterion.

SET UP: Rayleigh's criterion says that $\theta_{res} = 1.22\dfrac{\lambda}{D}$. $D = 4.00$ mm. $\theta_{res} = \dfrac{y}{s}$, where s is the altitude and $y = 65.0$ m.

EXECUTE: Combining two equations above gives $\dfrac{y}{s} = 1.22\dfrac{\lambda}{D}$.

$$s = \frac{yD}{1.22\lambda} = \frac{(65.0 \text{ m})(4.00 \times 10^{-3} \text{ m})}{1.22(550 \times 10^{-9} \text{ m})} = 3.87 \times 10^{5} \text{ m} = 387 \text{ km.}$$

EVALUATE: This is comparable to the altitude of the Hubble telescope.

36.71.   **IDENTIFY:**   The liquid reduces the wavlength of the light (compared to its value in air), and the scratch causes light passing through it to undergo single-slit diffraction.

SET UP: $\sin\theta = \dfrac{\lambda}{a}$, where λ is the wavelength in the liquid. $n = \dfrac{\lambda_{\text{air}}}{\lambda}$.

EXECUTE: $\tan\theta = \dfrac{(22.4/2) \text{ cm}}{30.0 \text{ cm}}$ and $\theta = 20.47°$.

$\lambda = a\sin\theta = (1.25 \times 10^{-6} \text{ m})\sin 20.47° = 4.372 \times 10^{-7} \text{ m} = 437.2 \text{ nm.}$ $n = \dfrac{\lambda_{\text{air}}}{\lambda} = \dfrac{612 \text{ nm}}{437.2 \text{ nm}} = 1.40.$

EVALUATE: $n > 1$, as it must be, and $n = 1.40$ is reasonable for many transparent films.

37

RELATIVITY

37.1. **IDENTIFY** and **SET UP:** Consider the distance A to O' and B to O' as observed by an observer on the ground (Figure 37.1).

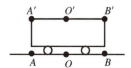

Figure 37.1

EXECUTE: Simultaneous to observer on train means light pulses from A' and B' arrive at O' at the same time. To observer at O light from A' has a longer distance to travel than light from B' so O will conclude that the pulse from $A(A')$ started before the pulse at $B(B')$. To observer at O bolt A appeared to strike first.

EVALUATE: Section 37.2 shows that if they are simultaneous to the observer on the ground then an observer on the train measures that the bolt at B' struck first.

37.5. **(a) IDENTIFY** and **SET UP:** $\Delta t_0 = 2.60 \times 10^{-8}$ s; $\Delta t = 4.20 \times 10^{-7}$ s. In the lab frame the pion is created and decays at different points, so this time is not the proper time.

EXECUTE: $\Delta t = \dfrac{\Delta t_0}{\sqrt{1 - u^2/c^2}}$ says $1 - \dfrac{u^2}{c^2} = \left(\dfrac{\Delta t_0}{\Delta t}\right)^2$

$\dfrac{u}{c} = \sqrt{1 - \left(\dfrac{\Delta t_0}{\Delta t}\right)^2} = \sqrt{1 - \left(\dfrac{2.60 \times 10^{-8} \text{ s}}{4.20 \times 10^{-7} \text{ s}}\right)^2} = 0.998; \ u = 0.998c$

EVALUATE: $u < c$, as it must be, but u/c is close to unity and the time dilation effects are large.

(b) IDENTIFY and **SET UP:** The speed in the laboratory frame is $u = 0.998c$; the time measured in this frame is Δt, so the distance as measured in this frame is $d = u\Delta t$.

EXECUTE: $d = (0.998)(2.998 \times 10^8 \text{ m/s})(4.20 \times 10^{-7} \text{ s}) = 126$ m

EVALUATE: The distance measured in the pion's frame will be different because the time measured in the pion's frame is different (shorter).

37.7. **IDENTIFY** and **SET UP:** A clock moving with respect to an observer appears to run more slowly than a clock at rest in the observer's frame. The clock in the spacecraft measurers the proper time Δt_0.

$\Delta t = 365$ days $= 8760$ hours.

EXECUTE: The clock on the moving spacecraft runs slow and shows the smaller elapsed time.

$\Delta t_0 = \Delta t \sqrt{1 - u^2/c^2} = (8760 \text{ h})\sqrt{1 - (4.80 \times 10^6 / 3.00 \times 10^8)^2} = 8758.88$ h. The difference in elapsed times is 8760 h $- 8758.88$ h $= 1.12$ h.

37.9. **IDENTIFY** and **SET UP:** $l = l_0 \sqrt{1 - u^2/c^2}$. The length measured when the spacecraft is moving is $l = 74.0$ m; l_0 is the length measured in a frame at rest relative to the spacecraft.

EXECUTE: $l_0 = \dfrac{l}{\sqrt{1-u^2/c^2}} = \dfrac{74.0 \text{ m}}{\sqrt{1-(0.600c/c)^2}} = 92.5 \text{ m}.$

EVALUATE: $l_0 > l.$ The moving spacecraft appears to an observer on the planet to be shortened along the direction of motion.

37.11.   **IDENTIFY and SET UP:** The 2.2 μs lifetime is Δt_0 and the observer on earth measures $\Delta t.$ The atmosphere is moving relative to the muon so in its frame the height of the atmosphere is l and l_0 is 10 km.

EXECUTE: (a) The greatest speed the muon can have is c, so the greatest distance it can travel in 2.2×10^{-6} s is $d = vt = (3.00 \times 10^8 \text{ m/s})(2.2 \times 10^{-6} \text{ s}) = 660 \text{ m} = 0.66 \text{ km}.$

(b) $\Delta t = \dfrac{\Delta t_0}{\sqrt{1-u^2/c^2}} = \dfrac{2.2 \times 10^{-6} \text{ s}}{\sqrt{1-(0.999)^2}} = 4.9 \times 10^{-5}$ s

$d = vt = (0.999)(3.00 \times 10^8 \text{ m/s})(4.9 \times 10^{-5} \text{ s}) = 15 \text{ km}$

In the frame of the earth the muon can travel 15 km in the atmosphere during its lifetime.

(c) $l = l_0 \sqrt{1-u^2/c^2} = (10 \text{ km})\sqrt{1-(0.999)^2} = 0.45 \text{ km}$

In the frame of the muon the height of the atmosphere is less than the distance it moves during its lifetime.

37.15.   **IDENTIFY:** Apply Eq. (37.23).

SET UP: The velocities $\vec{v}'$ and $\vec{v}$ are both in the +x-direction, so $v_x' = v'$ and $v_x = v.$

EXECUTE: (a) $v = \dfrac{v'+u}{1+uv'/c^2} = \dfrac{0.400c + 0.600c}{1+(0.400)(0.600)} = 0.806c$

(b) $v = \dfrac{v'+u}{1+uv'/c^2} = \dfrac{0.900c + 0.600c}{1+(0.900)(0.600)} = 0.974c$

(c) $v = \dfrac{v'+u}{1+uv'/c^2} = \dfrac{0.990c + 0.600c}{1+(0.990)(0.600)} = 0.997c.$

EVALUATE: Speed v is always less than c, even when $v'+u$ is greater than c.

37.17.   **IDENTIFY:** The relativistic velocity addition formulas apply since the speeds are close to that of light.

SET UP: The relativistic velocity addition formula is $v_x' = \dfrac{v_x - u}{1 - \dfrac{uv_x}{c^2}}.$

EXECUTE: (a) For the pursuit ship to catch the cruiser, the distance between them must be decreasing, so the velocity of the cruiser relative to the pursuit ship must be directed toward the pursuit ship.

(b) Let the unprimed frame be Tatooine and let the primed frame be the pursuit ship. We want the velocity v' of the cruiser knowing the velocity of the primed frame u and the velocity of the cruiser v in the unprimed frame (Tatooine).

$$v_x' = \dfrac{v_x - u}{1 - \dfrac{uv_x}{c^2}} = \dfrac{0.600c - 0.800c}{1-(0.600)(0.800)} = -0.385c$$

The result implies that the cruiser is moving toward the pursuit ship at $0.385c.$

EVALUATE: The nonrelativistic formula would have given $-0.200c$, which is considerably different from the correct result.

37.19.   **IDENTIFY and SET UP:** Reference frames S and S' are shown in Figure 37.19.

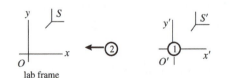

Frame S is at rest in the laboratory. Frame S' is attached to particle 1.

Figure 37.19

u is the speed of S' relative to S; this is the speed of particle 1 as measured in the laboratory. Thus $u = +0.650c$. The speed of particle 2 in S' is $0.950c$. Also, since the two particles move in opposite directions, 2 moves in the $-x'$-direction and $v'_x = -0.950c$. We want to calculate v_x, the speed of particle 2 in frame S; use Eq. (37.23).

EXECUTE: $v_x = \dfrac{v'_x + u}{1 + uv'_x/c^2} = \dfrac{-0.950c + 0.650c}{1 + (0.950c)(-0.650c)/c^2} = \dfrac{-0.300c}{1 - 0.6175} = -0.784c$. The speed of the second

particle, as measured in the laboratory, is $0.784c$.

EVALUATE: The incorrect Galilean expression for the relative velocity gives that the speed of the second particle in the lab frame is $0.300c$. The correct relativistic calculation gives a result more than twice this.

37.21. **IDENTIFY:** The relativistic velocity addition formulas apply since the speeds are close to that of light.

SET UP: The relativistic velocity addition formula is $v'_x = \dfrac{v_x - u}{1 - \dfrac{uv_x}{c^2}}$.

EXECUTE: In the relativistic velocity addition formula for this case, v'_x is the relative speed of particle 1 with respect to particle 2, v is the speed of particle 2 measured in the laboratory, and u is the speed of particle 1 measured in the laboratory, $u = -v$.

$v'_x = \dfrac{v - (-v)}{1 - (-v)v/c^2} = \dfrac{2v}{1 + v^2/c^2}$. $\dfrac{v'_x}{c^2}v^2 - 2v + v'_x = 0$ and $(0.890c)v^2 - 2c^2v + (0.890c^3) = 0$.

This is a quadratic equation with solution $v = 0.611c$ (v must be less than c).

EVALUATE: The nonrelativistic result would be $0.445c$, which is considerably different from this result.

37.23. **IDENTIFY** and **SET UP:** The reference frames are shown in Figure 37.23.

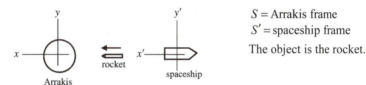

S = Arrakis frame
S' = spaceship frame
The object is the rocket.

Figure 37.23

u is the velocity of the spaceship relative to Arrakis.
$v_x = +0.360c$; $v'_x = +0.920c$
(In each frame the rocket is moving in the positive coordinate direction.)

Use the Lorentz velocity transformation equation, Eq. (37.22): $v'_x = \dfrac{v_x - u}{1 - uv_x/c^2}$.

EXECUTE: $v'_x = \dfrac{v_x - u}{1 - uv_x/c^2}$ so $v'_x - u\left(\dfrac{v_x v'_x}{c^2}\right) = v_x - u$ and $u\left(1 - \dfrac{v_x v'_x}{c^2}\right) = v_x - v'_x$

$u = \dfrac{v_x - v'_x}{1 - v_x v'_x/c^2} = \dfrac{0.360c - 0.920c}{1 - (0.360c)(0.920c)/c^2} = -\dfrac{0.560c}{0.6688} = -0.837c$

The speed of the spacecraft relative to Arrakis is $0.837c = 2.51 \times 10^8$ m/s. The minus sign in our result for u means that the spacecraft is moving in the $-x$-direction, so it is moving away from Arrakis.

EVALUATE: The incorrect Galilean expression also says that the spacecraft is moving away from Arrakis, but with speed $0.920c - 0.360c = 0.560c$.

37.25. **IDENTIFY** and **SET UP:** Source and observer are approaching, so use Eq. (37.25): $f = \sqrt{\dfrac{c + u}{c - u}}f_0$. Solve

for u, the speed of the light source relative to the observer.

(a) EXECUTE: $f^2 = \left(\dfrac{c + u}{c - u}\right)f_0^2$

$$(c-u)f^2 = (c+u)f_0^2 \text{ and } u = \frac{c(f^2 - f_0^2)}{f^2 + f_0^2} = c\left(\frac{(f/f_0)^2 - 1}{(f/f_0)^2 + 1}\right)$$

$\lambda_0 = 675 \text{ nm}, \quad \lambda = 575 \text{ nm}$

$$u = \left(\frac{(675 \text{ nm}/575 \text{ nm})^2 - 1}{(675 \text{ nm}/575 \text{ nm})^2 + 1}\right)c = 0.159c = (0.159)(2.998 \times 10^8 \text{ m/s}) = 4.77 \times 10^7 \text{ m/s; definitely speeding}$$

(b) $4.77 \times 10^7 \text{ m/s} = (4.77 \times 10^7 \text{ m/s})(1 \text{ km}/1000 \text{ m})(3600 \text{ s}/1 \text{ h}) = 1.72 \times 10^8 \text{ km/h}.$ Your fine would be

$\$1.72 \times 10^8$ (172 million dollars).

EVALUATE: The source and observer are approaching, so $f > f_0$ and $\lambda < \lambda_0$. Our result gives $u < c$, as it must.

37.29. **IDENTIFY:** Apply Eqs. (37.27) and (37.32).
SET UP: For a particle at rest (or with $v \ll c$), $a = F/m$.

EXECUTE: **(a)** $p = \dfrac{mv}{\sqrt{1 - v^2/c^2}} = 2mv.$

$$\Rightarrow 1 = 2\sqrt{1 - v^2/c^2} \Rightarrow \frac{1}{4} = 1 - \frac{v^2}{c^2} \Rightarrow v^2 = \frac{3}{4}c^2 \Rightarrow v = \frac{\sqrt{3}}{2}c = 0.866c.$$

(b) $F = \gamma^3 ma = 2ma \Rightarrow \gamma^3 = 2 \Rightarrow \gamma = (2)^{1/3}$ so $\dfrac{1}{1 - v^2/c^2} = 2^{2/3} \Rightarrow \dfrac{v}{c} = \sqrt{1 - 2^{-2/3}} = 0.608.$

EVALUATE: The momentum of a particle and the force required to give it a given acceleration both increase without bound as the speed of the particle approaches c.

37.31. **IDENTIFY:** When the speed of the electron is close to the speed of light, we must use the relativistic form of Newton's second law.

SET UP: When the force and velocity are parallel, as in part (b), $F = \dfrac{ma}{(1 - v^2/c^2)^{3/2}}.$ In part (a), $v \ll c$

so $F = ma.$

EXECUTE: **(a)** $a = \dfrac{F}{m} = \dfrac{5.00 \times 10^{-15} \text{ N}}{9.11 \times 10^{-31} \text{ kg}} = 5.49 \times 10^{15} \text{ m/s}^2.$

(b) $\gamma = \dfrac{1}{(1 - v^2/c^2)^{1/2}} = \dfrac{1}{(1 - [2.50 \times 10^8/3.00 \times 10^8]^2)^{1/2}} = 1.81.$

$a = \dfrac{F}{m\gamma^3} = \dfrac{5.49 \times 10^{15} \text{ m/s}^2}{(1.81)^3} = 9.26 \times 10^{14} \text{ m/s}^2.$

EVALUATE: The acceleration for low speeds is over 5 times greater than it is near the speed of light as in part (b).

37.35. **IDENTIFY and SET UP:** Use Eqs. (37.38) and (37.39).
EXECUTE: **(a)** $E = mc^2 + K$, so $E = 4.00mc^2$ means $K = 3.00mc^2 = 4.50 \times 10^{-10} \text{ J}$

(b) $E^2 = (mc^2)^2 + (pc)^2;$ $E = 4.00mc^2$, so $15.0(mc^2)^2 = (pc)^2$

$p = \sqrt{15}mc = 1.94 \times 10^{-18} \text{ kg} \cdot \text{m/s}$

(c) $E = mc^2/\sqrt{1 - v^2/c^2}$

$E = 4.00mc^2$ gives $1 - v^2/c^2 = 1/16$ and $v = \sqrt{15/16}c = 0.968c$

EVALUATE: The speed is close to c since the kinetic energy is greater than the rest energy. Nonrelativistic expressions relating E, K, p and v will be very inaccurate.

37.37. **IDENTIFY:** Use $E = mc^2$ to relate the mass increase to the energy increase.
(a) SET UP: Your total energy E increases because your gravitational potential energy mgy increases.
EXECUTE: $\Delta E = mg\Delta y$

$\Delta E = (\Delta m)c^2$ so $\Delta m = \Delta E/c^2 = mg(\Delta y)/c^2$

$\Delta m/m = (g\Delta y)/c^2 = (9.80 \text{ m/s}^2)(30 \text{ m})/(2.998 \times 10^8 \text{ m/s})^2 = 3.3 \times 10^{-13}\%$

This increase is much, much too small to be noticed.

(b) SET UP: The energy increases because potential energy is stored in the compressed spring.

EXECUTE: $\Delta E = \Delta U = \frac{1}{2}kx^2 = \frac{1}{2}(2.00 \times 10^4 \text{ N/m})(0.060 \text{ m})^2 = 36.0 \text{ J}$

$\Delta m = (\Delta E)/c^2 = 4.0 \times 10^{-16} \text{ kg}$

Energy increases so mass increases. The mass increase is much, much too small to be noticed.

EVALUATE: In both cases the energy increase corresponds to a mass increase. But since c^2 is a very large number the mass increase is very small.

37.41. **IDENTIFY and SET UP:** The total energy is given in terms of the momentum by Eq. (37.39). In terms of the total energy E, the kinetic energy K is $K = E - mc^2$ (from Eq. 37.38). The rest energy is mc^2.

EXECUTE: **(a)** $E = \sqrt{(mc^2)^2 + (pc)^2} = \sqrt{[(6.64 \times 10^{-27})(2.998 \times 10^8)^2]^2 + [(2.10 \times 10^{-18})(2.998 \times 10^8)]^2}$ J

$E = 8.67 \times 10^{-10}$ J

(b) $mc^2 = (6.64 \times 10^{-27} \text{ kg})(2.998 \times 10^8 \text{ m/s})^2 = 5.97 \times 10^{-10}$ J

$K = E - mc^2 = 8.67 \times 10^{-10} \text{ J} - 5.97 \times 10^{-10} \text{ J} = 2.70 \times 10^{-10}$ J

(c) $\dfrac{K}{mc^2} = \dfrac{2.70 \times 10^{-10} \text{ J}}{5.97 \times 10^{-10} \text{ J}} = 0.452$

EVALUATE: The incorrect nonrelativistic expressions for K and p give $K = p^2/2m = 3.3 \times 10^{-10}$ J; the correct relativistic value is less than this.

37.45. **IDENTIFY and SET UP:** Use Eq. (23.12) and conservation of energy to relate the potential difference to the kinetic energy gained by the electron. Use Eq. (37.36) to calculate the kinetic energy from the speed.

EXECUTE: **(a)** $K = q\Delta V = e\Delta V$

$K = mc^2 \left(\dfrac{1}{\sqrt{1 - v^2/c^2}} - 1 \right) = 4.025 mc^2 = 3.295 \times 10^{-13} \text{ J} = 2.06 \text{ MeV}$

$\Delta V = K/e = 2.06 \times 10^6$ V

(b) From part (a), $K = 3.30 \times 10^{-13}$ J $= 2.06$ MeV

EVALUATE: The speed is close to c and the kinetic energy is four times the rest mass.

37.49. **(a) IDENTIFY and SET UP:** $\Delta t_0 = 2.60 \times 10^{-8}$ s is the proper time, measured in the pion's frame. The time measured in the lab must satisfy $d = c\Delta t$, where $u \approx c$. Calculate Δt and then use Eq. (37.6) to calculate u.

EXECUTE: $\Delta t = \dfrac{d}{c} = \dfrac{1.90 \times 10^3 \text{ m}}{2.998 \times 10^8 \text{ m/s}} = 6.3376 \times 10^{-6}$ s. $\Delta t = \dfrac{\Delta t_0}{\sqrt{1 - u^2/c^2}}$ so $(1 - u^2/c^2)^{1/2} = \dfrac{\Delta t_0}{\Delta t}$ and

$(1 - u^2/c^2) = \left(\dfrac{\Delta t_0}{\Delta t} \right)^2$. Write $u = (1 - \Delta)c$ so that $(u/c)^2 = (1 - \Delta)^2 = 1 - 2\Delta + \Delta^2 \approx 1 - 2\Delta$ since Δ is small.

Using this in the above gives $1 - (1 - 2\Delta) = \left(\dfrac{\Delta t_0}{\Delta t} \right)^2$. $\Delta = \dfrac{1}{2} \left(\dfrac{\Delta t_0}{\Delta t} \right)^2 = \dfrac{1}{2} \left(\dfrac{2.60 \times 10^{-8} \text{ s}}{6.3376 \times 10^{-6} \text{ s}} \right)^2 = 8.42 \times 10^{-6}$.

EVALUATE: An alternative calculation is to say that the length of the tube must contract relative to the moving pion so that the pion travels that length before decaying. The contracted length must be

$l = c\Delta t_0 = (2.998 \times 10^8 \text{ m/s})(2.60 \times 10^{-8} \text{ s}) = 7.7948 \text{ m}$. $l = l_0 \sqrt{1 - u^2/c^2}$ so $1 - u^2/c^2 = \left(\dfrac{l}{l_0} \right)^2$. Then

$u = (1 - \Delta)c$ gives $\Delta = \dfrac{1}{2} \left(\dfrac{l}{l_0} \right)^2 = \dfrac{1}{2} \left(\dfrac{7.7948 \text{ m}}{1.90 \times 10^3 \text{ m}} \right)^2 = 8.42 \times 10^{-6}$, which checks.

(b) IDENTIFY and SET UP: $E = \gamma mc^2$ Eq. (37.38).

EXECUTE: $\gamma = \dfrac{1}{\sqrt{1-u^2/c^2}} = \dfrac{1}{\sqrt{2\Delta}} = \dfrac{1}{\sqrt{2(8.42\times10^{-6})}} = 244.$

$E = (244)(139.6 \text{ MeV}) = 3.40\times10^4 \text{ MeV} = 34.0 \text{ GeV.}$

EVALUATE: The total energy is 244 times the rest energy.

37.51.   **IDENTIFY and SET UP:**   There must be a length contraction such that the length a becomes the same as b; $l_0 = a$, $l = b$. l_0 is the distance measured by an observer at rest relative to the spacecraft. Use Eq. (37.16) and solve for u.

EXECUTE: $\dfrac{l}{l_0} = \sqrt{1-u^2/c^2}$ so $\dfrac{b}{a} = \sqrt{1-u^2/c^2}$;

$a = 1.40b$ gives $b/1.40b = \sqrt{1-u^2/c^2}$ and thus $1-u^2/c^2 = 1/(1.40)^2$

$u = \sqrt{1-1/(1.40)^2}\,c = 0.700c = 2.10\times10^8 \text{ m/s}$

EVALUATE: A length on the spacecraft in the direction of the motion is shortened. A length perpendicular to the motion is unchanged.

37.55.   **IDENTIFY:**   Since the speed is very close to the speed of light, we must use the relativistic formula for kinetic energy.

SET UP: The relativistic formula for kinetic energy is $K = mc^2\left(\dfrac{1}{\sqrt{1-v^2/c^2}} - 1\right)$ and the relativistic mass

is $m_{\text{rel}} = \dfrac{m}{\sqrt{1-v^2/c^2}}$.

EXECUTE: **(a)** $K = 7\times10^{12} \text{ eV} = 1.12\times10^{-6} \text{ J.}$ Using this value in the relativistic kinetic energy formula

and substituting the mass of the proton for m, we get $K = mc^2\left(\dfrac{1}{\sqrt{1-v^2/c^2}} - 1\right)$ which gives

$\dfrac{1}{\sqrt{1-v^2/c^2}} = 7.45\times10^3$ and $1-\dfrac{v^2}{c^2} = \dfrac{1}{(7.45\times10^3)^2}$. Solving for v gives $1-\dfrac{v^2}{c^2} = \dfrac{(c+v)(c-v)}{c^2} = \dfrac{2(c-v)}{c}$,

since $c + v \approx 2c$. Substituting $v = (1-\Delta)c$, we have $1-\dfrac{v^2}{c^2} = \dfrac{2(c-v)}{c} = \dfrac{2[c-(1-\Delta)c]}{c} = 2\Delta$. Solving for Δ

gives $\Delta = \dfrac{1-v^2/c^2}{2} = \dfrac{\dfrac{1}{(7.45\times10^3)^2}}{2} = 9\times10^{-9}$, to one significant digit.

(b) Using the relativistic mass formula and the result that $\dfrac{1}{\sqrt{1-v^2/c^2}} = 7.45\times10^3$, we have

$m_{\text{rel}} = \dfrac{m}{\sqrt{1-v^2/c^2}} = m\left(\dfrac{1}{\sqrt{1-v^2/c^2}}\right) = (7\times10^3)m$, to one significant digit.

EVALUATE: At such high speeds, the proton's mass is over 7000 times as great as its rest mass.

37.59.   **IDENTIFY and SET UP:**   Let S be the lab frame and S' be the frame of the proton that is moving in the $+x$-direction, so $u = +c/2$. The reference frames and moving particles are shown in Figure 37.59. The other proton moves in the $-x$-direction in the lab frame, so $v = -c/2$. A proton has rest mass $m_p = 1.67\times10^{-27} \text{ kg}$ and rest energy $m_pc^2 = 938 \text{ MeV.}$

EXECUTE: **(a)** $v' = \dfrac{v-u}{1-uv/c^2} = \dfrac{-c/2 - c/2}{1-(c/2)(-c/2)/c^2} = -\dfrac{4c}{5}$

The speed of each proton relative to the other is $\dfrac{4}{5}c$.

(b) In nonrelativistic mechanics the speeds just add and the speed of each relative to the other is c.

(c) $K = \dfrac{mc^2}{\sqrt{1 - v^2/c^2}} - mc^2$

(i) Relative to the lab frame each proton has speed $v = c/2$. The total kinetic energy of each proton is

$K = \dfrac{938 \text{ MeV}}{\sqrt{1 - \left(\dfrac{1}{2}\right)^2}} - (938 \text{ MeV}) = 145 \text{ MeV}.$

(ii) In its rest frame one proton has zero speed and zero kinetic energy and the other has speed $\dfrac{4}{5}c$. In this

frame the kinetic energy of the moving proton is $K = \dfrac{938 \text{ MeV}}{\sqrt{1 - \left(\dfrac{4}{5}\right)^2}} - (938 \text{ MeV}) = 625 \text{ MeV}.$

(d) (i) Each proton has speed $v = c/2$ and kinetic energy

$K = \dfrac{1}{2}mv^2 = \left(\dfrac{1}{2}m\right)(c/2)^2 = \dfrac{mc^2}{8} = \dfrac{938 \text{ MeV}}{8} = 117 \text{ MeV}.$

(ii) One proton has speed $v = 0$ and the other has speed c. The kinetic energy of the moving proton is

$K = \dfrac{1}{2}mc^2 = \dfrac{938 \text{ MeV}}{2} = 469 \text{ MeV}.$

EVALUATE: The relativistic expression for K gives a larger value than the nonrelativistic expression. The kinetic energy of the system is different in different frames.

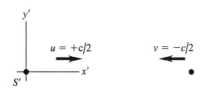

Figure 37.59

37.67. **IDENTIFY** and **SET UP:** An increase in wavelength corresponds to a decrease in frequency $(f = c/\lambda)$, so

the atoms are moving away from the earth. Receding, so use Eq. (37.26): $f = \sqrt{\dfrac{c-u}{c+u}}f_0$

EXECUTE: Solve for u: $(f/f_0)^2(c+u) = c - u$ and $u = c\left(\dfrac{1 - (f/f_0)^2}{1 + (f/f_0)^2}\right)$

$f = c/\lambda, \; f_0 = c/\lambda_0 \;$ so $\; f/f_0 = \lambda_0/\lambda$

$u = c\left(\dfrac{1 - (\lambda_0/\lambda)^2}{1 + (\lambda_0/\lambda)^2}\right) = c\left(\dfrac{1 - (656.3/953.4)^2}{1 + (656.3/953.4)^2}\right) = 0.357c = 1.07 \times 10^8 \text{ m/s}$

EVALUATE: The relative speed is large, 36% of c. The cosmological implication of such observations will be discussed in Chapter 44.

37.71.   **IDENTIFY:**   We need to use the relativistic form of Newton's second law because the speed of the proton is close to the speed of light.

SET UP: $\vec{F}$ and $\vec{v}$ are perpendicular, so $F = \gamma ma = \gamma m \dfrac{v^2}{R}$. $\gamma = \dfrac{1}{\sqrt{1 - v^2/c^2}} = \dfrac{1}{\sqrt{1 - (0.750)^2}} = 1.512$.

EXECUTE: $F = (1.512)(1.67 \times 10^{-27} \text{ kg}) \dfrac{[(0.750)(3.00 \times 10^8 \text{ m/s}]^2}{628 \text{ m}} = 2.04 \times 10^{-13} \text{ N}.$

EVALUATE: If we ignored relativity, the force would be $F_{\text{rel}}/\gamma = \dfrac{2.04 \times 10^{-13} \text{ N}}{1.512} = 1.35 \times 10^{-13} \text{ N},$ which is substantially less than the relativistic force.

38

PHOTONS: LIGHT WAVES BEHAVING AS PARTICLES

38.5. **IDENTIFY** and **SET UP:** $c = f\lambda$. The source emits $(0.05)(75\text{ J}) = 3.75\text{ J}$ of energy as visible light each second. $E = hf$, with $h = 6.63\times10^{-34}$ J·s.

EXECUTE: **(a)** $f = \dfrac{c}{\lambda} = \dfrac{3.00\times10^8\text{ m/s}}{600\times10^{-9}\text{ m}} = 5.00\times10^{14}$ Hz

(b) $E = hf = (6.63\times10^{-34}\text{ J·s})(5.00\times10^{14}\text{ Hz}) = 3.32\times10^{-19}$ J. The number of photons emitted per second

is $\dfrac{3.75\text{ J}}{3.32\times10^{-19}\text{ J/photon}} = 1.13\times10^{19}$ photons.

EVALUATE: **(c)** No. The frequency of the light depends on the energy of each photon. The number of photons emitted per second is proportional to the power output of the source.

38.7. **IDENTIFY** and **SET UP:** The stopping potential V_0 is related to the frequency of the light by $V_0 = \dfrac{h}{e}f - \dfrac{\phi}{e}$.

The slope of V_0 versus f is h/e. The value f_{th} of f when $V_0 = 0$ is related to ϕ by $\phi = hf_{\text{th}}$.

EXECUTE: **(a)** From the graph, $f_{\text{th}} = 1.25\times10^{15}$ Hz. Therefore, with the value of h from part (b), $\phi = hf_{\text{th}} = 4.8$ eV.

(b) From the graph, the slope is 3.8×10^{-15} V·s.

$h = (e)(\text{slope}) = (1.60\times10^{-16}\text{ C})(3.8\times10^{-15}\text{ V·s}) = 6.1\times10^{-34}$ J·s

(c) No photoelectrons are produced for $f < f_{\text{th}}$.

(d) For a different metal f_{th} and ϕ are different. The slope is h/e so would be the same, but the graph would be shifted right or left so it has a different intercept with the horizontal axis.

EVALUATE: As the frequency f of the light is increased above f_{th} the energy of the photons in the light increases and more energetic photons are produced. The work function we calculated is similar to that for gold or nickel.

38.9. **IDENTIFY** and **SET UP:** Eq. (38.3): $\dfrac{1}{2}mv_{\text{max}}^2 = hf - \phi = \dfrac{hc}{\lambda} - \phi$. Take the work function ϕ from Table 38.1.

Solve for v_{max}. Note that we wrote f as c/λ.

EXECUTE: $\dfrac{1}{2}mv_{\text{max}}^2 = \dfrac{(6.626\times10^{-34}\text{ J·s})(2.998\times10^8\text{ m/s})}{235\times10^{-9}\text{ m}} - (5.1\text{ eV})(1.602\times10^{-19}\text{ J/1 eV})$

$\dfrac{1}{2}mv_{\text{max}}^2 = 8.453\times10^{-19}\text{ J} - 8.170\times10^{-19}\text{ J} = 2.83\times10^{-20}$ J

$v_{\text{max}} = \sqrt{\dfrac{2(2.83\times10^{-20}\text{ J})}{9.109\times10^{-31}\text{ kg}}} = 2.49\times10^5$ m/s

EVALUATE: The work function in eV was converted to joules for use in Eq. (38.3). A photon with $\lambda = 235$ nm has energy greater then the work function for the surface.

38.11. **IDENTIFY:** The photoelectric effect occurs. The kinetic energy of the photoelectron is the difference between the initial energy of the photon and the work function of the metal.

SET UP: $\frac{1}{2}mv_{max}^2 = hf - \phi$, $E = hc/\lambda$.

EXECUTE: Use the data for the 400.0-nm light to calculate ϕ. Solving for ϕ gives $\phi = \frac{hc}{\lambda} - \frac{1}{2}mv_{max}^2 =$

$\frac{(4.136 \times 10^{-15} \text{ eV} \cdot \text{s})(3.00 \times 10^8 \text{ m/s})}{400.0 \times 10^{-9} \text{ m}} - 1.10 \text{ eV} = 3.10 \text{ eV} - 1.10 \text{ eV} = 2.00 \text{ eV}$. Then for 300.0 nm, we

have $\frac{1}{2}mv_{max}^2 = hf - \phi = \frac{hc}{\lambda} - \phi = \frac{(4.136 \times 10^{-15} \text{ eV} \cdot \text{s})(3.00 \times 10^8 \text{ m/s})}{300.0 \times 10^{-9} \text{ m}} - 2.00 \text{ eV}$, which gives

$\frac{1}{2}mv_{max}^2 = 4.14 \text{ eV} - 2.00 \text{ eV} = 2.14 \text{ eV}$.

EVALUATE: When the wavelength decreases the energy of the photons increases and the photoelectrons have a larger minimum kinetic energy.

38.15. **IDENTIFY:** Apply Eq. (38.6).

SET UP: For a 4.00-keV electron, $eV_{AC} = 4000$ eV.

EXECUTE: $eV_{AC} = hf_{max} = \frac{hc}{\lambda_{min}} \Rightarrow \lambda_{min} = \frac{hc}{eV_{AC}} = \frac{(6.63 \times 10^{-34} \text{ J} \cdot \text{s})(3.00 \times 10^8 \text{ m/s})}{(1.60 \times 10^{-19} \text{ C})(4000 \text{ V})} = 3.11 \times 10^{-10}$ m

EVALUATE: This is the same answer as would be obtained if electrons of this energy were used. Electron beams are much more easily produced and accelerated than proton beams.

38.17. **IDENTIFY:** Energy is conserved when the x ray collides with the stationary electron.

SET UP: $E = hc/\lambda$, and energy conservation gives $\frac{hc}{\lambda} = \frac{hc}{\lambda'} + K_e$.

EXECUTE: Solving for K_e gives $K_e = hc\left(\frac{1}{\lambda} - \frac{1}{\lambda'}\right) =$

$(6.63 \times 10^{-34} \text{ J} \cdot \text{s})(3.00 \times 10^8 \text{ m/s})\left(\frac{1}{0.100 \times 10^{-9} \text{ m}} - \frac{1}{0.110 \times 10^{-9} \text{ m}}\right)$. $K_e = 1.81 \times 10^{-16}$ J $= 1.13$ keV.

EVALUATE: The electron does not get all the energy of the incident photon.

38.19. **IDENTIFY:** Apply Eq. (38.7): $\lambda' - \lambda = \frac{h}{mc}(1 - \cos\phi) = \lambda_C(1 - \cos\phi)$

SET UP: Solve for λ': $\lambda' = \lambda + \lambda_C(1 - \cos\phi)$.

The largest λ' corresponds to $\phi = 180°$, so $\cos\phi = -1$.

EXECUTE: $\lambda' = \lambda + 2\lambda_C = 0.0665 \times 10^{-9}$ m $+ 2(2.426 \times 10^{-12}$ m$) = 7.135 \times 10^{-11}$ m $= 0.0714$ nm. This wavelength occurs at a scattering angle of $\phi = 180°$.

EVALUATE: The incident photon transfers some of its energy and momentum to the electron from which it scatters. Since the photon loses energy its wavelength increases, $\lambda' > \lambda$.

38.21. **IDENTIFY and SET UP:** The shift in wavelength of the photon is $\lambda' - \lambda = \frac{h}{mc}(1 - \cos\phi)$ where λ' is the

wavelength after the scattering and $\frac{h}{mc} = \lambda_C = 2.426 \times 10^{-12}$ m. The energy of a photon of wavelength λ

is $E = \frac{hc}{\lambda} = \frac{1.24 \times 10^{-6} \text{ eV} \cdot \text{m}}{\lambda}$. Conservation of energy applies to the collision, so the energy lost by the

photon equals the energy gained by the electron.

EXECUTE: **(a)** $\lambda' - \lambda = \lambda_C(1 - \cos\phi) = (2.426 \times 10^{-12} \text{ m})(1 - \cos 35.0°) = 4.39 \times 10^{-13}$ m $= 4.39 \times 10^{-4}$ nm.

(b) $\lambda' = \lambda + 4.39 \times 10^{-4}$ nm $= 0.04250$ nm $+ 4.39 \times 10^{-4}$ nm $= 0.04294$ nm.

(c) $E_\lambda = \frac{hc}{\lambda} = 2.918 \times 10^4$ eV and $E_{\lambda'} = \frac{hc}{\lambda'} = 2.888 \times 10^4$ eV so the photon loses 300 eV of energy.

(d) Energy conservation says the electron gains 300 eV of energy.

EVALUATE: The photon transfers energy to the electron. Since the photon loses energy, its wavelength increases.

38.23. **IDENTIFY:** During the Compton scattering, the wavelength of the x ray increases by 1.0%, which means that the x ray loses energy to the electron.

SET UP: $\Delta\lambda = \dfrac{h}{mc}(1 - \cos\phi)$ and $\dfrac{h}{mc} = 2.426 \times 10^{-12}$ m. $\lambda' = 1.010\lambda$ so $\Delta\lambda = 0.010\lambda$.

EXECUTE: $\cos\phi = 1 - \dfrac{\Delta\lambda}{h/mc} = 1 - \dfrac{(0.010)(0.900 \times 10^{-10}\text{ m})}{2.426 \times 10^{-12}\text{ m}} = 0.629$, so $\phi = 51.0°$.

EVALUATE: The scattering angle is less than 90°, so the x ray still has some forward momentum after scattering.

38.25. **(a) IDENTIFY and SET UP:** Use Eq. (37.36) to calculate the kinetic energy K.

EXECUTE: $K = mc^2\left(\dfrac{1}{\sqrt{1 - v^2/c^2}} - 1\right) = 0.1547mc^2$

$m = 9.109 \times 10^{-31}$ kg, so $K = 1.27 \times 10^{-14}$ J

(b) IDENTIFY and SET UP: The total energy of the particles equals the sum of the energies of the two photons. Linear momentum must also be conserved.

EXECUTE: The total energy of each electron or positron is $E = K + mc^2 = 1.1547mc^2 = 9.46 \times 10^{-13}$ J. The total energy of the electron and positron is converted into the total energy of the two photons. The initial momentum of the system in the lab frame is zero (since the equal-mass particles have equal speeds in opposite directions), so the final momentum must also be zero. The photons must have equal wavelengths and must be traveling in opposite directions. Equal λ means equal energy, so each photon has energy 9.46×10^{-14} J.

(c) IDENTIFY and SET UP: Use Eq. (38.2) to relate the photon energy to the photon wavelength.

EXECUTE: $E = hc/\lambda$ so $\lambda = hc/E = hc/(9.46 \times 10^{-14}\text{ J}) = 2.10$ pm

EVALUATE: When the particles also have kinetic energy, the energy of each photon is greater, so its wavelength is less.

38.27. **IDENTIFY:** The wavelength of the pulse tells us the momentum of the photon. The uncertainty in the momentum is determined by the uncertainty principle.

SET UP: $p = \dfrac{h}{\lambda}$ and $\Delta x \Delta p_x = \dfrac{\hbar}{2}$.

EXECUTE: $p = \dfrac{h}{\lambda} = \dfrac{6.626 \times 10^{-34}\text{ J}\cdot\text{s}}{556 \times 10^{-9}\text{ m}} = 1.19 \times 10^{-27}$ kg·m/s. The spatial length of the pulse is

$\Delta x = c\Delta t = (2.998 \times 10^8\text{ m/s})(9.00 \times 10^{-15}\text{ s}) = 2.698 \times 10^{-6}$ m. The uncertainty principle gives $\Delta x \Delta p_x = \dfrac{\hbar}{2}$.

Solving for the uncertainty in the momentum, we have $\Delta p_x = \dfrac{\hbar}{2\Delta x} = \dfrac{1.055 \times 10^{-34}\text{ J}\cdot\text{s}}{2(2.698 \times 10^{-6}\text{ m})} = 1.96 \times 10^{-29}$ kg·m/s.

EVALUATE: This is 1.6% of the average momentum.

38.33. **IDENTIFY and SET UP:** The energy added to mass m of the blood to heat it to $T_f = 100°C$ and to vaporize it is $Q = mc(T_f - T_i) + mL_v$, with $c = 4190$ J/kg·K and $L_v = 2.256 \times 10^6$ J/kg. The energy of one photon is $E = \dfrac{hc}{\lambda} = \dfrac{1.99 \times 10^{-25}\text{ J}\cdot\text{m}}{\lambda}$.

EXECUTE: **(a)** $Q = (2.0 \times 10^{-9}\text{ kg})(4190\text{ J/kg}\cdot\text{K})(100°C - 33°C) + (2.0 \times 10^{-9}\text{ kg})(2.256 \times 10^6\text{ J/kg}) = 5.07 \times 10^{-3}$ J. The pulse must deliver 5.07 mJ of energy.

(b) $P = \dfrac{\text{energy}}{t} = \dfrac{5.07 \times 10^{-3}\text{ J}}{450 \times 10^{-6}\text{ s}} = 11.3$ W

(c) One photon has energy $E = \dfrac{hc}{\lambda} = \dfrac{1.99 \times 10^{-25}\text{ J}\cdot\text{m}}{585 \times 10^{-9}\text{ m}} = 3.40 \times 10^{-19}$ J. The number N of photons per pulse is the energy per pulse divided by the energy of one photon: $N = \dfrac{5.07 \times 10^{-3}\text{ J}}{3.40 \times 10^{-19}\text{ J/photon}} = 1.49 \times 10^{16}$ photons.

EVALUATE: The power output of the laser is small but it is focused on a small area, so the laser intensity is large.

38.35.   IDENTIFY and SET UP: $\lambda' = \lambda + \dfrac{h}{mc}(1 - \cos\phi)$

$\phi = 180°$ so $\lambda' = \lambda + \dfrac{2h}{mc} = 0.09485$ nm. Use Eq. (38.5) to calculate the momentum of the scattered photon.

Apply conservation of energy to the collision to calculate the kinetic energy of the electron after the scattering. The energy of the photon is given by Eq. (38.2).

EXECUTE: (a) $p' = h/\lambda' = 6.99 \times 10^{-24}$ kg·m/s.

(b) $E = E' + E_e$; $hc/\lambda = hc/\lambda' + E_e$

$E_e = hc\left(\dfrac{1}{\lambda} - \dfrac{1}{\lambda'}\right) = (hc)\dfrac{\lambda' - \lambda}{\lambda\lambda'} = 1.129 \times 10^{-16}$ J $= 705$ eV

EVALUATE: The energy of the incident photon is 13.8 keV, so only about 5% of its energy is transferred to the electron. This corresponds to a fractional shift in the photon's wavelength that is also 5%.

38.37.   IDENTIFY: Compton scattering occurs, and we know the angle of scattering and the initial wavelength (and hence momentum) of the incident photon.

SET UP: $\lambda' - \lambda = \left(\dfrac{h}{mc}\right)(1 - \cos\phi)$ and $p = h/\lambda$. Let $+x$ be the direction of propagation of the incident

photon and let the scattered photon be moving at $30.0°$ clockwise from the $+y$ axis.

EXECUTE: $\lambda' - \lambda = \left(\dfrac{h}{mc}\right)(1 - \cos\phi) = 0.1050 \times 10^{-9}$ m $+ (2.426 \times 10^{-12}$ m$)(1 - \cos 60.0°) = 0.1062 \times 10^{-9}$ m.

$P_{ix} = P_{fx}$. $\dfrac{h}{\lambda} = \dfrac{h}{\lambda'}\cos 60.0° + p_{ex}$.

$p_{ex} = \dfrac{h}{\lambda} - \dfrac{h}{2\lambda'} = h\dfrac{2\lambda' - \lambda}{(2\lambda')(\lambda)} = (6.626 \times 10^{-34}$ J·s$)\dfrac{2.1243 \times 10^{-10}$ m $- 1.050 \times 10^{-10}$ m}{(2.1243 \times 10^{-10}$ m$)(1.050 \times 10^{-10}$ m$)}$.

$p_{ex} = 3.191 \times 10^{-24}$ kg·m/s. $P_{iy} = P_{fy}$. $0 = \dfrac{h}{\lambda'}\sin 60.0° + p_{ey}$.

$p_{ey} = -\dfrac{(6.626 \times 10^{-34}$ J·s$)\sin 60.0°}{0.1062 \times 10^{-9}$ m$} = -5.403 \times 10^{-24}$ kg·m/s. $p_e = \sqrt{p_{ex}^2 + p_{ey}^2} = 6.28 \times 10^{-24}$ kg·m/s.

$\tan\theta = \dfrac{p_{ey}}{p_{ex}} = \dfrac{-5.403}{3.191}$ and $\theta = -59.4°$.

EVALUATE: The electron gets only part of the momentum of the incident photon.

38.43.   IDENTIFY: Apply the Compton scattering formula $\lambda' - \lambda = \Delta\lambda = \dfrac{h}{mc}(1 - \cos\phi) = \lambda_C(1 - \cos\phi)$

(a) SET UP: Largest $\Delta\lambda$ is for $\phi = 180°$.

EXECUTE: For $\phi = 180°$, $\Delta\lambda = 2\lambda_C = 2(2.426$ pm$) = 4.85$ pm.

(b) SET UP: $\lambda' - \lambda = \lambda_C(1 - \cos\phi)$

Wavelength doubles implies $\lambda' = 2\lambda$ so $\lambda' - \lambda = \lambda$. Thus $\lambda = \lambda_C(1 - \cos\phi)$. λ is related to E by Eq. (38.2).

EXECUTE: $E = hc/\lambda$, so smallest energy photon means largest wavelength photon, so $\phi = 180°$ and

$\lambda = 2\lambda_C = 4.85$ pm. Then

$E = \dfrac{hc}{\lambda} = \dfrac{(6.626 \times 10^{-34}$ J·s$)(2.998 \times 10^8$ m/s$)}{4.85 \times 10^{-12}$ m$} = 4.096 \times 10^{-14}$ J$(1$ eV$/1.602 \times 10^{-19}$ J$) = 0.256$ MeV.

EVALUATE: Any photon Compton scattered at $\phi = 180°$ has a wavelength increase of $2\lambda_C = 4.85$ pm. 4.85 pm is near the short-wavelength end of the range of x-ray wavelengths.

PARTICLES BEHAVING AS WAVES

39.3. **IDENTIFY:** For a particle with mass, $\lambda = \dfrac{h}{p}$ and $K = \dfrac{p^2}{2m}$.

SET UP: $1\,\text{eV} = 1.60 \times 10^{-19}\,\text{J}$

EXECUTE: **(a)** $\lambda = \dfrac{h}{p} \Rightarrow p = \dfrac{h}{\lambda} = \dfrac{(6.63 \times 10^{-34}\,\text{J} \cdot \text{s})}{(2.80 \times 10^{-10}\,\text{m})} = 2.37 \times 10^{-24}\,\text{kg} \cdot \text{m/s}.$

(b) $K = \dfrac{p^2}{2m} = \dfrac{(2.37 \times 10^{-24}\,\text{kg} \cdot \text{m/s})^2}{2(9.11 \times 10^{-31}\,\text{kg})} = 3.08 \times 10^{-18}\,\text{J} = 19.3\,\text{eV}.$

EVALUATE: This wavelength is on the order of the size of an atom. This energy is on the order of the energy of an electron in an atom.

39.5. **IDENTIFY and SET UP:** The de Broglie wavelength is $\lambda = \dfrac{h}{p} = \dfrac{h}{mv}$. In the Bohr model, $mvr_n = n(h/2\pi)$,

so $mv = nh/(2\pi r_n)$. Combine these two expressions and obtain an equation for λ in terms of n. Then

$\lambda = h\left(\dfrac{2\pi r_n}{nh}\right) = \dfrac{2\pi r_n}{n}.$

EXECUTE: **(a)** For $n = 1$, $\lambda = 2\pi r_1$ with $r_1 = a_0 = 0.529 \times 10^{-10}\,\text{m}$, so

$\lambda = 2\pi(0.529 \times 10^{-10}\,\text{m}) = 3.32 \times 10^{-10}\,\text{m}.$

$\lambda = 2\pi r_1$; the de Broglie wavelength equals the circumference of the orbit.

(b) For $n = 4$, $\lambda = 2\pi r_4/4$.

$r_n = n^2 a_0$ so $r_4 = 16a_0$.

$\lambda = 2\pi(16a_0)/4 = 4(2\pi a_0) = 4(3.32 \times 10^{-10}\,\text{m}) = 1.33 \times 10^{-9}\,\text{m}$

$\lambda = 2\pi r_4/4$; the de Broglie wavelength is $\dfrac{1}{n} = \dfrac{1}{4}$ times the circumference of the orbit.

EVALUATE: As n increases the momentum of the electron increases and its de Broglie wavelength decreases. For any n, the circumference of the orbits equals an integer number of de Broglie wavelengths.

39.9. **IDENTIFY and SET UP:** A photon has zero mass and its energy and wavelength are related by Eq. (38.2). An electron has mass. Its energy is related to its momentum by $E = p^2/2m$ and its wavelength is related to its momentum by Eq. (39.1).

EXECUTE: **(a)** <u>photon:</u> $E = \dfrac{hc}{\lambda}$ so $\lambda = \dfrac{hc}{E} = \dfrac{(6.626 \times 10^{-34}\,\text{J} \cdot \text{s})(2.998 \times 10^8\,\text{m/s})}{(20.0\,\text{eV})(1.602 \times 10^{-19}\,\text{J/eV})} = 62.0\,\text{nm}.$

<u>electron:</u> $E = p^2/(2m)$ so $p = \sqrt{2mE} =$

$\sqrt{2(9.109 \times 10^{-31}\,\text{kg})(20.0\,\text{eV})(1.602 \times 10^{-19}\,\text{J/eV})} = 2.416 \times 10^{-24}\,\text{kg} \cdot \text{m/s}.$ $\lambda = h/p = 0.274\,\text{nm}.$

(b) <u>photon</u>: $E = hc/R = 7.946 \times 10^{-19}$ J $= 4.96$ eV.

<u>electron</u>: $\lambda = h/p$ so $p = h/\lambda = 2.650 \times 10^{-27}$ kg $\cdot$ m/s.

$E = p^2/(2m) = 3.856 \times 10^{-24}$ J $= 2.41 \times 10^{-5}$ eV.

(c) **EVALUATE:** You should use a probe of wavelength approximately 250 nm. An electron with $\lambda = 250$ nm has much less energy than a photon with $\lambda = 250$ nm, so is less likely to damage the molecule. Note that $\lambda = h/p$ applies to all particles, those with mass and those with zero mass.

$E = hf = hc/\lambda$ applies only to photons and $E = p^2/2m$ applies only to particles with mass.

39.13. **IDENTIFY:** The acceleration gives momentum to the electrons. We can use this momentum to calculate their de Broglie wavelength.

SET UP: The kinetic energy K of the electron is related to the accelerating voltage V by $K = eV$. For an electron $E = \frac{1}{2}mv^2 = \frac{p^2}{2m}$ and $\lambda = \frac{h}{p}$. For a photon $E = \frac{hc}{\lambda}$.

EXECUTE: **(a)** For an electron $p = \dfrac{h}{\lambda} = \dfrac{6.63 \times 10^{-34} \text{ J} \cdot \text{s}}{5.00 \times 10^{-9} \text{ m}} = 1.33 \times 10^{-25}$ kg $\cdot$ m/s and

$E = \dfrac{p^2}{2m} = \dfrac{(1.33 \times 10^{-25} \text{ kg} \cdot \text{m/s})^2}{2(9.11 \times 10^{-31} \text{ kg})} = 9.71 \times 10^{-21}$ J. $V = \dfrac{K}{e} = \dfrac{9.71 \times 10^{-21} \text{ J}}{1.60 \times 10^{-19} \text{ C}} = 0.0607$ V. The electrons would have kinetic energy 0.0607 eV.

(b) $E = \dfrac{hc}{\lambda} = \dfrac{1.24 \times 10^{-6} \text{ eV} \cdot \text{m}}{5.00 \times 10^{-9} \text{ m}} = 248$ eV.

(c) $E = 9.71 \times 10^{-21}$ J

so $\lambda = \dfrac{hc}{E} = \dfrac{(6.63 \times 10^{-34} \text{ J} \cdot \text{s})(3.00 \times 10^{8} \text{ m/s})}{9.71 \times 10^{-21} \text{ J}} = 20.5 \ \mu\text{m}$.

EVALUATE: If they have the same wavelength, the photon has vastly more energy than the electron.

39.17. **IDENTIFY:** The intensity maxima are located by Eq. (39.4). Use $\lambda = \dfrac{h}{p}$ for the wavelength of the neutrons. For a particle, $p = \sqrt{2mE}$.

SET UP: For a neutron, $m = 1.67 \times 10^{-27}$ kg.

EXECUTE: For $m = 1$, $\lambda = d\sin\theta = \dfrac{h}{\sqrt{2mE}}$.

$E = \dfrac{h^2}{2md^2\sin^2\theta} = \dfrac{(6.63 \times 10^{-34} \text{ J} \cdot \text{s})^2}{2(1.675 \times 10^{-27} \text{ kg})(9.10 \times 10^{-11} \text{ m})^2 \sin^2(28.6°)} = 6.91 \times 10^{-20}$ J $= 0.432$ eV.

EVALUATE: The neutrons have $\lambda = 0.0436$ nm, comparable to the atomic spacing.

39.19. **IDENTIFY:** The condition for a maximum is $d\sin\theta = m\lambda$. $\lambda = \dfrac{h}{p} = \dfrac{h}{Mv}$, so $\theta = \arcsin\left(\dfrac{mh}{dMv}\right)$.

SET UP: Here m is the order of the maximum, whereas M is the incoming particle mass.

EXECUTE: **(a)** $m = 1 \Rightarrow \theta_1 = \arcsin\left(\dfrac{h}{dMv}\right)$

$= \arcsin\left(\dfrac{6.63 \times 10^{-34} \text{ J} \cdot \text{s}}{(1.60 \times 10^{-6} \text{ m})(9.11 \times 10^{-31} \text{ kg})(1.26 \times 10^{4} \text{ m/s})}\right) = 2.07°$.

$m = 2 \Rightarrow \theta_2 = \arcsin\left(\dfrac{(2)(6.63 \times 10^{-34} \text{ J} \cdot \text{s})}{(1.60 \times 10^{-6} \text{ m})(9.11 \times 10^{-31} \text{ kg})(1.26 \times 10^{4} \text{ m/s})}\right) = 4.14°$.

(b) For small angles (in radians!) $y \cong D\theta$, so $y_1 \approx (50.0 \text{ cm})(2.07°)\left(\dfrac{\pi \text{ radians}}{180°}\right) = 1.81 \text{ cm}$,

$y_2 \approx (50.0 \text{ cm})(4.14°)\left(\dfrac{\pi \text{ radians}}{180°}\right) = 3.61 \text{ cm}$ and $y_2 - y_1 = 3.61 \text{ cm} - 1.81 \text{ cm} = 1.80 \text{ cm}$.

EVALUATE: For these electrons, $\lambda = \dfrac{h}{mv} = 0.0577 \ \mu\text{m}$. λ is much less than d and the intensity maxima occur at small angles.

39.21. **IDENTIFY** and **SET UP:** For a photon $E_{\text{ph}} = \dfrac{hc}{\lambda} = \dfrac{1.99 \times 10^{-25} \text{ J} \cdot \text{m}}{\lambda}$. For an electron

$E_{\text{e}} = \dfrac{p^2}{2m} = \dfrac{1}{2m}\left(\dfrac{h}{\lambda}\right)^2 = \dfrac{h^2}{2m\lambda^2}$.

EXECUTE: **(a)** <u>photon</u> $E_{\text{ph}} = \dfrac{1.99 \times 10^{-25} \text{ J} \cdot \text{m}}{10.0 \times 10^{-9} \text{ m}} = 1.99 \times 10^{-17} \text{ J}$

<u>electron</u> $E_{\text{e}} = \dfrac{(6.63 \times 10^{-34} \text{ J} \cdot \text{s})^2}{2(9.11 \times 10^{-31} \text{ kg})(10.0 \times 10^{-9} \text{ m})^2} = 2.41 \times 10^{-21} \text{ J}$

$\dfrac{E_{\text{ph}}}{E_{\text{e}}} = \dfrac{1.99 \times 10^{-17} \text{ J}}{2.41 \times 10^{-21} \text{ J}} = 8.26 \times 10^3$

(b) The electron has much less energy so would be less damaging.

EVALUATE: For a particle with mass, such as an electron, $E \sim \lambda^{-2}$. For a massless photon $E \sim \lambda^{-1}$.

39.25. **IDENTIFY** and **SET UP:** Use the energy to calculate n for this state. Then use the Bohr equation, Eq. (39.6), to calculate L.

EXECUTE: $E_n = -(13.6 \text{ eV})/n^2$, so this state has $n = \sqrt{13.6/1.51} = 3$. In the Bohr model, $L = n\hbar$ so for this state $L = 3\hbar = 3.16 \times 10^{-34} \text{ kg} \cdot \text{m}^2/\text{s}$.

EVALUATE: We will find in Section 41.1 that the modern quantum mechanical description gives a different result.

39.27. **IDENTIFY:** The force between the electron and the nucleus in Be^{3+} is $F = \dfrac{1}{4\pi\epsilon_0}\dfrac{Ze^2}{r^2}$, where $Z = 4$ is the nuclear charge. All the equations for the hydrogen atom apply to Be^{3+} if we replace e^2 by Ze^2.

(a) **SET UP:** Modify Eq. (39.14).

EXECUTE: $E_n = -\dfrac{1}{\epsilon_0^2}\dfrac{me^4}{8n^2h^2}$ (hydrogen) becomes

$E_n = -\dfrac{1}{\epsilon_0^2}\dfrac{m(Ze^2)^2}{8n^2h^2} = Z^2\left(-\dfrac{1}{\epsilon_0^2}\dfrac{me^4}{8n^2h^2}\right) = Z^2\left(-\dfrac{13.60 \text{ eV}}{n^2}\right)$ (for Be^{3+})

The ground-level energy of Be^{3+} is $E_1 = 16\left(-\dfrac{13.60 \text{ eV}}{1^2}\right) = -218 \text{ eV}$.

EVALUATE: The ground-level energy of Be^{3+} is $Z^2 = 16$ times the ground-level energy of H.

(b) **SET UP:** The ionization energy is the energy difference between the $n \to \infty$ level energy and the $n = 1$ level energy.

EXECUTE: The $n \to \infty$ level energy is zero, so the ionization energy of Be^{3+} is 218 eV.

EVALUATE: This is 16 times the ionization energy of hydrogen.

(c) **SET UP:** $\dfrac{1}{\lambda} = R\left(\dfrac{1}{n_1^2} - \dfrac{1}{n_2^2}\right)$ just as for hydrogen but now R has a different value.

EXECUTE: $R_{\mathrm{H}} = \dfrac{me^4}{8\epsilon_0^2 h^3 c} = 1.097 \times 10^7 \text{ m}^{-1}$ for hydrogen becomes

$$R_{\mathrm{Be}} = Z^2 \dfrac{me^4}{8\epsilon_0^2 h^3 c} = 16(1.097 \times 10^7 \text{ m}^{-1}) = 1.755 \times 10^8 \text{ m}^{-1} \text{ for } \mathrm{Be}^{3+}.$$

For $n = 2$ to $n = 1$, $\dfrac{1}{\lambda} = R_{\mathrm{Be}}\left(\dfrac{1}{1^2} - \dfrac{1}{2^2}\right) = 3R_{\mathrm{Be}}/4.$

$\lambda = 4/(3R_{\mathrm{Be}}) = 4/(3(1.755 \times 10^8 \text{ m}^{-1})) = 7.60 \times 10^{-9} \text{ m} = 7.60 \text{ nm}.$

EVALUATE: This wavelength is smaller by a factor of 16 compared to the wavelength for the corresponding transition in the hydrogen atom.

(d) SET UP: Modify Eq. (39.8): $r_n = \epsilon_0 \dfrac{n^2 h^2}{\pi m e^2}$ (hydrogen).

EXECUTE: $r_n = \epsilon_0 \dfrac{n^2 h^2}{\pi m (Z e^2)}$ (Be^{3+}).

EVALUATE: For a given n the orbit radius for Be^{3+} is smaller by a factor of $Z = 4$ compared to the corresponding radius for hydrogen.

39.31.   IDENTIFY and SET UP: The wavelength of the photon is related to the transition energy $E_i - E_f$ of the atom by $E_i - E_f = \dfrac{hc}{\lambda}$ where $hc = 1.240 \times 10^{-6} \text{ eV} \cdot \text{m}.$

EXECUTE: (a) The minimum energy to ionize an atom is when the upper state in the transition has $E = 0$, so $E_1 = -17.50 \text{ eV}$. For $n = 5 \rightarrow n = 1$, $\lambda = 73.86 \text{ nm}$ and $E_5 - E_1 = \dfrac{1.240 \times 10^{-6} \text{ eV} \cdot \text{m}}{73.86 \times 10^{-9} \text{ m}} = 16.79 \text{ eV}.$

$E_5 = -17.50 \text{ eV} + 16.79 \text{ eV} = -0.71 \text{ eV}$. For $n = 4 \rightarrow n = 1$, $\lambda = 75.63 \text{ nm}$ and $E_4 = -1.10 \text{ eV}$. For $n = 3 \rightarrow n = 1$, $\lambda = 79.76 \text{ nm}$ and $E_3 = -1.95 \text{ eV}$. For $n = 2 \rightarrow n = 1$, $\lambda = 94.54 \text{ nm}$ and $E_2 = -4.38 \text{ eV}.$

(b) $E_i - E_f = E_4 - E_2 = -1.10 \text{ eV} - (-4.38 \text{ eV}) = 3.28 \text{ eV}$ and $\lambda = \dfrac{hc}{E_i - E_f} = \dfrac{1.240 \times 10^{-6} \text{ eV} \cdot \text{m}}{3.28 \text{ eV}} = 378 \text{ nm}$

EVALUATE: The $n = 4 \rightarrow n = 2$ transition energy is smaller than the $n = 4 \rightarrow n = 1$ transition energy so the wavelength is longer. In fact, this wavelength is longer than for any transition that ends in the $n = 1$ state.

39.33.   IDENTIFY: Apply conservation of energy to the system of atom and photon.

SET UP: The energy of a photon is $E_\gamma = \dfrac{hc}{\lambda}.$

EXECUTE: (a) $E_\gamma = \dfrac{hc}{\lambda} = \dfrac{(6.63 \times 10^{-34} \text{ J} \cdot \text{s})(3.00 \times 10^8 \text{ m/s})}{8.60 \times 10^{-7} \text{ m}} = 2.31 \times 10^{-19} \text{ J} = 1.44 \text{ eV}$. So the internal energy of the atom increases by 1.44 eV to $E = -6.52 \text{ eV} + 1.44 \text{ eV} = -5.08 \text{ eV}.$

(b) $E_\gamma = \dfrac{hc}{\lambda} = \dfrac{(6.63 \times 10^{-34} \text{ J} \cdot \text{s})(3.00 \times 10^8 \text{ m/s})}{4.20 \times 10^{-7} \text{ m}} = 4.74 \times 10^{-19} \text{ J} = 2.96 \text{ eV}$. So the final internal energy of the atom decreases to $E = -2.68 \text{ eV} - 2.96 \text{ eV} = -5.64 \text{ eV}.$

EVALUATE: When an atom absorbs a photon the energy of the atom increases. When an atom emits a photon the energy of the atom decreases.

39.35.   IDENTIFY: We know the power of the laser beam, so we know the energy per second that it delivers. The wavelength of the light tells us the energy of each photon, so we can use that to calculate the number of photons delivered per second.

SET UP: The energy of each photon is $E = hf = \dfrac{hc}{\lambda} = \dfrac{1.99 \times 10^{-25} \text{ J} \cdot \text{m}}{\lambda}$. The power is the total energy per second and the total energy E_{tot} is the number of photons N times the energy E of each photon.

EXECUTE: $\lambda = 10.6 \times 10^{-6}$ m, so $E = 1.88 \times 10^{-20}$ J. $P = \dfrac{E_{\text{tot}}}{t} = \dfrac{NE}{t}$ so

$$\frac{N}{t} = \frac{P}{E} = \frac{0.100 \times 10^3 \text{ W}}{1.88 \times 10^{-20} \text{ J}} = 5.32 \times 10^{21} \text{ photons/s.}$$

EVALUATE: At over 10^{21} photons per second, we can see why we do not detect individual photons.

39.39. **IDENTIFY:** Apply Eq. (39.18): $\dfrac{n_{5s}}{n_{3p}} = e^{-(E_{5s} - E_{3p})/kT}$

SET UP: $E_{5s} = 20.66$ eV and $E_{3p} = 18.70$ eV

EXECUTE: $E_{5s} - E_{3p} = 20.66$ eV $- 18.70$ eV $= 1.96$ eV$(1.602 \times 10^{-19}$ J/1 eV$) = 3.140 \times 10^{-19}$ J

(a) $\dfrac{n_{5s}}{n_{3p}} = e^{-(3.140 \times 10^{-19} \text{ J})/[(1.38 \times 10^{-23} \text{ J/K})(300 \text{ K})]} = e^{-75.79} = 1.2 \times 10^{-33}$

(b) $\dfrac{n_{5s}}{n_{3p}} = e^{-(3.140 \times 10^{-19} \text{ J})/[(1.38 \times 10^{-23} \text{ J/K})(600 \text{ K})]} = e^{-37.90} = 3.5 \times 10^{-17}$

(c) $\dfrac{n_{5s}}{n_{3p}} = e^{-(3.140 \times 10^{-19} \text{ J})/[(1.38 \times 10^{-23} \text{ J/K})(1200 \text{ K})]} = e^{-18.95} = 5.9 \times 10^{-9}$

(d) EVALUATE: At each of these temperatures the number of atoms in the $5s$ excited state, the initial state for the transition that emits 632.8 nm radiation, is quite small. The ratio increases as the temperature increases.

39.41. **IDENTIFY:** Energy radiates at the rate $H = Ae\sigma T^4$.
SET UP: The surface area of a cylinder of radius r and length l is $A = 2\pi rl$.

EXECUTE: **(a)** $T = \left(\dfrac{H}{Ae\sigma}\right)^{1/4} = \left(\dfrac{100 \text{ W}}{2\pi(0.20 \times 10^{-3} \text{ m})(0.30 \text{ m})(0.26)(5.671 \times 10^{-8} \text{ W/m}^2 \cdot \text{K}^4)}\right)^{1/4}.$

$T = 2.06 \times 10^3$ K.

(b) $\lambda_{\text{m}} T = 2.90 \times 10^{-3}$ m $\cdot$ K; $\lambda_{\text{m}} = 1410$ nm.

EVALUATE: **(c)** λ_{m} is in the infrared. The incandescent bulb is not a very efficient source of visible light because much of the emitted radiation is in the infrared.

39.45. **IDENTIFY:** Since the stars radiate as blackbodies, they obey the Stefan-Boltzmann law and Wien's displacement law.

SET UP: The Stefan-Boltzmann law says that the intensity of the radiation is $I = \sigma T^4$, so the total radiated power is $P = \sigma A T^4$. Wien's displacement law tells us that the peak-intensity wavelength is $\lambda_{\text{m}} = (\text{constant})/T$.

EXECUTE: **(a)** The hot and cool stars radiate the same total power, so the Stefan-Boltzmann law gives $\sigma A_{\text{h}} T_{\text{h}}^4 = \sigma A_{\text{c}} T_{\text{c}}^4 \Rightarrow 4\pi R_{\text{h}}^2 T_{\text{h}}^4 = 4\pi R_{\text{c}}^2 T_{\text{c}}^4 = 4\pi (3R_{\text{h}})^2 T_{\text{c}}^4 \Rightarrow T_{\text{h}}^4 = 9T^4 \Rightarrow T_{\text{h}} = T\sqrt{3} = 1.7T$, rounded to two significant digits.
(b) Using Wien's law, we take the ratio of the wavelengths, giving

$$\frac{\lambda_{\text{m}}(\text{hot})}{\lambda_{\text{m}}(\text{cool})} = \frac{T_{\text{c}}}{T_{\text{h}}} = \frac{T}{T\sqrt{3}} = \frac{1}{\sqrt{3}} = 0.58, \text{ rounded to two significant digits.}$$

EVALUATE: Although the hot star has only $1/9$ the surface area of the cool star, its absolute temperature has to be only 1.7 times as great to radiate the same amount of energy.

39.47. **IDENTIFY:** Apply the Wien displacement law to relate λ_{m} and T. Apply the Stefan-Boltzmann law to relate the power output of the star to its surface area and therefore to its radius.
SET UP: For a sphere $A = 4\pi r^2$. Since we assume a blackbody, $e = 1$.

EXECUTE: **(a)** Wien's law: $\lambda_m = \dfrac{k}{T}$. $\lambda_m = \dfrac{2.90 \times 10^{-3} \text{ K} \cdot \text{m}}{30,000 \text{ K}} = 9.7 \times 10^{-8} \text{ m} = 97 \text{ nm}$. This peak is in the

ultraviolet region, which is *not* visible. The star is blue because the largest part of the visible light radiated
is in the blue/violet part of the visible spectrum.

(b) $P = \sigma A T^4$ (Stefan-Boltzmann law)

$$(100,000)(3.86 \times 10^{26} \text{ W}) = \left(5.67 \times 10^{-8} \frac{\text{W}}{\text{m}^2 \text{K}^4}\right)(4\pi R^2)(30,000 \text{ K})^4$$

$$R = 8.2 \times 10^9 \text{ m}$$

$$R_{star}/R_{sun} = \frac{8.2 \times 10^9 \text{ m}}{6.96 \times 10^8 \text{ m}} = 12$$

EVALUATE: **(c)** The visual luminosity is proportional to the power radiated at visible wavelengths. Much
of the power is radiated nonvisible wavelengths, which does not contribute to the visible luminosity.

39.49.   **(a) IDENTIFY and SET UP:**   Use $\Delta x \Delta p_x \geq \hbar/2$ to calculate Δp_x and obtain Δv_x from this.

EXECUTE: $\Delta p_x \geq \dfrac{\hbar}{2\Delta x} = \dfrac{1.055 \times 10^{-34} \text{ J} \cdot \text{s}}{2(1.00 \times 10^{-6} \text{ m})} = 5.725 \times 10^{-29} \text{ kg} \cdot \text{m/s}.$

$\Delta v_x = \dfrac{\Delta p_x}{m} = \dfrac{5.275 \times 10^{-29} \text{ kg} \cdot \text{m/s}}{1200 \text{ kg}} = 4.40 \times 10^{-32} \text{ m/s}.$

(b) EVALUATE: Even for this very small Δx the minimum Δv_x required by the Heisenberg uncertainty
principle is very small. The uncertainty principle does not impose any practical limit on the simultaneous
measurements of the positions and velocities of ordinary objects.

39.53.   **IDENTIFY:**   Apply the Heisenberg Uncertainty Principle in the form $\Delta E \Delta t = \hbar/2$.

SET UP: Let $\Delta t = 5.2 \times 10^{-3}$ s, the lifetime of the state of the atom, and let ΔE be the uncertainty in the
energy of the state.

EXECUTE: $\Delta E > \dfrac{\hbar}{2\Delta t} = \dfrac{(1.055 \times 10^{-34} \text{ J} \cdot \text{s})}{2(5.2 \times 10^{-3} \text{ s})} = 1.01 \times 10^{-32} \text{ J} = 6.34 \times 10^{-14} \text{ eV}.$

EVALUATE: The uncertainty in the energy is a very small fraction of the typical energy of atomic states,
which is on the order of 1 eV.

39.55.   **(a) IDENTIFY and SET UP:**   Apply Eq. (39.17): $m_r = \dfrac{m_1 m_2}{m_1 + m_2} = \dfrac{207 m_e m_p}{207 m_e + m_p}$

EXECUTE: $m_r = \dfrac{207(9.109 \times 10^{-31} \text{ kg})(1.673 \times 10^{-27} \text{ kg})}{207(9.109 \times 10^{-31} \text{ kg}) + 1.673 \times 10^{-27} \text{ kg}} = 1.69 \times 10^{-28} \text{ kg}$

We have used m_e to denote the electron mass.

(b) IDENTIFY: In Eq. (39.14) replace $m = m_e$ by m_r: $E_n = -\dfrac{1}{\epsilon_0^2} \dfrac{m_r e^4}{8n^2 h^2}$.

SET UP: Write as $E_n = \left(\dfrac{m_r}{m_H}\right)\left(-\dfrac{1}{\epsilon_0^2} \dfrac{m_H e^4}{8n^2 h^2}\right)$, since we know that $\dfrac{1}{\epsilon_0^2} \dfrac{m_H e^4}{8 h^2} = 13.60 \text{ eV}$. Here m_H denotes

the reduced mass for the hydrogen atom; $m_H = 0.99946(9.109 \times 10^{-31} \text{ kg}) = 9.104 \times 10^{-31} \text{ kg}$.

EXECUTE: $E_n = \left(\dfrac{m_r}{m_H}\right)\left(-\dfrac{13.60 \text{ eV}}{n^2}\right)$

$E_1 = \dfrac{1.69 \times 10^{-28} \text{ kg}}{9.109 \times 10^{-31} \text{ kg}}(-13.60 \text{ eV}) = 186(-13.60 \text{ eV}) = -2.53 \text{ keV}$

(c) **Set Up:** From part (b), $E_n = \left(\dfrac{m_r}{m_H}\right)\left(-\dfrac{R_H ch}{n^2}\right)$, where $R_H = 1.097 \times 10^7 \text{ m}^{-1}$ is the Rydberg constant

for the hydrogen atom. Use this result in $\dfrac{hc}{\lambda} = E_i - E_f$ to find an expression for $1/\lambda$. The initial level for

the transition is the $n_i = 2$ level and the final level is the $n_f = 1$ level.

Execute: $\dfrac{hc}{\lambda} = \dfrac{m_r}{m_H}\left(-\dfrac{R_H ch}{n_i^2} - \left(-\dfrac{R_H ch}{n_f^2}\right)\right)$

$\dfrac{1}{\lambda} = \dfrac{m_r}{m_H} R_H \left(\dfrac{1}{n_f^2} - \dfrac{1}{n_i^2}\right)$

$\dfrac{1}{\lambda} = \dfrac{1.69 \times 10^{-28} \text{ kg}}{9.109 \times 10^{-31} \text{ kg}} (1.097 \times 10^7 \text{ m}^{-1})\left(\dfrac{1}{1^2} - \dfrac{1}{2^2}\right) = 1.527 \times 10^9 \text{ m}^{-1}$

$\lambda = 0.655 \text{ nm}$

Evaluate: From Example 39.6 the wavelength of the radiation emitted in this transition in hydrogen is

122 nm. The wavelength for muonium is $\dfrac{m_H}{m_r} = 5.39 \times 10^{-3}$ times this. The reduced mass for hydrogen is

very close to the electron mass because the electron mass is much less then the proton mass: $m_p/m_e = 1836$.

The muon mass is $207 m_e = 1.886 \times 10^{-28}$ kg. The proton is only about 10 times more massive than the

muon, so the reduced mass is somewhat smaller than the muon mass. The muon-proton atom has much

more strongly bound energy levels and much shorter wavelengths in its spectrum than for hydrogen.

39.57. **Identify and Set Up:** The H_α line in the Balmer series corresponds to the $n = 3$ to $n = 2$ transition.

$E_n = -\dfrac{13.6 \text{ eV}}{n^2}$. $\dfrac{hc}{\lambda} = \Delta E$.

Execute: **(a)** The atom must be given an amount of energy $E_3 - E_1 = -(13.6 \text{ eV})\left(\dfrac{1}{3^2} - \dfrac{1}{1^2}\right) = 12.1 \text{ eV}$.

(b) There are three possible transitions. $n = 3 \rightarrow n = 1$: $\Delta E = 12.1 \text{ eV}$ and $\lambda = \dfrac{hc}{\Delta E} = 103 \text{ nm}$;

$n = 3 \rightarrow n = 2$: $\Delta E = -(13.6 \text{ eV})\left(\dfrac{1}{3^2} - \dfrac{1}{2^2}\right) = 1.89 \text{ eV}$ and $\lambda = 657 \text{ nm}$; $n = 2 \rightarrow n = 1$:

$\Delta E = -(13.6 \text{ eV})\left(\dfrac{1}{2^2} - \dfrac{1}{1^2}\right) = 10.2 \text{ eV}$ and $\lambda = 122 \text{ nm}$.

Evaluate: The larger the transition energy for the atom, the shorter the wavelength.

39.59. **(a)** **Identify and Set Up:** The photon energy is given to the electron in the atom. Some of this energy
overcomes the binding energy of the atom and what is left appears as kinetic energy of the free electron.
Apply $hf = E_f - E_i$, the energy given to the electron in the atom when a photon is absorbed.

Execute: The energy of one photon is $\dfrac{hc}{\lambda} = \dfrac{(6.626 \times 10^{-34} \text{ J} \cdot \text{s})(2.998 \times 10^8 \text{ m/s})}{85.5 \times 10^{-9} \text{ m}}$

$\dfrac{hc}{\lambda} = 2.323 \times 10^{-18} \text{ J}(1 \text{ eV}/1.602 \times 10^{-19} \text{ J}) = 14.50 \text{ eV}$.

The final energy of the electron is $E_f = E_i + hf$. In the ground state of the hydrogen atom the energy of the

electron is $E_i = -13.60 \text{ eV}$. Thus $E_f = -13.60 \text{ eV} + 14.50 \text{ eV} = 0.90 \text{ eV}$.

(b) **Evaluate:** At thermal equilibrium a few atoms will be in the $n = 2$ excited levels, which have an

energy of $-13.6 \text{ eV}/4 = -3.40 \text{ eV}, 10.2 \text{ eV}$ greater than the energy of the ground state. If an electron with

$E = -3.40 \text{ eV}$ gains 14.5 eV from the absorbed photon, it will end up with $14.5 \text{ eV} - 3.4 \text{ eV} = 11.1 \text{ eV}$ of

kinetic energy.

39.63. **IDENTIFY:** The energy of the peak-intensity photons must be equal to the energy difference between the $n = 1$ and the $n = 4$ states. Wien's law allows us to calculate what the temperature of the blackbody must be for it to radiate with its peak intensity at this wavelength.

SET UP: In the Bohr model, the energy of an electron in shell n is $E_n = -\dfrac{13.6 \text{ eV}}{n^2}$, and Wien's

displacement law is $\lambda_m = \dfrac{2.90 \times 10^{-3} \text{ m} \cdot \text{K}}{T}$. The energy of a photon is $E = hf = hc/\lambda$.

EXECUTE: First find the energy (ΔE) that a photon would need to excite the atom. The ground state of the atom is $n = 1$ and the third excited state is $n = 4$. This energy is the *difference* between the two energy

levels. Therefore $\Delta E = (-13.6 \text{ eV})\left(\dfrac{1}{4^2} - \dfrac{1}{1^2}\right) = 12.8 \text{ eV}$. Now find the wavelength of the photon having

this amount of energy. $hc/\lambda = 12.8 \text{ eV}$ and

$$\lambda = (4.136 \times 10^{-15} \text{ eV} \cdot \text{s})(3.00 \times 10^8 \text{ m/s})/(12.8 \text{ eV}) = 9.73 \times 10^{-8} \text{ m}$$

Now use Wien's law to find the temperature. $T = (0.00290 \text{ m} \cdot \text{K})/(9.73 \times 10^{-8} \text{ m}) = 2.98 \times 10^4 \text{ K}$.

EVALUATE: This temperature is well above ordinary room temperatures, which is why hydrogen atoms are not in excited states during everyday conditions.

39.65. **IDENTIFY:** Apply conservation of energy and conservation of linear momentum to the system of atom plus photon.

(a) SET UP: Let E_{tr} be the transition energy, E_{ph} be the energy of the photon with wavelength λ', and E_r be the kinetic energy of the recoiling atom. Conservation of energy gives $E_{ph} + E_r = E_{tr}$.

$E_{ph} = \dfrac{hc}{\lambda'}$ so $\dfrac{hc}{\lambda'} = E_{tr} - E_r$ and $\lambda' = \dfrac{hc}{E_{tr} - E_r}$.

EXECUTE: If the recoil energy is neglected then the photon wavelength is $\lambda = hc/E_{tr}$.

$$\Delta\lambda = \lambda' - \lambda = hc\left(\dfrac{1}{E_{tr} - E_r} - \dfrac{1}{E_{tr}}\right) = \left(\dfrac{hc}{E_{tr}}\right)\left(\dfrac{1}{1 - E_r/E_{tr}} - 1\right)$$

$$\dfrac{1}{1 - E_r/E_{tr}} = \left(1 - \dfrac{E_r}{E_{tr}}\right)^{-1} \approx 1 + \dfrac{E_r}{E_{tr}} \text{ since } \dfrac{E_r}{E_{tr}} \ll 1$$

(We have used the binomial theorem, Appendix B.)

Thus $\Delta\lambda = \dfrac{hc}{E_{tr}}\left(\dfrac{E_r}{E_{tr}}\right)$, or since $E_{tr} = hc/\lambda$, $\Delta\lambda = \left(\dfrac{E_r}{hc}\right)\lambda^2$.

SET UP: Use conservation of linear momentum to find E_r: Assuming that the atom is initially at rest, the momentum p_r of the recoiling atom must be equal in magnitude and opposite in direction to the momentum $p_{ph} = h/\lambda$ of the emitted photon: $h/\lambda = p_r$.

EXECUTE: $E_r = \dfrac{p_r^2}{2m}$, where m is the mass of the atom, so $E_r = \dfrac{h^2}{2m\lambda^2}$.

Use this result in the above equation: $\Delta\lambda = \left(\dfrac{E_r}{hc}\right)\lambda^2 = \left(\dfrac{h^2}{2m\lambda^2}\right)\left(\dfrac{\lambda^2}{hc}\right) = \dfrac{h}{2mc}$;

note that this result for $\Delta\lambda$ is independent of the atomic transition energy.

(b) For a hydrogen atom $m = m_p$ and $\Delta\lambda = \dfrac{h}{2m_p c} = \dfrac{6.626 \times 10^{-34} \text{ J} \cdot \text{s}}{2(1.673 \times 10^{-27} \text{ kg})(2.998 \times 10^8 \text{ m/s})} = 6.61 \times 10^{-16} \text{ m}$

EVALUATE: The correction is independent of n. The wavelengths of photons emitted in hydrogen atom transitions are on the order of $100 \text{ nm} = 10^{-7} \text{ m}$, so the recoil correction is exceedingly small.

39.71. **IDENTIFY:** The electrons behave like waves and produce a double-slit interference pattern after passing through the slits.

SET UP: The first angle at which destructive interference occurs is given by $d \sin\theta = \lambda/2$. The de Broglie wavelength of each of the electrons is $\lambda = h/mv$.

EXECUTE: **(a)** First find the wavelength of the electrons. For the first dark fringe, we have $d \sin\theta = \lambda/2$, which gives $(1.25 \text{ nm})(\sin 18.0°) = \lambda/2$, and $\lambda = 0.7725 \text{ nm}$. Now solve the de Broglie wavelength equation for the speed of the electron:

$$v = \frac{h}{m\lambda} = \frac{6.626 \times 10^{-34} \text{ J} \cdot \text{s}}{(9.11 \times 10^{-31} \text{ kg})(0.7725 \times 10^{-9} \text{ m})} = 9.42 \times 10^5 \text{ m/s}$$

which is about 0.3% the speed of light, so they are *nonrelativistic*.

(b) Energy conservation gives $eV = \frac{1}{2}mv^2$ and

$$V = mv^2/2e = (9.11 \times 10^{-31} \text{ kg})(9.42 \times 10^5 \text{ m/s})^2/[2(1.60 \times 10^{-19} \text{ C})] = 2.52 \text{ V}$$

EVALUATE: The hole must be much smaller than the wavelength of visible light for the electrons to show diffraction.

39.73. **IDENTIFY:** Both the electrons and photons behave like waves and exhibit single-slit diffraction after passing through their respective slits.

SET UP: The energy of the photon is $E = hc/\lambda$ and the de Broglie wavelength of the electron is $\lambda = h/mv = h/p$. Destructive interference for a single slit first occurs when $a \sin\theta = \lambda$.

EXECUTE: **(a)** For the photon: $\lambda = hc/E$ and $a \sin\theta = \lambda$. Since the a and θ are the same for the photons and electrons, they must both have the same wavelength. Equating these two expressions for λ gives $a \sin\theta = hc/E$. For the electron, $\lambda = h/p = \dfrac{h}{\sqrt{2mK}}$ and $a \sin\theta = \lambda$. Equating these two expressions for λ gives $a \sin\theta = \dfrac{h}{\sqrt{2mK}}$. Equating the two expressions for $a\sin\theta$ gives $hc/E = \dfrac{h}{\sqrt{2mK}}$, which gives $E = c\sqrt{2mK} = (4.05 \times 10^{-7} \text{ J}^{1/2})\sqrt{K}$.

(b) $\dfrac{E}{K} = \dfrac{c\sqrt{2mK}}{K} = \sqrt{\dfrac{2mc^2}{K}}$. Since $v \ll c$, $mc^2 > K$, so the square root is > 1. Therefore $E/K > 1$, meaning that the photon has more energy than the electron.

EVALUATE: When a photon and a particle have the same wavelength, the photon has more energy than the particle.

39.79. **IDENTIFY and SET UP:** Combining the two equations in the hint gives $pc = \sqrt{K(K + 2mc^2)}$ and

$$\lambda = \frac{hc}{\sqrt{K(K + 2mc^2)}}.$$

EXECUTE: **(a)** With $K = 3mc^2$ this becomes $\lambda = \dfrac{hc}{\sqrt{3mc^2(3mc^2 + 2mc^2)}} = \dfrac{h}{\sqrt{15}mc}$.

(b) (i) $K = 3mc^2 = 3(9.109 \times 10^{-31} \text{ kg})(2.998 \times 10^8 \text{ m/s})^2 = 2.456 \times 10^{13} \text{ J} = 1.53 \text{ MeV}$

$$\lambda = \frac{h}{\sqrt{15}mc} = \frac{6.626 \times 10^{-34} \text{ J} \cdot \text{s}}{\sqrt{15}(9.109 \times 10^{-31} \text{ kg})(2.998 \times 10^8 \text{ m/s})} = 6.26 \times 10^{-13} \text{ m}$$

(ii) K is proportional to m, so for a proton $K = (m_p/m_e)(1.53 \text{ MeV}) = 1836(1.53 \text{ MeV}) = 2810 \text{ MeV}$

λ is proportional to $1/m$, so for a proton

$$\lambda = (m_e/m_p)(6.26 \times 10^{-13} \text{ m}) = (1/1836)(6.26 \times 10^{-13} \text{ m}) = 3.41 \times 10^{-16} \text{ m}.$$

EVALUATE: The proton has a larger rest mass energy so its kinetic energy is larger when $K = 3mc^2$. The proton also has larger momentum so has a smaller λ.

39.81.   **(a) IDENTIFY and SET UP:** $\Delta x \Delta p_x \geq \hbar/2$. Estimate Δx as $\Delta x \approx 5.0 \times 10^{-15}$ m.

EXECUTE: Then the minimum allowed Δp_x is $\Delta p_x \approx \dfrac{\hbar}{2\Delta x} = \dfrac{1.055 \times 10^{-34} \text{ J} \cdot \text{s}}{2(5.0 \times 10^{-15} \text{ m})} = 1.1 \times 10^{-20}$ kg $\cdot$ m/s.

(b) IDENTIFY and SET UP: Assume $p \approx 1.1 \times 10^{-20}$ kg $\cdot$ m/s. Use Eq. (37.39) to calculate E, and then $K = E - mc^2$.

EXECUTE: $E = \sqrt{(mc^2)^2 + (pc)^2}$. $mc^2 = (9.109 \times 10^{-31} \text{ kg})(2.998 \times 10^8 \text{ m/s})^2 = 8.187 \times 10^{-14}$ J.

$pc = (1.1 \times 10^{-20} \text{ kg} \cdot \text{m/s})(2.998 \times 10^8 \text{ m/s}) = 3.165 \times 10^{-12}$ J.

$E = \sqrt{(8.187 \times 10^{-14} \text{ J})^2 + (3.165 \times 10^{-12} \text{ J})^2} = 3.166 \times 10^{-12}$ J.

$K = E - mc^2 = 3.166 \times 10^{-12} \text{ J} - 8.187 \times 10^{-14} \text{ J} = 3.084 \times 10^{-12} \text{ J} \times (1 \text{ eV}/1.602 \times 10^{-19} \text{ J}) = 19$ MeV.

(c) IDENTIFY and SET UP: The Coulomb potential energy for a pair of point charges is given by Eq. (23.9). The proton has charge $+e$ and the electron has charge $-e$.

EXECUTE: $U = -\dfrac{ke^2}{r} = -\dfrac{(8.988 \times 10^9 \text{ N} \cdot \text{m}^2/\text{C}^2)(1.602 \times 10^{-19} \text{ C})^2}{5.0 \times 10^{-15} \text{ m}} = -4.6 \times 10^{-14} \text{ J} = -0.29$ MeV.

EVALUATE: The kinetic energy of the electron required by the uncertainty principle would be much larger than the magnitude of the negative Coulomb potential energy. The total energy of the electron would be large and positive and the electron could not be bound within the nucleus.

39.87.   **IDENTIFY:**   The electrons behave as waves whose wavelength is equal to the de Broglie wavelength.

SET UP: The de Broglie wavelength is $\lambda = h/mv$, and the energy of a photon is $E = hf = hc/\lambda$.

EXECUTE: **(a)** Use the de Broglie wavelength to find the speed of the electron.

$$v = \frac{h}{m\lambda} = \frac{6.626 \times 10^{-34} \text{ J} \cdot \text{s}}{(9.11 \times 10^{-31} \text{ kg})(1.00 \times 10^{-9} \text{ m})} = 7.27 \times 10^5 \text{ m/s}$$

which is much less than the speed of light, so it is nonrelativistic.

(b) Energy conservation gives $eV = \frac{1}{2}mv^2$.

$V = mv^2/2e = (9.11 \times 10^{-31} \text{ kg})(7.27 \times 10^5 \text{ m/s})^2/[2(1.60 \times 10^{-19} \text{ C})] = 1.51$ V

(c) $K = eV = e(1.51 \text{ V}) = 1.51$ eV, which is about ¼ the potential energy of the NaCl molecule, so the electron would not be too damaging.

(d) $E = hc/\lambda = (4.136 \times 10^{-15} \text{ eV s})(3.00 \times 10^8 \text{ m/s})/(1.00 \times 10^{-9} \text{ m}) = 1240$ eV

which would certainly destroy the molecules under study.

EVALUATE: As we have seen in Problems 39.73 and 39.76, when a particle and a photon have the same wavelength, the photon has much more energy.

39.91.   **IDENTIFY:**   The wave (light or electron matter wave) having less energy will cause less damage to the virus.

SET UP: For a photon $E_{ph} = \dfrac{hc}{\lambda} = \dfrac{1.24 \times 10^{-6} \text{ eV} \cdot \text{m}}{\lambda}$. For an electron $E_e = \dfrac{p^2}{2m} = \dfrac{h^2}{2m\lambda^2}$.

EXECUTE: **(a)** $E = \dfrac{hc}{\lambda} = \dfrac{1.24 \times 10^{-6} \text{ eV} \cdot \text{m}}{5.00 \times 10^{-9} \text{ m}} = 248$ eV.

(b) $E_e = \dfrac{h^2}{2m\lambda^2} = \dfrac{(6.63 \times 10^{-34} \text{ J} \cdot \text{s})^2}{2(9.11 \times 10^{-31} \text{ kg})(5.00 \times 10^{-9} \text{ m})^2} = 9.65 \times 10^{-21} \text{ J} = 0.0603$ eV.

EVALUATE: The electron has much less energy than a photon of the same wavelength and therefore would cause much less damage to the virus.

40

QUANTUM MECHANICS

40.3. **IDENTIFY:** Use the wave function from Example 40.1.

SET UP: $|\Psi(x,t)|^2 = 2|A|^2 \{1 + \cos[(k_2 - k_1)x - (\omega_2 - \omega_1)t]\}$. $k_2 = 3k_1 = 3k$. $\omega = \dfrac{\hbar k^2}{2m}$, so $\omega_2 = 9\omega_1 = 9\omega$.

$|\Psi(x,t)|^2 = 2|A|^2 \{1 + \cos(2kx - 8\omega t)\}$.

EXECUTE: **(a)** At $t = 2\pi/\omega$, $|\Psi(x,t)|^2 = 2|A|^2 \{1 + \cos(2kx - 16\pi)\}$. $|\Psi(x,t)|^2$ is maximum for

$\cos(2kx - 16\pi) = 1$. This happens for $2kx - 16\pi = 0, 2\pi, \ldots$. Smallest positive x where $|\Psi(x,t)|^2$ is a

maximum is $x = \dfrac{8\pi}{k}$.

(b) From the result of part (a), $v_{av} = \dfrac{8\pi/k}{2\pi/\omega} = \dfrac{4\omega}{k}$. $v_{av} = \dfrac{\omega_2 - \omega_1}{k_2 - k_1} = \dfrac{8\omega}{2k} = \dfrac{4\omega}{k}$.

EVALUATE: The two expressions agree.

40.5. **IDENTIFY and SET UP:** $\psi(x) = A\sin kx$. The position probability density is given by $|\psi(x)|^2 = A^2 \sin^2 kx$.

EXECUTE: **(a)** The probability is highest where $\sin kx = 1$ so $kx = 2\pi x/\lambda = n\pi/2$, $n = 1, 3, 5, \ldots$

$x = n\lambda/4$, $n = 1, 3, 5, \ldots$ so $x = \lambda/4, 3\lambda/4, 5\lambda/4, \ldots$

(b) The probability of finding the particle is zero where $|\psi|^2 = 0$, which occurs where $\sin kx = 0$ and

$kx = 2\pi x/\lambda = n\pi$, $n = 0, 1, 2, \ldots$

$x = n\lambda/2$, $n = 0, 1, 2, \ldots$ so $x = 0, \lambda/2, \lambda, 3\lambda/2, \ldots$

EVALUATE: The situation is analogous to a standing wave, with the probability analogous to the square of the amplitude of the standing wave.

40.11. **IDENTIFY and SET UP:** The energy levels for a particle in a box are given by $E_n = \dfrac{n^2 h^2}{8mL^2}$.

EXECUTE: **(a)** The lowest level is for $n = 1$, and $E_1 = \dfrac{(1)(6.626 \times 10^{-34}\ \text{J} \cdot \text{s})^2}{8(0.20\ \text{kg})(1.3\ \text{m})^2} = 1.6 \times 10^{-67}$ J.

(b) $E = \dfrac{1}{2}mv^2$ so $v = \sqrt{\dfrac{2E}{m}} = \sqrt{\dfrac{2(1.2 \times 10^{-67}\ \text{J})}{0.20\ \text{kg}}} = 1.3 \times 10^{-33}$ m/s. If the ball has this speed the time it

would take it to travel from one side of the table to the other is

$t = \dfrac{1.3\ \text{m}}{1.3 \times 10^{-33}\ \text{m/s}} = 1.0 \times 10^{33}$ s.

(c) $E_1 = \dfrac{h^2}{8mL^2}$, $E_2 = 4E_1$, so $\Delta E = E_2 - E_1 = 3E_1 = 3(1.6 \times 10^{-67}\ \text{J}) = 4.9 \times 10^{-67}$ J.

(d) EVALUATE: No, quantum mechanical effects are not important for the game of billiards. The discrete, quantized nature of the energy levels is completely unobservable.

40.13. **IDENTIFY:** An electron in the lowest energy state in this box must have the same energy as it would in the ground state of hydrogen.

SET UP: The energy of the nth level of an electron in a box is $E_n = \dfrac{nh^2}{8mL^2}$.

EXECUTE: An electron in the ground state of hydrogen has an energy of -13.6 eV, so find the width corresponding to an energy of $E_1 = 13.6$ eV. Solving for L gives

$$L = \frac{h}{\sqrt{8mE_1}} = \frac{(6.626 \times 10^{-34} \text{ J} \cdot \text{s})}{\sqrt{8(9.11 \times 10^{-31} \text{ kg})(13.6 \text{ eV})(1.602 \times 10^{-19} \text{ J/eV})}} = 1.66 \times 10^{-10} \text{ m}.$$

EVALUATE: This width is of the same order of magnitude as the diameter of a Bohr atom with the electron in the K shell.

40.15. **IDENTIFY and SET UP:** Eq. (40.31) gives the energy levels. Use this to obtain an expression for $E_2 - E_1$ and use the value given for this energy difference to solve for L.

EXECUTE: Ground state energy is $E_1 = \dfrac{h^2}{8mL^2}$; first excited state energy is $E_2 = \dfrac{4h^2}{8mL^2}$. The energy

separation between these two levels is $\Delta E = E_2 - E_1 = \dfrac{3h^2}{8mL^2}$. This gives $L = h\sqrt{\dfrac{3}{8m\Delta E}} =$

$$L = 6.626 \times 10^{-34} \text{ J} \cdot \text{s} \sqrt{\frac{3}{8(9.109 \times 10^{-31} \text{ kg})(3.0 \text{ eV})(1.602 \times 10^{-19} \text{ J/1 eV})}} = 6.1 \times 10^{-10} \text{ m} = 0.61 \text{ nm}.$$

EVALUATE: This energy difference is typical for an atom and L is comparable to the size of an atom.

40.19. **IDENTIFY and SET UP:** For the $n = 2$ first excited state the normalized wave function is given by

Eq. (40.35). $\psi_2(x) = \sqrt{\dfrac{2}{L}} \sin\left(\dfrac{2\pi x}{L}\right)$. $|\psi_2(x)|^2 dx = \dfrac{2}{L} \sin^2\left(\dfrac{2\pi x}{L}\right) dx$. Examine $|\psi_2(x)|^2 dx$ and find where it is zero and where it is maximum.

EXECUTE: **(a)** $|\psi_2|^2 dx = 0$ implies $\sin\left(\dfrac{2\pi x}{L}\right) = 0$

$\dfrac{2\pi x}{L} = m\pi$, $m = 0, 1, 2, \ldots$; $x = m(L/2)$

For $m = 0$, $x = 0$; for $m = 1$, $x = L/2$; for $m = 2$, $x = L$

The probability of finding the particle is zero at $x = 0$, $L/2$, and L.

(b) $|\psi_2|^2 dx$ is maximum when $\sin\left(\dfrac{2\pi x}{L}\right) = \pm 1$

$\dfrac{2\pi x}{L} = m(\pi/2)$, $m = 1, 3, 5, \ldots$; $x = m(L/4)$

For $m = 1$, $x = L/4$; for $m = 3$, $x = 3L/4$

The probability of finding the particle is largest at $x = L/4$ and $3L/4$.

(c) EVALUATE: The answers to part (a) correspond to the zeros of $|\psi|^2$ shown in Figure 40.12 in the textbook and the answers to part (b) correspond to the two values of x where $|\psi|^2$ in the figure is maximum.

40.23. **IDENTIFY and SET UP:** $\lambda = \dfrac{h}{p} = \dfrac{h}{\sqrt{2mE}}$. The energy of the electron in level n is given by Eq. (40.31).

EXECUTE: **(a)** $E_1 = \dfrac{h^2}{8mL^2} \Rightarrow \lambda_1 = \dfrac{h}{\sqrt{2mh^2/8mL^2}} = 2L = 2(3.0 \times 10^{-10} \text{ m}) = 6.0 \times 10^{-10} \text{ m}$. The wavelength

is twice the width of the box. $p_1 = \dfrac{h}{\lambda_1} = \dfrac{(6.63 \times 10^{-34} \text{ J} \cdot \text{s})}{6.0 \times 10^{-10} \text{ m}} = 1.1 \times 10^{-24} \text{ kg} \cdot \text{m/s}$.

(b) $E_2 = \dfrac{4h^2}{8mL^2} \Rightarrow \lambda_2 = L = 3.0 \times 10^{-10}$ m. The wavelength is the same as the width of the box.

$p_2 = \dfrac{h}{\lambda_2} = 2p_1 = 2.2 \times 10^{-24}$ kg·m/s.

(c) $E_3 = \dfrac{9h^2}{8mL^2} \Rightarrow \lambda_3 = \dfrac{2}{3}L = 2.0 \times 10^{-10}$ m. The wavelength is two-thirds the width of the box.

$p_3 = 3p_1 = 3.3 \times 10^{-24}$ kg·m/s.

EVALUATE: In each case the wavelength is an integer multiple of $\lambda/2$. In the nth state, $p_n = np_1$.

40.27.   **IDENTIFY:** Figure 40.15b in the textbook gives values for the bound state energy of a square well for which $U_0 = 6E_{\text{1-1DW}}$.

SET UP: $E_{\text{1-1DW}} = \dfrac{\pi^2 \hbar^2}{2mL^2}$.

EXECUTE: $E_1 = 0.625E_{\text{1-1DW}} = 0.625\dfrac{\pi^2 \hbar^2}{2mL^2}$; $E_1 = 2.00$ eV $= 3.20 \times 10^{-19}$ J.

$L = \pi\hbar\left(\dfrac{0.625}{2(9.109 \times 10^{-31} \text{ kg})(3.20 \times 10^{-19} \text{ J})}\right)^{1/2} = 3.43 \times 10^{-10}$ m.

EVALUATE: As L increases the ground state energy decreases.

40.31.   **IDENTIFY:** Find the transition energy ΔE and set it equal to the energy of the absorbed photon. Use $E = hc/\lambda$, to find the wavelength of the photon.

SET UP: $U_0 = 6E_{\text{1-1DW}}$, as in Figure 40.15 in the textbook, so $E_1 = 0.625E_{\text{1-1DW}}$ and $E_3 = 5.09E_{\text{1-1DW}}$

with $E_{\text{1-1DW}} = \dfrac{\pi^2 \hbar^2}{2mL^2}$. In this problem the particle bound in the well is a proton, so $m = 1.673 \times 10^{-27}$ kg.

EXECUTE: $E_{\text{1-1DW}} = \dfrac{\pi^2 \hbar^2}{2mL^2} = \dfrac{\pi^2 (1.055 \times 10^{-34} \text{ J·s})^2}{2(1.673 \times 10^{-27} \text{ kg})(4.0 \times 10^{-15} \text{ m})^2} = 2.052 \times 10^{-12}$ J. The transition energy

is $\Delta E = E_3 - E_1 = (5.09 - 0.625)E_{\text{1-1DW}} = 4.465E_{\text{1-1DW}}$. $\Delta E = 4.465(2.052 \times 10^{-12} \text{ J}) = 9.162 \times 10^{-12}$ J

The wavelength of the photon that is absorbed is related to the transition energy by $\Delta E = hc/\lambda$, so

$\lambda = \dfrac{hc}{\Delta E} = \dfrac{(6.626 \times 10^{-34} \text{ J·s})(2.998 \times 10^8 \text{ m/s})}{9.162 \times 10^{-12} \text{ J}} = 2.2 \times 10^{-14}$ m $= 22$ fm.

EVALUATE: The wavelength of the photon is comparable to the size of the box.

40.33.   **IDENTIFY:** The tunneling probability is $T = 16\dfrac{E}{U_0}\left(1 - \dfrac{E}{U_0}\right)e^{-2L\sqrt{2m(U_0 - E)}/\hbar}$.

SET UP: $\dfrac{E}{U_0} = \dfrac{6.0 \text{ eV}}{11.0 \text{ eV}}$ and $E - U_0 = 5$ eV $= 8.0 \times 10^{-19}$ J.

EXECUTE: **(a)** $L = 0.80 \times 10^{-9}$ m:

$T = 16\left(\dfrac{6.0 \text{ eV}}{11.0 \text{ eV}}\right)\left(1 - \dfrac{6.0 \text{ ev}}{11.0 \text{ eV}}\right)e^{-2(0.80 \times 10^{-9} \text{ m})\sqrt{2(9.11 \times 10^{-31} \text{ kg})(8.0 \times 10^{-19} \text{ J})}/1.055 \times 10^{-34} \text{ J·s}} = 4.4 \times 10^{-8}$.

(b) $L = 0.40 \times 10^{-9}$ m: $T = 4.2 \times 10^{-4}$.

EVALUATE: The tunneling probability is less when the barrier is wider.

40.35.   **IDENTIFY and SET UP:** Use Eq. (39.1), where $K = p^2/2m$ and $E = K + U$.

EXECUTE: $\lambda = h/p = h/\sqrt{2mK}$, so $\lambda\sqrt{K}$ is constant. $\lambda_1\sqrt{K_1} = \lambda_2\sqrt{K_2}$; λ_1 and K_1 are for $x > L$ where $K_1 = 2U_0$ and λ_2 and K_2 are for $0 < x < L$ where $K_2 = E - U_0 = U_0$.

$$\frac{\lambda_1}{\lambda_2} = \sqrt{\frac{K_2}{K_1}} = \sqrt{\frac{U_0}{2U_0}} = \frac{1}{\sqrt{2}}$$

EVALUATE: When the particle is passing over the barrier its kinetic energy is less and its wavelength is larger.

40.37. **IDENTIFY and SET UP:** The probability is $T = Ae^{-2\kappa L}$, with $A = 16\frac{E}{U_0}\left(1 - \frac{E}{U_0}\right)$ and $\kappa = \frac{\sqrt{2m(U_0 - E)}}{\hbar}$.

$E = 32$ eV, $U_0 = 41$ eV, $L = 0.25 \times 10^{-9}$ m. Calculate T.

EXECUTE: **(a)** $A = 16\frac{E}{U_0}\left(1 - \frac{E}{U_0}\right) = 16\frac{32}{41}\left(1 - \frac{32}{41}\right) = 2.741$.

$$\kappa = \frac{\sqrt{2m(U_0 - E)}}{\hbar}$$

$$\kappa = \frac{\sqrt{2(9.109 \times 10^{-31} \text{ kg})(41 \text{ eV} - 32 \text{ eV})(1.602 \times 10^{-19} \text{ J/eV})}}{1.055 \times 10^{-34} \text{ J} \cdot \text{s}} = 1.536 \times 10^{10} \text{ m}^{-1}$$

$T = Ae^{-2\kappa L} = (2.741)e^{-2(1.536 \times 10^{10} \text{ m}^{-1})(0.25 \times 10^{-9} \text{ m})} = 2.741e^{-7.68} = 0.0013$

(b) The only change in the mass m, which appears in κ.

$$\kappa = \frac{\sqrt{2m(U_0 - E)}}{\hbar}$$

$$\kappa = \frac{\sqrt{2(1.673 \times 10^{-27} \text{ kg})(41 \text{ eV} - 32 \text{ eV})(1.602 \times 10^{-19} \text{ J/eV})}}{1.055 \times 10^{-34} \text{ J} \cdot \text{s}} = 6.584 \times 10^{11} \text{ m}^{-1}$$

Then $T = Ae^{-2\kappa L} = (2.741)e^{-2(6.584 \times 10^{11} \text{ m}^{-1})(0.25 \times 10^{-9} \text{ m})} = 2.741e^{-392.2} = 10^{-143}$

EVALUATE: The more massive proton has a much smaller probability of tunneling than the electron does.

40.39. **IDENTIFY and SET UP:** The energy levels are given by Eq. (40.46), where $\omega = \sqrt{\frac{k'}{m}}$.

EXECUTE: $\omega = \sqrt{\frac{k'}{m}} = \sqrt{\frac{110 \text{ N/m}}{0.250 \text{ kg}}} = 21.0$ rad/s

The ground state energy is given by Eq. (40.46):

$E_0 = \frac{1}{2}\hbar\omega = \frac{1}{2}(1.055 \times 10^{-34} \text{ J} \cdot \text{s})(21.0 \text{ rad/s}) = 1.11 \times 10^{-33} \text{ J}(1 \text{ eV}/1.602 \times 10^{-19} \text{ J}) = 6.93 \times 10^{-15}$ eV

$E_n = \left(n + \frac{1}{2}\right)\hbar\omega, \quad E_{(n+1)} = \left(n + 1 + \frac{1}{2}\right)\hbar\omega$

The energy separation between these adjacent levels is

$\Delta E = E_{n+1} - E_n = \hbar\omega = 2E_0 = 2(1.11 \times 10^{-33} \text{ J}) = 2.22 \times 10^{-33} \text{ J} = 1.39 \times 10^{-14}$ eV.

EVALUATE: These energies are extremely small; quantum effects are not important for this oscillator.

40.41. **IDENTIFY:** We can model the molecule as a harmonic oscillator. The energy of the photon is equal to the energy difference between the two levels of the oscillator.

SET UP: The energy of a photon is $E_\gamma = hf = hc/\lambda$, and the energy levels of a harmonic oscillator are

given by $E_n = \left(n + \frac{1}{2}\right)\hbar\sqrt{\frac{k'}{m}} = \left(n + \frac{1}{2}\right)\hbar\omega$.

EXECUTE: **(a)** The photon's energy is $E_\gamma = \frac{hc}{\lambda} = \frac{(6.63 \times 10^{-34} \text{ J} \cdot \text{s})(3.00 \times 10^8 \text{ m/s})}{5.8 \times 10^{-6} \text{ m}} = 0.21$ eV.

(b) The transition energy is $\Delta E = E_{n+1} - E_n = \hbar\omega = \hbar\sqrt{\frac{k'}{m}}$, which gives $\frac{2\pi\hbar c}{\lambda} = \hbar\sqrt{\frac{k'}{m}}$. Solving for k',

we get $k' = \frac{4\pi^2 c^2 m}{\lambda^2} = \frac{4\pi^2(3.00 \times 10^8 \text{ m/s})^2(5.6 \times 10^{-26} \text{ kg})}{(5.8 \times 10^{-6} \text{ m})^2} = 5,900$ N/m.

EVALUATE: This would be a rather strong spring in the physics lab.

40.45.   **IDENTIFY:**   We model the atomic vibration in the crystal as a harmonic oscillator.

SET UP: The energy levels of a harmonic oscillator are given by $E_n = \left(n + \dfrac{1}{2}\right)\hbar\sqrt{\dfrac{k'}{m}} = \left(n + \dfrac{1}{2}\right)\hbar\omega$.

EXECUTE: **(a)** The ground state energy of a simple harmonic oscillator is

$$E_0 = \frac{1}{2}\hbar\omega = \frac{1}{2}\hbar\sqrt{\frac{k'}{m}} = \frac{(1.055\times10^{-34}\ \text{J}\cdot\text{s})}{2}\sqrt{\frac{12.2\ \text{N/m}}{3.82\times10^{-26}\ \text{kg}}} = 9.43\times10^{-22}\ \text{J} = 5.89\times10^{-3}\ \text{eV}$$

(b) $E_4 - E_3 = \hbar\omega = 2E_0 = 0.0118$ eV, so $\lambda = \dfrac{hc}{E} = \dfrac{(6.63\times10^{-34}\ \text{J}\cdot\text{s})(3.00\times10^8\ \text{m/s})}{1.88\times10^{-21}\ \text{J}} = 106\ \mu\text{m}$

(c) $E_{n+1} - E_n = \hbar\omega = 2E_0 = 0.0118$ eV

EVALUATE: These energy differences are much smaller than those due to electron transitions in the hydrogen atom.

40.49.   **IDENTIFY:**   Evaluate $\psi(x) = \displaystyle\int_0^\infty B(k)\cos kx\ dk$ for the function $B(k)$ specified in the problem.

SET UP: $\displaystyle\int \cos kx\ dk = \frac{1}{x}\sin kx$.

EXECUTE: **(a)** $\psi(x) = \displaystyle\int_0^\infty B(k)\cos kx\,dk = \int_0^{k_0}\left(\frac{1}{k_0}\right)\cos kx\,dk = \left.\frac{\sin kx}{k_0 x}\right|_0^{k_0} = \frac{\sin k_0 x}{k_0 x}$

(b) $\psi(x)$ has a maximum value at the origin $x = 0$. $\psi(x_0) = 0$ when $k_0 x_0 = \pi$ so $x_0 = \dfrac{\pi}{k_0}$. Thus the width of

this function $w_x = 2x_0 = \dfrac{2\pi}{k_0}$. If $k_0 = \dfrac{2\pi}{L}$, $w_x = L$. $B(k)$ versus k is graphed in Figure 40.49a. The graph of $\psi(x)$ versus x is in Figure 40.49b.

(c) If $k_0 = \dfrac{\pi}{L}$, $w_x = 2L$.

EVALUATE: **(d)** $w_p w_x = \left(\dfrac{hw_k}{2\pi}\right)\left(\dfrac{2\pi}{k_0}\right) = \dfrac{hw_k}{k_0} = \dfrac{hk_0}{k_0} = h$. If $\Delta x = w_x$ and $\Delta p_x = w_p$, then the uncertainty

principle states that $w_p w_x \geq \dfrac{\hbar}{2}$. For us, no matter what k_0 is, $w_p w_x = h$, which is greater than $\hbar/2$.

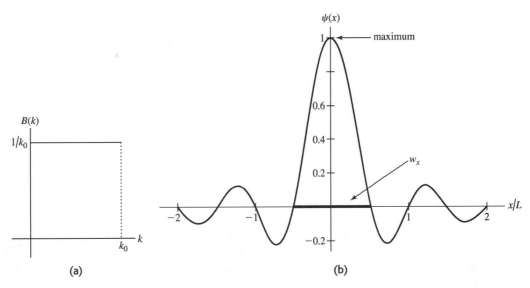

(a)   (b)

Figure 40.49

40.53.   **IDENTIFY** and **SET UP:**   The energy levels are given by Eq. (40.31): $E_n = \dfrac{n^2 h^2}{8mL^2}$. Calculate ΔE for the

transition and set $\Delta E = hc/\lambda$, the energy of the photon.

EXECUTE: **(a)** Ground level, $n = 1$, $E_1 = \dfrac{h^2}{8mL^2}$. First excited level, $n = 2$, $E_2 = \dfrac{4h^2}{8mL^2}$. The transition

energy is $\Delta E = E_2 - E_1 = \dfrac{3h^2}{8mL^2}$. Set the transition energy equal to the energy hc/λ of the emitted photon.

This gives $\dfrac{hc}{\lambda} = \dfrac{3h^2}{8mL^2}$. $\lambda = \dfrac{8mcL^2}{3h} = \dfrac{8(9.109 \times 10^{-31} \text{ kg})(2.998 \times 10^8 \text{ m/s})(4.18 \times 10^{-9} \text{ m})^2}{3(6.626 \times 10^{-34} \text{ J} \cdot \text{s})}$.

$\lambda = 1.92 \times 10^{-5}$ m $= 19.2 \ \mu$m.

(b) Second excited level has $n = 3$ and $E_3 = \dfrac{9h^2}{8mL^2}$ The transition energy is

$\Delta E = E_3 - E_2 = \dfrac{9h^2}{8mL^2} - \dfrac{4h^2}{8mL^2} = \dfrac{5h^2}{8mL^2}$. $\dfrac{hc}{\lambda} = \dfrac{5h^2}{8mL^2}$ so $\lambda = \dfrac{8mcL^2}{5h} = \dfrac{3}{5}(19.2 \ \mu m) = 11.5 \ \mu$m.

EVALUATE: The energy spacing between adjacent levels increases with n, and this corresponds to a
shorter wavelength and more energetic photon in part (b) than in part (a).

40.55.   **IDENTIFY:**   The probability of the particle being between x_1 and x_2 is $\int_{x_1}^{x_2} |\psi|^2 \, dx$, where ψ is the

normalized wave function for the particle.

(a) SET UP: The normalized wave function for the ground state is $\psi_1 = \sqrt{\dfrac{2}{L}} \sin\left(\dfrac{\pi x}{L}\right)$.

EXECUTE: The probability P of the particle being between $x = L/4$ and $x = 3L/4$ is

$P = \int_{L/4}^{3L/4} |\psi_1|^2 dx = \dfrac{2}{L} \int_{L/4}^{3L/4} \sin^2\left(\dfrac{\pi x}{L}\right) dx$. Let $y = \pi x / L$; $dx = (L/\pi) dy$ and the integration limits become

$\pi/4$ and $3\pi/4$.

$P = \dfrac{2}{L}\left(\dfrac{L}{\pi}\right) \int_{\pi/4}^{3\pi/4} \sin^2 y \, dy = \dfrac{2}{\pi}\left[\dfrac{1}{2}y - \dfrac{1}{4}\sin 2y\right]_{\pi/4}^{3\pi/4}$

$P = \dfrac{2}{\pi}\left[\dfrac{3\pi}{8} - \dfrac{\pi}{8} - \dfrac{1}{4}\sin\left(\dfrac{3\pi}{2}\right) + \dfrac{1}{4}\sin\left(\dfrac{\pi}{2}\right)\right]$

$P = \dfrac{2}{\pi}\left(\dfrac{\pi}{4} - \dfrac{1}{4}(-1) + \dfrac{1}{4}(1)\right) = \dfrac{1}{2} + \dfrac{1}{\pi} = 0.818$. (Note: The integral formula $\int \sin^2 y \, dy = \dfrac{1}{2}y - \dfrac{1}{4}\sin 2y$ was used.)

(b) SET UP: The normalized wave function for the first excited state is $\psi_2 = \sqrt{\dfrac{2}{L}} \sin\left(\dfrac{2\pi x}{L}\right)$.

EXECUTE: $P = \int_{L/4}^{3L/4} |\psi_2|^2 dx = \dfrac{2}{L} \int_{L/4}^{3L/4} \sin^2\left(\dfrac{2\pi x}{L}\right) dx$. Let $y = 2\pi x / L$; $dx = (L/2\pi) dy$ and the integration

limits become $\pi/2$ and $3\pi/2$.

$P = \dfrac{2}{L}\left(\dfrac{L}{2\pi}\right) \int_{\pi/2}^{3\pi/2} \sin^2 y \, dy = \dfrac{1}{\pi}\left[\dfrac{1}{2}y - \dfrac{1}{4}\sin 2y\right]_{\pi/2}^{3\pi/2} = \dfrac{1}{\pi}\left(\dfrac{3\pi}{4} - \dfrac{\pi}{4}\right) = 0.500$

(c) EVALUATE: These results are consistent with Figure 40.11b in the textbook. That figure shows that $|\psi|^2$
is more concentrated near the center of the box for the ground state than for the first excited state; this is
consistent with the answer to part (a) being larger than the answer to part (b). Also, this figure shows that for
the first excited state half the area under $|\psi|^2$ curve lies between $L/4$ and $3L/4$, consistent with our answer
to part (b).

40.65. **IDENTIFY** and **SET UP:** Calculate the angular frequency ω of the pendulum and apply Eq. (40.46) for the energy levels.

EXECUTE: $\omega = \dfrac{2\pi}{T} = \dfrac{2\pi}{0.500 \text{ s}} = 4\pi \text{ s}^{-1}$

The ground-state energy is $E_0 = \dfrac{1}{2}\hbar\omega = \dfrac{1}{2}(1.055 \times 10^{-34} \text{ J} \cdot \text{s})(4\pi \text{ s}^{-1}) = 6.63 \times 10^{-34}$ J.

$E_0 = 6.63 \times 10^{-34}$ J$(1 \text{ eV}/1.602 \times 10^{-19} \text{ J}) = 4.14 \times 10^{-15}$ eV

$E_n = \left(n + \dfrac{1}{2}\right)\hbar\omega$

$E_{n+1} = \left(n + 1 + \dfrac{1}{2}\right)\hbar\omega$

The energy difference between the adjacent energy levels is

$\Delta E = E_{n+1} - E_n = \hbar\omega = 2E_0 = 1.33 \times 10^{-33}$ J $= 8.30 \times 10^{-15}$ eV.

EVALUATE: These energies are much too small to detect. Quantum effects are not important for ordinary size objects.

40.69. **IDENTIFY:** For a standing wave in the box, there must be a node at each wall and $n\left(\dfrac{\lambda}{2}\right) = L$.

SET UP: $p = \dfrac{h}{\lambda}$ so $mv = \dfrac{h}{\lambda}$.

EXECUTE: **(a)** For a standing wave, $n\lambda = 2L$, and $E_n = \dfrac{p^2}{2m} = \dfrac{(h/\lambda)^2}{2m} = \dfrac{n^2 h^2}{8mL^2}$.

(b) With $L = a_0 = 0.5292 \times 10^{-10}$ m, $E_1 = 2.15 \times 10^{-17}$ J $= 134$ eV.

EVALUATE: For a hydrogen atom, E_n is proportional to $1/n^2$ so this is a very poor model for a hydrogen atom. In particular, it gives very inaccurate values for the separations between energy levels.

40.71. **IDENTIFY:** Require $\psi(-L/2) = \psi(L/2) = 0$.

SET UP: $k = \dfrac{2\pi}{\lambda}$, $p = \dfrac{h}{\lambda}$ and $E = \dfrac{p^2}{2m}$.

EXECUTE: **(a)** $\psi(x) = A \sin kx$ and $\psi(-L/2) = 0 = \psi(+L/2)$

$\Rightarrow 0 = A\sin\left(\dfrac{+kL}{2}\right) \Rightarrow \dfrac{+kL}{2} = n\pi \Rightarrow k = \dfrac{2n\pi}{L} = \dfrac{2\pi}{\lambda}$

$\Rightarrow \lambda = \dfrac{L}{n} \Rightarrow p_n = \dfrac{h}{\lambda} = \dfrac{nh}{L} \Rightarrow E_n = \dfrac{p^2}{2m} = \dfrac{n^2 h^2}{2mL^2} = \dfrac{(2n)^2 h^2}{8mL^2}$, where $n = 1, 2\ldots$

(b) $\psi(x) = A \cos kx$ and $\psi(-L/2) = 0 = \psi(+L/2)$

$\Rightarrow 0 = A\cos\left(\dfrac{kL}{2}\right) \Rightarrow \dfrac{kL}{2} = (2n+1)\dfrac{\pi}{2} \Rightarrow k = \dfrac{(2n+1)\pi}{L} = \dfrac{2\pi}{\lambda}$

$\Rightarrow \lambda = \dfrac{2L}{(2n+1)} \Rightarrow p_n = \dfrac{(2n+1)h}{2L}$

$\Rightarrow E_n = \dfrac{(2n+1)^2 h^2}{8mL^2}$ $n = 0, 1, 2\ldots$

(c) The combination of all the energies in parts (a) and (b) is the same energy levels as given in

Eq. (40.31), where $E_n = \dfrac{n^2 h^2}{8mL^2}$.

EVALUATE: **(d)** Part (a)'s wave functions are odd, and part (b)'s are even.

41

ATOMIC STRUCTURE

41.1. **IDENTIFY:** For a particle in a cubical box, different values of n_X, n_Y and n_Z can give the same energy.

SET UP: $E_{n_x, n_y, n_z} = \dfrac{(n_X^2 + n_Y^2 + n_Z^2)\pi^2 \hbar^2}{2mL^2}$.

EXECUTE: **(a)** $n_X^2 + n_Y^2 + n_Z^2 = 3$. This only occurs for $n_X = 1, n_Y = 1, n_Z = 1$ and the degeneracy is 1.

(b) $n_X^2 + n_Y^2 + n_Z^2 = 9$. Occurs for $n_X = 2, n_Y = 1, n_Z = 1$, for $n_X = 1, n_Y = 2, n_Z = 1$ and for $n_X = 1, n_Y = 1, n_Z = 2$. The degeneracy is 3.

EVALUATE: In the second case, three different states all have the same energy.

41.3. **IDENTIFY:** The energy of the photon is equal to the energy difference between the states. We can use this energy to calculate its wavelength.

SET UP: $E_{1,1,1} = \dfrac{3\pi^2 \hbar^2}{2mL^2}$. $E_{2,2,1} = \dfrac{9\pi^2 \hbar^2}{2mL^2}$. $\Delta E = \dfrac{3\pi^2 \hbar^2}{mL^2}$. $\Delta E = \dfrac{hc}{\lambda}$.

EXECUTE: $\Delta E = \dfrac{3\pi^2 (1.055 \times 10^{-34} \text{ J} \cdot \text{s})^2}{(9.109 \times 10^{-31} \text{ kg})(8.00 \times 10^{-11} \text{ m})^2} = 5.653 \times 10^{-17}$ J. $\Delta E = \dfrac{hc}{\lambda}$ gives

$\lambda = \dfrac{hc}{\Delta E} = \dfrac{(6.626 \times 10^{-34} \text{ J} \cdot \text{s})(2.998 \times 10^8 \text{ m/s})}{5.653 \times 10^{-17} \text{ J}} = 3.51 \times 10^{-9}$ m $= 3.51$ nm.

EVALUATE: This wavelength is much shorter than that of visible light.

41.5. **IDENTIFY:** A particle is in a three-dimensional box. At what planes is its probability function zero?

SET UP: $|\psi_{2,2,1}|^2 = \left(\dfrac{L}{2}\right)^3 \left(\sin^2 \dfrac{2\pi x}{L}\right)\left(\sin^2 \dfrac{2\pi y}{L}\right)\left(\sin^2 \dfrac{\pi z}{L}\right)$.

EXECUTE: $|\psi_{2,2,1}|^2 = 0$ for $\dfrac{2\pi x}{L} = 0, \pi, 2\pi, \ldots$ $x = 0$ and $x = L$ correspond to walls of the box. $x = \dfrac{L}{2}$

is the other plane where $|\psi_{2,2,1}|^2 = 0$. Similarly, $|\psi_{2,2,1}|^2 = 0$ on the plane $y = \dfrac{L}{2}$. The $\sin^2 \dfrac{\pi z}{L}$ factor is

zero only on the walls of the box. Therefore, for this state $|\psi_{2,2,1}|^2 = 0$ on the following two planes other

than walls of the box: $x = \dfrac{L}{2}$ and $y = \dfrac{L}{2}$.

$|\psi_{2,1,1}|^2 = \left(\dfrac{L}{2}\right)^3 \left(\sin^2 \dfrac{2\pi x}{L}\right)\left(\sin^2 \dfrac{\pi y}{L}\right)\left(\sin^2 \dfrac{\pi z}{L}\right)$ is zero only on one plane $(x = L/2)$ other than the walls

of the box.

$|\psi_{1,1,1}|^2 = \left(\dfrac{L}{2}\right)^3 \left(\sin^2 \dfrac{\pi x}{L}\right)\left(\sin^2 \dfrac{\pi y}{L}\right)\left(\sin^2 \dfrac{\pi z}{L}\right)$ is zero only on the walls of the box; for this state there are

zero additional planes.

EVALUATE: For comparison, (2,1,1) has two nodal planes, (2,1,1) has one nodal and (1,1,1) has no nodal planes. The number of nodal planes increases as the energy of the state increases.

41.7. **IDENTIFY:** The possible values of the angular momentum are limited by the value of n.

SET UP: For the N shell $n = 4$, $0 \leq l \leq n - 1$, $|m| \leq l$, $m_s = \pm\frac{1}{2}$.

EXECUTE: **(a)** The smallest l is $l = 0$. $L = \sqrt{l(l+1)}\hbar$, so $L_{\min} = 0$.

(b) The largest l is $n - 1 = 3$ so $L_{\max} = \sqrt{3(4)}\hbar = 2\sqrt{3}\hbar = 3.65 \times 10^{-34}$ kg·m^2/s.

(c) Let the chosen direction be the z-axis. The largest m is $m = l = 3$.

$L_{z,\max} = m\hbar = 3\hbar = 3.16 \times 10^{-34}$ kg·m^2/s.

(d) $S_z = \pm\frac{1}{2}\hbar$. The maximum value is $S_z = \hbar/2 = 5.27 \times 10^{-35}$ kg·m^2/s.

(e) $\dfrac{S_z}{L_z} = \dfrac{\frac{1}{2}\hbar}{3\hbar} = \dfrac{1}{6}$.

EVALUATE: The orbital and spin angular momenta are of comparable sizes.

41.11. **IDENTIFY and SET UP:** The angular momentum L is related to the quantum number l by Eq. (41.22), $L = \sqrt{l(l+1)}\hbar$. The maximum l, $l_{\max}$, for a given n is $l_{\max} = n - 1$.

EXECUTE: For $n = 2, l_{\max} = 1$ and $L = \sqrt{2}\hbar = 1.414\hbar$.

For $n = 20, l_{\max} = 19$ and $L = \sqrt{(19)(20)}\hbar = 19.49\hbar$.

For $n = 200, l_{\max} = 199$ and $L = \sqrt{(199)(200)}\hbar = 199.5\hbar$.

EVALUATE: As n increases, the maximum L gets closer to the value $n\hbar$ postulated in the Bohr model.

41.13. **IDENTIFY:** For the $5g$ state, $l = 4$, which limits the other quantum numbers.

SET UP: $m_l = 0, \pm1, \pm2, \ldots, \pm l$. g means $l = 4$. $\cos\theta = L_z/L$, with $L = \sqrt{l(l+1)}\,\hbar$ and $L_z = m_l\hbar$.

EXECUTE: **(a)** There are eighteen $5g$ states: $m_l = 0, \pm1, \pm2, \pm3, \pm4$, with $m_s = \pm\frac{1}{2}$ for each.

(b) The largest θ is for the most negative m_l. $L = 2\sqrt{5}\hbar$. The most negative L_z is $L_z = -4\hbar$.

$\cos\theta = \dfrac{-4\hbar}{2\sqrt{5}\hbar}$ and $\theta = 153.4°$.

(c) The smallest θ is for the largest positive m_l, which is $m_l = +4$. $\cos\theta = \dfrac{4\hbar}{2\sqrt{5}\hbar}$ and $\theta = 26.6°$.

EVALUATE: The minimum angle between $\vec{L}$ and the z-axis is for $m_l = +l$ and for that m_l, $\cos\theta = \dfrac{l}{\sqrt{l(l+1)}}$.

41.17. **IDENTIFY:** Apply $\Delta U = \mu_B B$.

SET UP: For a $3p$ state, $l = 1$ and $m_l = 0, \pm1$.

EXECUTE: **(a)** $B = \dfrac{U}{\mu_B} = \dfrac{(2.71 \times 10^{-5} \text{ eV})}{(5.79 \times 10^{-5} \text{ eV/T})} = 0.468$ T.

(b) Three: $m_l = 0, \pm1$.

EVALUATE: The $m_l = +1$ level will be highest in energy and the $m_l = -1$ level will be lowest. The $m_l = 0$ level is unaffected by the magnetic field.

41.19. **IDENTIFY and SET UP:** The interaction energy between an external magnetic field and the orbital angular momentum of the atom is given by Eq. (41.36). The energy depends on m_l with the most negative m_l value having the lowest energy.

EXECUTE: **(a)** For the $5g$ level, $l = 4$ and there are $2l + 1 = 9$ different m_l states. The $5g$ level is split into 9 levels by the magnetic field.

(b) Each m_l level is shifted in energy an amount given by $U = m_l\mu_B B$. Adjacent levels differ in m_l by one, so $\Delta U = \mu_B B$.

$$\mu_B = \frac{e\hbar}{2m} = \frac{(1.602 \times 10^{-19} \text{ C})(1.055 \times 10^{-34} \text{ J} \cdot \text{s})}{2(9.109 \times 10^{-31} \text{ kg})} = 9.277 \times 10^{-24} \text{ A} \cdot \text{m}^2$$

$$\Delta U = \mu_B B = (9.277 \times 10^{-24} \text{ A/m}^2)(0.600 \text{ T}) = 5.566 \times 10^{-24} \text{ J}(1 \text{ eV}/1.602 \times 10^{-19} \text{ J}) = 3.47 \times 10^{-5} \text{ eV}$$

(c) The level of highest energy is for the largest m_l, which is $m_l = l = 4$; $U_4 = 4\mu_B B$. The level of lowest energy is for the smallest m_l, which is $m_l = -l = -4$; $U_{-4} = -4\mu_B B$. The separation between these two levels is $U_4 - U_{-4} = 8\mu_B B = 8(3.47 \times 10^{-5} \text{ eV}) = 2.78 \times 10^{-4} \text{ eV}$.

EVALUATE: The energy separations are proportional to the magnetic field. The energy of the $n = 5$ level in the absence of the external magnetic field is $(-13.6 \text{ eV})/5^2 = -0.544 \text{ eV}$, so the interaction energy with the magnetic field is much less than the binding energy of the state.

41.21. **IDENTIFY** and **SET UP:** For a classical particle $L = I\omega$. For a uniform sphere with mass m and radius R, $I = \frac{2}{5}mR^2$, so $L = \left(\frac{2}{5}mR^2\right)\omega$. Solve for ω and then use $v = r\omega$ to solve for v.

EXECUTE: **(a)** $L = \sqrt{\frac{3}{4}}\hbar$ so $\frac{2}{5}mR^2\omega = \sqrt{\frac{3}{4}}\hbar$

$$\omega = \frac{5\sqrt{3/4}\hbar}{2mR^2} = \frac{5\sqrt{3/4}(1.055 \times 10^{-34} \text{ J} \cdot \text{s})}{2(9.109 \times 10^{-31} \text{ kg})(1.0 \times 10^{-17} \text{ m})^2} = 2.5 \times 10^{30} \text{ rad/s}$$

(b) $v = r\omega = (1.0 \times 10^{-17} \text{ m})(2.5 \times 10^{30} \text{ rad/s}) = 2.5 \times 10^{13} \text{ m/s}$

EVALUATE: This is much greater than the speed of light c, so the model cannot be valid.

41.23. **IDENTIFY** and **SET UP:** The interaction energy is $U = -\vec{\mu} \cdot \vec{B}$, with μ_z given by Eq. (41.40).

EXECUTE: $U = -\vec{\mu} \cdot \vec{B} = +\mu_z B$, since the magnetic field is in the negative z-direction.

$$\mu_z = -(2.00232)\left(\frac{e}{2m}\right)S_z, \text{ so } U = -(2.00232)\left(\frac{e}{2m}\right)S_z B$$

$$S_z = m_s\hbar, \text{ so } U = -2.00232\left(\frac{e\hbar}{2m}\right)m_s B$$

$$\frac{e\hbar}{2m} = \mu_B = 5.788 \times 10^{-5} \text{ eV/T}$$

$$U = -2.00232\mu_B m_s B$$

The $m_s = +\frac{1}{2}$ level has lower energy.

$$\Delta U = U\left(m_s = -\frac{1}{2}\right) - U\left(m_s = +\frac{1}{2}\right) = -2.00232\,\mu_B B\left(-\frac{1}{2} - \left(+\frac{1}{2}\right)\right) = +2.00232\,\mu_B B$$

$$\Delta U = +2.00232(5.788 \times 10^{-5} \text{ eV/T})(1.45 \text{ T}) = 1.68 \times 10^{-4} \text{ eV}$$

EVALUATE: The interaction energy with the electron spin is the same order of magnitude as the interaction energy with the orbital angular momentum for states with $m_l \neq 0$. But a $1s$ state has $l = 0$ and $m_l = 0$, so there is no orbital magnetic interaction.

41.25. **IDENTIFY** and **SET UP:** j can have the values $l + 1/2$ and $l - 1/2$.
EXECUTE: If j takes the values $7/2$ and $9/2$ it must be that $l - 1/2 = 7/2$ and $l = 8/2 = 4$. The letter that labels this l is g.
EVALUATE: l must be an integer.

41.31. **IDENTIFY** and **SET UP:** The energy of an atomic level is given in terms of n and Z_{eff} by Eq. (41.45),

$$E_n = -\left(\frac{Z_{\text{eff}}^2}{n^2}\right)(13.6 \text{ eV}). \text{ The ionization energy for a level with energy } -E_n \text{ is } +E_n.$$

EXECUTE: $n = 5$ and $Z_{eff} = 2.771$ gives $E_5 = -\dfrac{(2.771)^2}{5^2}(13.6 \text{ eV}) = -4.18 \text{ eV}$

The ionization energy is 4.18 eV.

EVALUATE: The energy of an atomic state is proportional to Z_{eff}^2.

41.33.   **IDENTIFY** and **SET UP:**   Use the exclusion principle to determine the ground-state electron configuration, as in Table 41.3. Estimate the energy by estimating Z_{eff}, taking into account the electron screening of the nucleus.

EXECUTE: **(a)** $Z = 7$ for nitrogen so a nitrogen atom has 7 electrons. N^{2+} has 5 electrons: $1s^2 2s^2 2p$.

(b) $Z_{eff} = 7 - 4 = 3$ for the $2p$ level.

$$E_n = -\left(\frac{Z_{eff}^2}{n^2}\right)(13.6 \text{ eV}) = -\frac{3^2}{2^2}(13.6 \text{ eV}) = -30.6 \text{ eV}$$

(c) $Z = 15$ for phosphorus so a phosphorus atom has 15 electrons.

P^{2+} has 13 electrons: $1s^2 2s^2 2p^6 3s^2 3p$

(d) $Z_{eff} = 15 - 12 = 3$ for the $3p$ level.

$$E_n = -\left(\frac{Z_{eff}^2}{n^2}\right)(13.6 \text{ eV}) = -\frac{3^2}{3^2}(13.6 \text{ eV}) = -13.6 \text{ eV}$$

EVALUATE: In these ions there is one electron outside filled subshells, so it is a reasonable approximation to assume full screening by these inner-subshell electrons.

41.35.   **IDENTIFY** and **SET UP:**   Estimate Z_{eff} by considering electron screening and use Eq. (41.45) to calculate the energy. Z_{eff} is calculated as in Example 41.9.

EXECUTE: **(a)** The element Be has nuclear charge $Z = 4$. The ion Be^+ has 3 electrons. The outermost electron sees the nuclear charge screened by the other two electrons so $Z_{eff} = 4 - 2 = 2$.

$$E_n = -\left(\frac{Z_{eff}^2}{n^2}\right)(13.6 \text{ eV}) \text{ so } E_2 = -\frac{2^2}{2^2}(13.6 \text{ eV}) = -13.6 \text{ eV}$$

(b) The outermost electron in Ca^+ sees a $Z_{eff} = 2$. $E_4 = -\dfrac{2^2}{4^2}(13.6 \text{ eV}) = -3.4 \text{ eV}$

EVALUATE: For the electron in the highest l-state it is reasonable to assume full screening by the other electrons, as in Example 41.9. The highest l-states of Be^+, Mg^+, Ca^+, etc. all have a $Z_{eff} = 2$. But the energies are different because for each ion the outermost sublevel has a different n quantum number.

41.37.   **IDENTIFY** and **SET UP:**   Apply Eq. (41.47). $E = hf$ and $c = f\lambda$.

EXECUTE: **(a)** $Z = 20$: $f = (2.48 \times 10^{15} \text{ Hz})(20 - 1)^2 = 8.95 \times 10^{17} \text{ Hz}$.

$E = hf = (4.14 \times 10^{-15} \text{ eV} \cdot \text{s})(8.95 \times 10^{17} \text{ Hz}) = 3.71 \text{ keV}$. $\lambda = \dfrac{c}{f} = \dfrac{3.00 \times 10^8 \text{ m/s}}{8.95 \times 10^{17} \text{ Hz}} = 3.35 \times 10^{-10} \text{ m}$.

(b) $Z = 27$: $f = 1.68 \times 10^{18} \text{ Hz}$. $E = 6.96 \text{ keV}$. $\lambda = 1.79 \times 10^{-10} \text{ m}$.

(c) $Z = 48$: $f = 5.48 \times 10^{18} \text{ Hz}$, $E = 22.7 \text{ keV}$, $\lambda = 5.47 \times 10^{-11} \text{ m}$.

EVALUATE: f and E increase and λ decreases as Z increases.

41.39.   **IDENTIFY:**   The electrons cannot all be in the same state in a cubical box.
SET UP and **EXECUTE:**   The ground state can hold 2 electrons, the first excited state can hold 6 electrons and the second excited state can hold 6. Therefore, two electrons will be in the second excited state, which has energy $3E_{1,1,1}$.

EVALUATE: The second excited state is the third state, which has energy $3E_{1,1,1}$, as shown in Figure 41.4.

41.41. **IDENTIFY:** Calculate the probability of finding a particle in a given region within a cubical box.

(a) SET UP and **EXECUTE:** The box has volume L^3. The specified cubical space has volume $(L/4)^3$. Its fraction of the total volume is $\dfrac{1}{64} = 0.0156$.

(b) SET UP and **EXECUTE:** $P = \left(\dfrac{2}{L}\right)^3 \left[\displaystyle\int_0^{L/4} \sin^2 \dfrac{\pi x}{L}\, dx\right]\left[\displaystyle\int_0^{L/4} \sin^2 \dfrac{\pi y}{L}\, dy\right]\left[\displaystyle\int_0^{L/4} \sin^2 \dfrac{\pi z}{L}\, dz\right].$

From Example 41.1, each of the three integrals equals $\dfrac{L}{8} - \dfrac{L}{4\pi} = \dfrac{1}{2}\left(\dfrac{L}{2}\right)\left(\dfrac{1}{2} - \dfrac{1}{\pi}\right).$

$P = \left(\dfrac{2}{L}\right)^3 \left(\dfrac{L}{2}\right)^3 \left(\dfrac{1}{2}\right)^3 \left(\dfrac{1}{2} - \dfrac{1}{\pi}\right)^3 = 7.50 \times 10^{-4}.$

EVALUATE: Note that this is the cube of the probability of finding the particle anywhere between $x = 0$ and $x = L/4$. This probability is much less that the fraction of the total volume that this space represents. In this quantum state the probability distribution function is much larger near the center of the box than near its walls.

(c) SET UP and **EXECUTE:** $|\psi_{2,1,1}|^2 = \left(\dfrac{L}{2}\right)^3 \left(\sin^2 \dfrac{2\pi x}{L}\right)\left(\sin^2 \dfrac{\pi y}{L}\right)\left(\sin^2 \dfrac{\pi z}{L}\right).$

$P = \left(\dfrac{2}{L}\right)^3 \left[\displaystyle\int_0^{L/4} \sin^2 \dfrac{2\pi x}{L}\, dx\right]\left[\displaystyle\int_0^{L/4} \sin^2 \dfrac{\pi y}{L}\, dy\right]\left[\displaystyle\int_0^{L/4} \sin^2 \dfrac{\pi z}{L}\, dz\right].$

$\left[\displaystyle\int_0^{L/4} \sin^2 \dfrac{\pi y}{L}\, dy\right] = \left[\displaystyle\int_0^{L/4} \sin^2 \dfrac{\pi z}{L}\, dz\right] = \dfrac{L}{2}\left(\dfrac{1}{2}\right)\left(\dfrac{1}{2} - \dfrac{1}{\pi}\right). \quad \displaystyle\int_0^{L/4} \sin^2 \dfrac{2\pi x}{L}\, dx = \dfrac{L}{8}.$

$P = \left(\dfrac{2}{L}\right)^3 \left(\dfrac{L}{2}\right)^2 \left(\dfrac{1}{2}\right)^2 \left(\dfrac{1}{2} - \dfrac{1}{\pi}\right)^2 \left(\dfrac{L}{8}\right) = 2.06 \times 10^{-3}.$

EVALUATE: This is about a factor of three larger than the probability when the particle is in the ground state.

41.49. **IDENTIFY:** The total energy determines what shell the electron is in, which limits its angular momentum.

SET UP: The electron's orbital angular momentum is given by $L = \sqrt{l(l+1)}\hbar$, and its total energy in the nth shell is $E_n = -(13.6\text{ eV})/n^2$.

EXECUTE: **(a)** First find n: $E_n = -(13.6\text{eV})/n^2 = -0.5440$ eV which gives $n = 5$, so $l = 4, 3, 2, 1, 0$. Therefore the possible values of L are given by $L = \sqrt{l(l+1)}\hbar$, giving $L = 0, \sqrt{2}\hbar, \sqrt{6}\hbar, \sqrt{12}\hbar, \sqrt{20}\hbar$.

(b) $E_6 = -(13.6\text{ eV})/6^2 = -0.3778$ eV. $\Delta E = E_6 - E_5 = -0.3778$ eV $- (-0.5440$ eV$) = +0.1662$ eV
This must be the energy of the photon, so $\Delta E = hc/\lambda$, which gives
$\lambda = hc/\Delta E = (4.136 \times 10^{-15}\text{eV} \cdot \text{s})(3.00 \times 10^8\text{ m/s})/(0.1662\text{ eV}) = 7.47 \times 10^{-6}\text{ m} = 7470$ nm, which is in the infrared and hence not visible.

EVALUATE: The electron can have any of the five possible values for its angular momentum, but it cannot have any others.

41.51. **IDENTIFY:** The inner electrons shield part of the nuclear charge from the outer electron.

SET UP: The electron's energy in the nth shell, due to shielding, is $E_n = -\dfrac{Z_{\text{eff}}^2}{n^2}(13.6\text{ eV})$, where $Z_{\text{eff}}e$ is the effective charge that the electron "sees" for the nucleus.

EXECUTE: **(a)** $E_n = -\dfrac{Z_{\text{eff}}^2}{n^2}(13.6\text{ eV})$ and $n = 4$ for the $4s$ state. Solving for Z_{eff} gives

$Z_{\text{eff}} = \sqrt{-\dfrac{(4^2)(-1.947\text{ eV})}{13.6\text{ eV}}} = 1.51.$ The nucleus contains a charge of $+11e$, so the average number of electrons that screen this nucleus must be $11 - 1.51 = 9.49$ electrons.

(b) (i) The charge of the nucleus is $+19e$, but $17.2e$ is screened by the electrons, so the outer electron "sees" $19e - 17.2e = 1.8e$ and $Z_{\text{eff}} = 1.8$.

(ii) $E_n = -\dfrac{Z_{\text{eff}}^2}{n^2}(13.6 \text{ eV}) = -\dfrac{(1.8)^2}{4^2}(13.6 \text{ eV}) = -2.75 \text{ eV}$

EVALUATE: Sodium has 11 protons, so the inner 10 electrons shield a large portion of this charge from the outer electron. But they don't shield 10 of the protons, since the inner electrons are not totally equivalent to a uniform spherical shell. (They are lumpy.)

41.53. **(a) IDENTIFY** and **SET UP:** The energy is given by Eq. (39.14), which is identical to Eq. (41.21). The potential energy is given by Eq. (23.9), with $q = +Ze$ and $q_0 = -e$.

EXECUTE: $E_{1s} = -\dfrac{1}{(4\pi\epsilon_0)^2}\dfrac{me^4}{2\hbar^2}$; $U(r) = -\dfrac{1}{4\pi\epsilon_0}\dfrac{e^2}{r}$

$E_{1s} = U(r)$ gives $-\dfrac{1}{(4\pi\epsilon_0)^2}\dfrac{me^4}{2\hbar^2} = -\dfrac{1}{4\pi\epsilon_0}\dfrac{e^2}{r}$

$r = \dfrac{(4\pi\epsilon_0)2\hbar^2}{me^2} = 2a$

EVALUATE: The turning point is twice the Bohr radius.

(b) IDENTIFY and **SET UP:** For the $1s$ state the probability that the electron is in the classically forbidden region is $P(r > 2a) = \int_{2a}^{\infty}|\psi_{1s}|^2\, dV = 4\pi\int_{2a}^{\infty}|\psi_{1s}|^2\, r^2 dr$. The normalized wave function of the $1s$ state of hydrogen is given in Example 41.4: $\psi_{1s}(r) = \dfrac{1}{\sqrt{\pi a^3}}e^{-r/a}$. Evaluate the integral; the integrand is the same as in Example 41.4.

EXECUTE: $P(r > 2a) = 4\pi\left(\dfrac{1}{\pi a^3}\right)\int_{2a}^{\infty} r^2 e^{-2r/a}\, dr$

Use the integral formula $\int r^2 e^{-\alpha r}\, dr = -e^{-\alpha r}\left(\dfrac{r^2}{\alpha} + \dfrac{2r}{\alpha^2} + \dfrac{2}{\alpha^3}\right)$, with $\alpha = 2/a$.

$P(r > 2a) = -\dfrac{4}{a^3}\left[e^{-2r/a}\left(\dfrac{ar^2}{2} + \dfrac{a^2 r}{2} + \dfrac{a^3}{4}\right)\right]_{2a}^{\infty} = +\dfrac{4}{a^3}e^{-4}(2a^3 + a^3 + a^3/4)$

$P(r > 2a) = 4e^{-4}(13/4) = 13e^{-4} = 0.238$.

EVALUATE: These is a 23.8% probability of the electron being found in the classically forbidden region, where classically its kinetic energy would be negative.

41.55. $\psi_{2s}(r) = \dfrac{1}{\sqrt{32\pi a^3}}\left(2 - \dfrac{r}{a}\right)e^{-r/2a}$

(a) IDENTIFY and **SET UP:** Let $I = \int_0^{\infty}|\psi_{2s}|^2\, dV = 4\pi|\psi_{2s}|^2\, r^2 dr$. If ψ_{2s} is normalized then we will find that $I = 1$.

EXECUTE: $I = 4\pi\left(\dfrac{1}{32\pi a^3}\right)\int_0^{\infty}\left(2 - \dfrac{r}{a}\right)^2 e^{-r/a}r^2\, dr = \dfrac{1}{8a^3}\int_0^{\infty}\left(4r^2 - \dfrac{4r^3}{a} + \dfrac{r^4}{a^2}\right)e^{-r/a}\, dr$

Use the integral formula $\int_0^{\infty} x^n e^{-\alpha x}\, dx = \dfrac{n!}{\alpha^{n+1}}$, with $\alpha = 1/a$.

$I = \dfrac{1}{8a^3}\left(4(2!)(a^3) - \dfrac{4}{a}(3!)(a)^4 + \dfrac{1}{a^2}(4!)(a)^5\right) = \dfrac{1}{8}(8 - 24 + 24) = 1$; this ψ_{2s} is normalized.

(b) SET UP: For a spherically symmetric state such as the $2s$, the probability that the electron will be found at $r < 4a$ is $P(r < 4a) = \int_0^{4a}|\psi_{2s}|^2\, dV = 4\pi\int_0^{4a}|\psi_{2s}|^2\, r^2 dr$.

EXECUTE: $P(r < 4a) = \dfrac{1}{8a^3} \displaystyle\int_0^{4a} \left(4r^2 - \dfrac{4r^3}{a} + \dfrac{r^4}{a^2} \right) e^{-r/a} dr$

Let $P(r < 4a) = \dfrac{1}{8a^3}(I_1 + I_2 + I_3).$

$I_1 = 4 \displaystyle\int_0^{4a} r^2 e^{-r/a} dr$

Use the integral formula $\displaystyle\int r^2 e^{-\alpha r} dr = -e^{-\alpha r} \left(\dfrac{r^2}{\alpha} + \dfrac{2r}{\alpha^2} + \dfrac{2}{\alpha^3} \right)$ with $\alpha = 1/a.$

$I_1 = -4 \left[e^{-r/a}(r^2 a + 2ra^2 + 2a^3) \right]_0^{4a} = (-104e^{-4} + 8)a^3.$

$I_2 = -\dfrac{4}{a} \displaystyle\int_0^{4a} r^3 e^{-r/a} dr$

Use the integral formula $\displaystyle\int r^3 e^{-\alpha r} dr = -e^{-\alpha r} \left(\dfrac{r^3}{\alpha} + \dfrac{3r^2}{\alpha^2} + \dfrac{6r}{\alpha^3} + \dfrac{6}{\alpha^4} \right)$ with $\alpha = 1/a.$

$I_2 = \dfrac{4}{a} \left[e^{-r/a}(r^3 a + 3r^2 a^2 + 6ra^3 + 6a^4) \right]_0^{4a} = (568e^{-4} - 24)a^3.$

$I_3 = \dfrac{1}{a^2} \displaystyle\int_0^{4a} r^4 e^{-r/a} dr$

Use the integral formula $\displaystyle\int r^4 e^{-\alpha r} dr = -e^{-\alpha r} \left(\dfrac{r^4}{\alpha} + \dfrac{4r^3}{\alpha^2} + \dfrac{12r^2}{\alpha^3} + \dfrac{24r}{\alpha^4} + \dfrac{24}{\alpha^5} \right)$ with $\alpha = 1/a.$

$I_3 = -\dfrac{1}{a^2} \left[e^{-r/a}(r^4 a + 4r^3 a^2 + 12r^2 a^3 + 24ra^4 + 24a^5) \right]_0^{4a} = (-824e^{-4} + 24)a^3.$

Thus $P(r < 4a) = \dfrac{1}{8a^3}(I_1 + I_2 + I_3) = \dfrac{1}{8a^3} a^3([8 - 24 + 24] + e^{-4}[-104 + 568 - 824])$

$P(r < 4a) = \dfrac{1}{8}(8 - 360e^{-4}) = 1 - 45e^{-4} = 0.176.$

EVALUATE: There is an 82.4% probability that the electron will be found at $r > 4a$. In the Bohr model the electron is for certain at $r = 4a$; this is a poor description of the radial probability distribution for this state.

41.57.   **IDENTIFY:**   Use Figure 41.6 in the textbook to relate θ_L to L_z and L: $\cos\theta_L = \dfrac{L_z}{L}$ so $\theta_L = \arccos\left(\dfrac{L_z}{L} \right).$

(a) **SET UP:**   The smallest angle $(\theta_L)_{\min}$ is for the state with the largest L and the largest L_z. This is the state with $l = n-1$ and $m_l = l = n-1.$

EXECUTE: $L_z = m_l \hbar = (n-1)\hbar$

$L = \sqrt{l(l+1)}\hbar = \sqrt{(n-1)n}\hbar$

$(\theta_L)_{\min} = \arccos\left(\dfrac{(n-1)\hbar}{\sqrt{(n-1)n}\hbar} \right) = \arccos\left(\dfrac{(n-1)}{\sqrt{(n-1)n}} \right) = \arccos\left(\sqrt{\dfrac{n-1}{n}} \right) = \arccos(\sqrt{(1-1)/n}).$

EVALUATE: Note that $(\theta_L)_{\min}$ approaches $0°$ as $n \to \infty.$

(b) **SET UP:**   The largest angle $(\theta_L)_{\max}$ is for $l = n-1$ and $m_l = -l = -(n-1).$

EXECUTE: A similar calculation to part (a) yields $(\theta_L)_{\max} = \arccos\left(-\sqrt{1-1/n} \right)$

EVALUATE: Note that $(\theta_L)_{\max}$ approaches $180°$ as $n \to \infty.$

41.61. **IDENTIFY:** Apply Eq. (41.36).

SET UP: Decay from a $3d$ to $2p$ state in hydrogen means that $n = 3 \to n = 2$ and $m_l = \pm 2, \pm 1, 0 \to m_l = \pm 1, 0$. However, selection rules limit the possibilities for decay. The emitted photon carries off one unit of angular momentum so l must change by 1 and hence m_l must change by 0 or ± 1.

EXECUTE: The shift in the transition energy from the zero field value is

$U = (m_{l_3} - m_{l_2})\mu_B B = \dfrac{e\hbar B}{2m}(m_{l_3} - m_{l_2})$, where m_{l_3} is the $3d$ m_l value and m_{l_2} is the $2p$ m_l value. Thus there are only three different energy shifts. The shifts and the transitions that have them, labeled by the m_l values, are:

$$\frac{e\hbar B}{2m}: 2 \to 1, 1 \to 0, 0 \to -1. \quad 0: 1 \to 1, \ 0 \to 0, -1 \to -1. \quad -\frac{e\hbar B}{2m}: 0 \to 1, -1 \to 0, -2 \to -1.$$

EVALUATE: Our results are consistent with Figure 41.15 in the textbook.

41.63. **IDENTIFY:** The presence of an external magnetic field shifts the energy levels up or down, depending upon the value of m_l.

SET UP: The energy difference due to the magnetic field is $\Delta E = \mu_B B$ and the energy of a photon is $E = hc/\lambda$.

EXECUTE: For the p state, $m_l = 0$ or ± 1, and for the s state $m_l = 0$. Between any two adjacent lines, $\Delta E = \mu_B B$. Since the change in the wavelength $(\Delta \lambda)$ is very small, the energy change (ΔE) is also very small, so we can use differentials. $E = hc/\lambda$. $|dE| = \dfrac{hc}{\lambda^2}d\lambda$ and $\Delta E = \dfrac{hc\Delta\lambda}{\lambda^2}$. Since $\Delta E = \mu_B B$, we get

$\mu_B B = \dfrac{hc\Delta\lambda}{\lambda^2}$ and $B = \dfrac{hc\Delta\lambda}{\mu_B \lambda^2}$.

$B = (4.136 \times 10^{-15} \text{ eV} \cdot \text{s})(3.00 \times 10^8 \text{ m/s})(0.0462 \text{ nm})/(5.788 \times 10^{-5} \text{ eV/T})(575.050 \text{ nm})^2 = 3.00 \text{ T}$

EVALUATE: Even a strong magnetic field produces small changes in the energy levels, and hence in the wavelengths of the emitted light.

41.65. **IDENTIFY:** The ratio according to the Boltzmann distribution is given by Eq. (39.18): $\dfrac{n_1}{n_0} = e^{-(E_1 - E_0)/kT}$, where 1 is the higher energy state and 0 is the lower energy state.

SET UP: The interaction energy with the magnetic field is $U = -\mu_z B = 2.00232\left(\dfrac{e\hbar}{2m}\right)m_s B$ (Example 41.6.).

The energy of the $m_s = +\dfrac{1}{2}$ level is increased and the energy of the $m_s = -\dfrac{1}{2}$ level is decreased.

$\dfrac{n_{1/2}}{n_{-1/2}} = e^{-(U_{1/2} - U_{-1/2})/kT}$

EXECUTE: $U_{1/2} - U_{-1/2} = 2.00232\left(\dfrac{e\hbar}{2m}\right)B\left(\dfrac{1}{2} - \left(-\dfrac{1}{2}\right)\right) = 2.00232\left(\dfrac{e\hbar}{2m}\right)B = 2.00232\mu_B B$

$\dfrac{n_{1/2}}{n_{-1/2}} = e^{-(2.00232)\mu_B B/kT}$

(a) $B = 5.00 \times 10^{-5} \text{ T}$

$\dfrac{n_{1/2}}{n_{-1/2}} = e^{-2.00232(9.274 \times 10^{-24} \text{ A/m}^2)(5.00 \times 10^{-5} \text{ T})/([1.381 \times 10^{-23} \text{ J/K}][300 \text{ K}])}$

$\dfrac{n_{1/2}}{n_{-1/2}} = e^{-2.24 \times 10^{-7}} = 0.99999978 = 1 - 2.2 \times 10^{-7}$

(b) $B = 5.00 \times 10^{-5} \text{ T}$, $\dfrac{n_{1/2}}{n_{-1/2}} = e^{-2.24 \times 10^{-3}} = 0.9978$

(c) $B = 5.00 \times 10^{-5}$ T, $\dfrac{n_{1/2}}{n_{-1/2}} = e^{-2.24 \times 10^{-2}} = 0.978$

EVALUATE: For small fields the energy separation between the two spin states is much less than kT for $T = 300$ K and the states are equally populated. For $B = 5.00$ T the energy spacing is large enough for there to be a small excess of atoms in the lower state.

41.67. **IDENTIFY and SET UP:** m_s can take on 4 different values: $m_s = -\dfrac{3}{2}, -\dfrac{1}{2}, +\dfrac{1}{2}, +\dfrac{3}{2}$. Each nlm_l state can have 4 electrons, each with one of the four different m_s values. Apply the exclusion principle to determine the electron configurations.

EXECUTE: **(a)** For a filled $n = 1$ shell, the electron configuration would be $1s^4$; four electrons and $Z = 4$. For a filled $n = 2$ shell, the electron configuration would be $1s^4 2s^4 2p^{12}$; twenty electrons and $Z = 20$.

(b) Sodium has $Z = 11$; 11 electrons. The ground-state electron configuration would be $1s^4 2s^4 2p^3$.

EVALUATE: The chemical properties of each element would be very different.

41.69. **(a) IDENTIFY and SET UP:** The energy of the photon equals the transition energy of the atom: $\Delta E = hc/\lambda$. The energies of the states are given by Eq. (41.21).

EXECUTE: $E_n = -\dfrac{13.60 \text{ eV}}{n^2}$ so $E_2 = -\dfrac{13.60 \ eV}{4}$ and $E_1 = -\dfrac{13.60 \text{ eV}}{1}$

$\Delta E = E_2 - E_1 = 13.60 \text{ eV}\left(-\dfrac{1}{4} + 1\right) = \dfrac{3}{4}(13.60 \text{ eV}) = 10.20 \text{ eV} = (10.20 \text{ eV})(1.602 \times 10^{-19} \text{ J/eV}) = 1.634 \times 10^{-18} \text{ J}$

$\lambda = \dfrac{hc}{\Delta E} = \dfrac{(6.626 \times 10^{-34} \text{ J} \cdot \text{s})(2.998 \times 10^8 \text{ m/s})}{1.634 \times 10^{-18} \text{ J}} = 1.22 \times 10^{-7} \text{ m} = 122 \text{ nm}$

(b) IDENTIFY and SET UP: Calculate the change in ΔE due to the orbital magnetic interaction energy, Eq. (41.36), and relate this to the shift $\Delta \lambda$ in the photon wavelength.

EXECUTE: The shift of a level due to the energy of interaction with the magnetic field in the z-direction is $U = m_l \mu_B B$. The ground state has $m_l = 0$ so is unaffected by the magnetic field. The $n = 2$ initial state has $m_l = -1$ so its energy is shifted downward an amount $U = m_l \mu_B B = (-1)(9.274 \times 10^{-24} \text{ A/m}^2)(2.20 \text{ T}) = (-2.040 \times 10^{-23} \text{ J})(1 \text{ eV}/1.602 \times 10^{-19} \text{ J}) = 1.273 \times 10^{-4}$ eV.

Note that the shift in energy due to the magnetic field is a very small fraction of the 10.2 eV transition energy. Problem 39.86c shows that in this situation $|\Delta\lambda/\lambda| = |\Delta E/E|$. This gives

$|\Delta\lambda| = \lambda |\Delta E/E| = 122 \text{ nm} \left(\dfrac{1.273 \times 10^{-4} \text{ eV}}{10.2 \text{ eV}}\right) = 1.52 \times 10^{-3} \text{ nm} = 1.52 \text{ pm}.$

EVALUATE: The upper level in the transition is lowered in energy so the transition energy is decreased. A smaller ΔE means a larger λ; the magnetic field increases the wavelength. The fractional shift in wavelength, $\Delta\lambda/\lambda$ is small, only 1.2×10^{-5}.

41.71. **IDENTIFY:** Estimate the atomic transition energy and use Eq. (39.5) to relate this to the photon wavelength.

(a) SET UP: vanadium, $Z = 23$

minimum wavelength; corresponds to largest transition energy

EXECUTE: The highest occupied shell is the N shell ($n = 4$). The highest energy transition is $N \rightarrow K$, with transition energy $\Delta E = E_N - E_K$. Since the shell energies scale like $1/n^2$ neglect E_N relative to E_K, so $\Delta E = E_K = (Z - 1)^2(13.6 \text{ eV}) = (23 - 1)^2(13.6 \text{ eV}) = 6.582 \times 10^3 \text{ eV} = 1.055 \times 10^{-15} \text{ J}$. The energy of the emitted photon equals this transition energy, so the photon's wavelength is given by $\Delta E = hc/\lambda$ so $\lambda = hc/\Delta E$.

$\lambda = \dfrac{(6.626 \times 10^{-34} \text{ J} \cdot \text{s})(2.998 \times 10^8 \text{ m/s})}{1.055 \times 10^{-15} \text{ J}} = 1.88 \times 10^{-10} \text{ m} = 0.188 \text{ nm}.$

SET UP: maximum wavelength; corresponds to smallest transition energy, so for the K_α transition

EXECUTE: The frequency of the photon emitted in this transition is given by Moseley's law (Eq. 41.47):

$$f = (2.48 \times 10^{15} \text{ Hz})(Z-1)^2 = (2.48 \times 10^{15} \text{ Hz})(23-1)^2 = 1.200 \times 10^{18} \text{ Hz}$$

$$\lambda = \frac{c}{f} = \frac{2.998 \times 10^8 \text{ m/s}}{1.200 \times 10^{18} \text{ Hz}} = 2.50 \times 10^{-10} \text{ m} = 0.250 \text{ nm}$$

(b) rhenium, $Z = 45$

Apply the analysis of part (a), just with this different value of Z.

minimum wavelength

$$\Delta E = E_K = (Z-1)^2 (13.6 \text{ eV}) = (45-1)^2 (13.6 \text{ eV}) = 2.633 \times 10^4 \text{ eV} = 4.218 \times 10^{-15} \text{ J}.$$

$$\lambda = hc/\Delta E = \frac{(6.626 \times 10^{-34} \text{ J} \cdot \text{s})(2.998 \times 10^8 \text{ m/s})}{4.218 \times 10^{-15} \text{ J}} = 4.71 \times 10^{-11} \text{ m} = 0.0471 \text{ nm}.$$

maximum wavelength

$$f = (2.48 \times 10^{15} \text{ Hz})(Z-1)^2 = (2.48 \times 10^{15} \text{ Hz})(45-1)^2 = 4.801 \times 10^{18} \text{ Hz}$$

$$\lambda = \frac{c}{f} = \frac{2.998 \times 10^8 \text{ m/s}}{4.801 \times 10^{18} \text{ Hz}} = 6.24 \times 10^{-11} \text{ m} = 0.0624 \text{ nm}$$

EVALUATE: Our calculated wavelengths have values corresponding to x rays. The transition energies increase when Z increases and the photon wavelengths decrease.

42

MOLECULES AND CONDENSED MATTER

42.3. **IDENTIFY:** Set $\frac{3}{2}kT$ equal to the specified bond energy E.

SET UP: $k = 1.38 \times 10^{-23}$ J/K.

EXECUTE: **(a)** $E = \frac{3}{2}kT \Rightarrow T = \frac{2E}{3k} = \frac{2(7.9 \times 10^{-4}\text{ eV})(1.60 \times 10^{-19}\text{ J/eV})}{3(1.38 \times 10^{-23}\text{ J/K})} = 6.1$ K.

(b) $T = \frac{2(4.48\text{ eV})(1.60 \times 10^{-19}\text{ J/eV})}{3(1.38 \times 10^{-23}\text{ J/K})} = 34,600$ K.

EVALUATE: **(c)** The thermal energy associated with room temperature (300 K) is much greater than the bond energy of He_2 (calculated in part (a)), so the typical collision at room temperature will be more than enough to break up He_2. However, the thermal energy at 300 K is much less than the bond energy of H_2, so we would expect it to remain intact at room temperature.

42.5. **IDENTIFY:** The energy of the photon is equal to the energy difference between the $l = 1$ and $l = 2$ states. This energy determines its wavelength.

SET UP: The reduced mass of the molecule is $m_r = \frac{m_H m_H}{m_H + m_H} = \frac{1}{2}m_H$, its moment of inertia is $I = m_r r_0^2$,

the photon energy is $\Delta E = \frac{hc}{\lambda}$, and the energy of the state l is $E_l = l(l+1)\frac{\hbar^2}{2I}$.

EXECUTE: $I = m_r r_0^2 = \frac{1}{2}(1.67 \times 10^{-27}\text{ kg})(0.074 \times 10^{-9}\text{ m})^2 = 4.57 \times 10^{-48}\text{ kg} \cdot \text{m}^2$. Using $E_l = l(l+1)\frac{\hbar^2}{2I}$,

the energy levels are $E_2 = 6\frac{\hbar^2}{2I} = 6\frac{(1.055 \times 10^{-34}\text{ J} \cdot \text{s})^2}{2(4.57 \times 10^{-48}\text{ kg} \cdot \text{m}^2)} = 6(1.218 \times 10^{-21}\text{ J}) = 7.307 \times 10^{-21}\text{ J}$ and

$E_{1'} = 2\frac{\hbar^2}{2I} = 2(1.218 \times 10^{-21}\text{ J}) = 2.436 \times 10^{-21}\text{ J}$. $\Delta E = E_2 - E_1 = 4.87 \times 10^{-21}\text{ J}$. Using $\Delta E = \frac{hc}{\lambda}$ gives

$\lambda = \frac{hc}{\Delta E} = \frac{(6.626 \times 10^{-34}\text{ J} \cdot \text{s})(2.998 \times 10^8\text{ m/s})}{4.871 \times 10^{-21}\text{ J}} = 4.08 \times 10^{-5}\text{ m} = 40.8\ \mu\text{m}$.

EVALUATE: This wavelength is much longer than that of visible light.

42.7. **IDENTIFY:** The energy given to the photon comes from a transition between rotational states.

SET UP: The rotational energy of a molecule is $E = l(l+1)\frac{\hbar^2}{2I}$ and the energy of the photon is $E = hc/\lambda$.

EXECUTE: Use the energy formula, the energy difference between the $l = 3$ and $l = 1$ rotational levels of

the molecule is $\Delta E = \frac{\hbar^2}{2I}[3(3+1) - 1(1+1)] = \frac{5\hbar^2}{I}$. Since $\Delta E = hc/\lambda$, we get $hc/\lambda = 5\hbar^2/I$. Solving for I gives

$$I = \frac{5\hbar\lambda}{2\pi c} = \frac{5(1.055 \times 10^{-34}\text{ J} \cdot \text{s})(1.780\text{ nm})}{2\pi(3.00 \times 10^8\text{ m/s})} = 4.981 \times 10^{-52}\text{ kg} \cdot \text{m}^2.$$

Using $I = m_r r_0^2$, we can solve for r_0:

$$r_0 = \sqrt{\frac{I(m_N + m_H)}{m_N m_H}} = \sqrt{\frac{(4.981 \times 10^{-52} \text{ kg} \cdot \text{m}^2)(2.33 \times 10^{-26} \text{ kg} + 1.67 \times 10^{-27} \text{ kg})}{(2.33 \times 10^{-26} \text{ kg})(1.67 \times 10^{-27} \text{ kg})}} \quad r_0 = 5.65 \times 10^{-13} \text{ m}$$

EVALUATE: This separation is much smaller than the diameter of a typical atom and is not very realistic. But we are treating a *hypothetical* NH molecule.

42.11. **IDENTIFY** and **SET UP:** Set $K = E_1$ from Example 42.2. Use $K = \frac{1}{2} I \omega^2$ to solve for ω and $v = r\omega$ to solve for v.

EXECUTE: **(a)** From Example 42.2, $E_1 = 0.479$ meV $= 7.674 \times 10^{-23}$ J and $I = 1.449 \times 10^{-46}$ kg $\cdot$ m^2

$K = \frac{1}{2} I \omega^2$ and $K = E$ gives $\omega = \sqrt{2E_1 / I} = 1.03 \times 10^{12}$ rad/s

(b) $v_1 = r_1 \omega_1 = (0.0644 \times 10^{-9} \text{ m})(1.03 \times 10^{12} \text{ rad/s}) = 66.3$ m/s (carbon)

$v_2 = r_2 \omega_2 = (0.0484 \times 10^{-9} \text{ m})(1.03 \times 10^{12} \text{ rad/s}) = 49.8$ m/s (oxygen)

(c) $T = 2\pi / \omega = 6.10 \times 10^{-12}$ s

EVALUATE: Even for fast rotation rates, $v \ll c$.

42.13. **IDENTIFY** and **SET UP:** The energy of a rotational level with quantum number l is $E_l = l(l+1)\hbar^2 / 2I$ (Eq. (42.3)). $I = m_r r^2$, with the reduced mass m_r given by Eq. (42.4). Calculate I and ΔE and then use $\Delta E = hc / \lambda$ to find λ.

EXECUTE: **(a)** $m_r = \dfrac{m_1 m_2}{m_1 + m_2} = \dfrac{m_{Li} m_H}{m_{Li} + m_H} = \dfrac{(1.17 \times 10^{-26} \text{ kg})(1.67 \times 10^{-27} \text{ kg})}{1.17 \times 10^{-26} \text{ kg} + 1.67 \times 10^{-27} \text{ kg}} = 1.461 \times 10^{-27}$ .kg

$I = m_r r^2 = (1.461 \times 10^{-27} \text{ kg})(0.159 \times 10^{-9} \text{ m})^2 = 3.694 \times 10^{-47}$ kg $\cdot$ m^2

$l = 3 : E = 3(4)\left(\dfrac{\hbar^2}{2I}\right) = 6\left(\dfrac{\hbar^2}{I}\right)$

$l = 4 : E = 4(5)\left(\dfrac{\hbar^2}{2I}\right) = 10\left(\dfrac{\hbar^2}{I}\right)$

$\Delta E = E_4 - E_3 = 4\left(\dfrac{\hbar^2}{I}\right) = 4\left(\dfrac{(1.055 \times 10^{-34} \text{ J} \cdot \text{s})^2}{3.694 \times 10^{-47} \text{ kg} \cdot \text{m}^2}\right) = 1.20 \times 10^{-21}$ J $= 7.49 \times 10^{-3}$ eV

(b) $\Delta E = hc / \lambda$ so $\lambda = \dfrac{hc}{\Delta E} = \dfrac{(4.136 \times 10^{-15} \text{ eV})(2.998 \times 10^8 \text{ m/s})}{7.49 \times 10^{-3} \text{ eV}} = 166 \, \mu\text{m}$

EVALUATE: LiH has a smaller reduced mass than CO and λ is somewhat smaller here than the λ calculated for CO in Example 42.2

42.19. **IDENTIFY:** The energy gap is the energy of the maximum-wavelength photon.
SET UP: The energy difference is equal to the energy of the photon, so $\Delta E = hc / \lambda$.
EXECUTE: **(a)** Using the photon wavelength to find the energy difference gives

$$\Delta E = hc / \lambda = (4.136 \times 10^{-15} \text{ eV} \cdot \text{s})(3.00 \times 10^8 \text{ m/s}) / (1.11 \times 10^{-6} \text{ m}) = 1.12 \text{ eV}$$

(b) A wavelength of $1.11 \, \mu$m $= 1110$ nm is in the infrared, shorter than that of visible light.

EVALUATE: Since visible photons have more than enough energy to excite electrons from the valence to the conduction band, visible light will be absorbed, which makes silicon opaque.

42.21. **IDENTIFY** and **SET UP:** The energy ΔE deposited when a photon with wavelength λ is absorbed is

$$\Delta E = \frac{hc}{\lambda}.$$

EXECUTE: $\Delta E = \dfrac{hc}{\lambda} = \dfrac{(6.63 \times 10^{-34} \text{ J} \cdot \text{s})(3.00 \times 10^{8} \text{ m/s})}{9.31 \times 10^{-13} \text{ m}} = 2.14 \times 10^{-13} \text{ J} = 1.34 \times 10^{6} \text{ eV}$. So the number

of electrons that can be excited to the conduction band is $n = \dfrac{1.34 \times 10^{6} \text{ eV}}{1.12 \text{ eV}} = 1.20 \times 10^{6}$ electrons.

EVALUATE: A photon of wavelength

$\lambda = \dfrac{hc}{\Delta E} = \dfrac{(4.13 \times 10^{-15} \text{ eV} \cdot \text{s})(3.00 \times 10^{8} \text{ m/s})}{1.12 \text{ eV}} = 1.11 \times 10^{-6} \text{ m} = 1110 \text{ nm}$ can excite one electron. This

photon is in the infrared.

42.23.   **IDENTIFY:** $g(E)$ is given by Eq. (42.10).

SET UP: $m = 9.11 \times 10^{-31}$ kg, the mass of an electron.

EXECUTE: $g(E) = \dfrac{(2m)^{3/2}V}{2\pi^{2}\hbar^{3}} E^{1/2} = \dfrac{(2(9.11 \times 10^{-31} \text{ kg}))^{3/2}(1.0 \times 10^{-6} \text{ m}^{3})(5.0 \text{ eV})^{1/2}(1.60 \times 10^{-19} \text{ J/eV})^{1/2}}{2\pi^{2}(1.054 \times 10^{-34} \text{ J} \cdot \text{s})^{3}}$.

$g(E) = (9.5 \times 10^{40} \text{ states/J})(1.60 \times 10^{-19} \text{ J/eV}) = 1.5 \times 10^{22}$ states/eV.

EVALUATE: For a metal the density of states expressed as states/eV is very large.

42.25.   **(a) IDENTIFY** and **SET UP:** The electron contribution to the molar heat capacity at constant volume of a

metal is $C_V = \left(\dfrac{\pi^{2}KT}{2E_{\text{F}}} \right) R$.

EXECUTE: $C_V = \dfrac{\pi^{2}(1.381 \times 10^{-23} \text{ J/K})(300 \text{ K})}{2(5.48 \text{ eV})(1.602 \times 10^{-19} \text{ J/eV})} R = 0.0233R$.

(b) EVALUATE: The electron contribution found in part (a) is $0.0233R = 0.194$ J/mol · K. This is

$0.194/25.3 = 7.67 \times 10^{-3} = 0.767\%$ of the total C_V.

(c) Only a small fraction of C_V is due to the electrons. Most of C_V is due to the vibrational motion of

the ions.

42.27.   **IDENTIFY:** The probability is given by the Fermi-Dirac distribution.

SET UP: The Fermi-Dirac distribution is $f(E) = \dfrac{1}{e^{(E-E_{\text{F}})/kT} + 1}$.

EXECUTE: We calculate the value of $f(E)$, where $E = 8.520$ eV, $E_{\text{F}} = 8.500$ eV,

$k = 1.38 \times 10^{-23}$ J/K $= 8.625 \times 10^{-5}$ eV/K, and $T = 20°$C $= 293$ K. The result is $f(E) = 0.312 = 31.2\%$.

EVALUATE: Since the energy is close to the Fermi energy, the probability is quite high that the state is

occupied by an electron.

42.29.   **IDENTIFY:** Use Eq. (42.16), $f(E) = \dfrac{1}{e^{(E-E_{\text{F}})/kT} + 1}$. Solve for $E - E_{\text{F}}$.

SET UP: $e^{(E-E_{\text{F}})/kT} = \dfrac{1}{f(E)} - 1$

The problem states that $f(E) = 4.4 \times 10^{-4}$ for E at the bottom of the conduction band.

EXECUTE: $e^{(E-E_{\text{F}})/kT} = \dfrac{1}{4.4 \times 10^{-4}} - 1 = 2.272 \times 10^{3}$.

$E - E_{\text{F}} = kT \ln(2.272 \times 10^{3}) = (1.3807 \times 10^{-23} \text{ J/T})(300 \text{ K})\ln(2.272 \times 10^{3}) = 3.201 \times 10^{-20} \text{ J} = 0.20 \text{ eV}$

$E_{\text{F}} = E - 0.20$ eV; the Fermi level is 0.20 eV below the bottom of the conduction band.

EVALUATE: The energy gap between the Fermi level and bottom of the conduction band is large

compared to kT at $T = 300$ K and as a result $f(E)$ is small.

42.31.   **IDENTIFY:** Knowing the saturation current of a *p-n* junction at a given temperature, we want to find the

current at that temperature for various voltages.

SET UP: $I = I_S(e^{eV/kT} - 1)$.

EXECUTE: **(a)** (i) For $V = 1.00$ mV, $\dfrac{eV}{kT} = \dfrac{(1.602 \times 10^{-19} \text{ C})(1.00 \times 10^{-3} \text{ V})}{(1.381 \times 10^{-23} \text{ J/K})(290 \text{ K})} = 0.0400$.

$I = (0.500 \text{ mA})(e^{0.0400} - 1) = 0.0204$ mA.

(ii) For $V = -1.00$ mV, $\dfrac{eV}{kT} = -0.0400$. $I = (0.500 \text{ mA})(e^{-0.0400} - 1) = -0.0196$ mA.

(iii) For $V = 100$ mV, $\dfrac{eV}{kT} = 4.00$. $I = (0.500 \text{ mA})(e^{4.00} - 1) = 26.8$ mA.

(iv) For $V = -100$ mV, $\dfrac{eV}{kT} = -4.00$. $I = (0.500 \text{ mA})(e^{-4.00} - 1) = -0.491$ mA.

EXECUTE: **(b)** For small V, between ± 1.00 mV, $R = V/I$ is approximately constant and the diode obeys Ohm's law to a good approximation. For larger V the deviation from Ohm's law is substantial.

42.33.   **IDENTIFY and SET UP:** The voltage-current relation is given by Eq. (42.22): $I = I_s(e^{eV/kT} - 1)$. Use the current for $V = +15.0$ mV to solve for the constant I_s.

EXECUTE: **(a)** Find I_s: $V = +15.0 \times 10^{-3}$ V gives $I = 9.25 \times 10^{-3}$ A

$\dfrac{eV}{kT} = \dfrac{(1.602 \times 10^{-19} \text{ C})(15.0 \times 10^{-3} \text{ V})}{(1.381 \times 10^{-23} \text{ J/K})(300 \text{ K})} = 0.5800$

$I_s = \dfrac{I}{e^{eV/kT} - 1} = \dfrac{9.25 \times 10^{-3} \text{ A}}{e^{0.5800} - 1} = 1.177 \times 10^{-2} = 11.77$ mA

Then can calculate I for $V = 10.0$ mV: $\dfrac{eV}{kT} = \dfrac{(1.602 \times 10^{-19} \text{ C})(10.0 \times 10^{-3} \text{ V})}{(1.381 \times 10^{-23} \text{ J/K})(300 \text{ K})} = 0.3867$

$I = I_s(e^{eV/kT} - 1) = (11.77 \text{ mA})(e^{0.3867} - 1) = 5.56$ mA

(b) $\dfrac{eV}{kT}$ has the same magnitude as in part (a) but not V is negative so $\dfrac{eV}{kT}$ is negative.

$V = -15.0$ mV : $\dfrac{eV}{kT} = -0.5800$ and $I = I_s(e^{eV/kT} - 1) = (11.77 \text{ mA})(e^{-0.5800} - 1) = -5.18$ mA

$V = -10.0$ mV : $\dfrac{eV}{kT} = -0.3867$ and $I = I_s(e^{eV/kT} - 1) = (11.77 \text{ mA})(e^{-0.3867} - 1) = -3.77$ mA

EVALUATE: There is a directional asymmetry in the current, with a forward-bias voltage producing more current than a reverse-bias voltage of the same magnitude, but the voltage is small enough for the asymmetry not be pronounced.

42.35.   **IDENTIFY:** During the transition, the molecule emits a photon of light having energy equal to the energy difference between the two vibrational states of the molecule.

SET UP: The vibrational energy is $E_n = \left(n + \dfrac{1}{2}\right)\hbar\omega = \left(n + \dfrac{1}{2}\right)\hbar\sqrt{\dfrac{k'}{m_r}}$.

EXECUTE: **(a)** The energy difference between two adjacent energy states is $\Delta E = \hbar\sqrt{\dfrac{k'}{m_r}}$, and this is the energy of the photon, so $\Delta E = hc/\lambda$. Equating these two expressions for ΔE and solving for k', we have

$k' = m_r \left(\dfrac{\Delta E}{\hbar}\right)^2 = \dfrac{m_H m_O}{m_H + m_O}\left(\dfrac{\Delta E}{\hbar}\right)^2$, and using $\dfrac{\Delta E}{\hbar} = \dfrac{hc/\lambda}{\hbar} = \dfrac{2\pi c}{\lambda}$ with the appropriate numbers gives us

$k' = \dfrac{(1.67 \times 10^{-27} \text{ kg})(2.656 \times 10^{-26} \text{ kg})}{1.67 \times 10^{-27} \text{ kg} + 2.656 \times 10^{-26} \text{ kg}}\left[\dfrac{2\pi(3.00 \times 10^8 \text{ m/s})}{2.39 \times 10^{-6} \text{ m}}\right]^2 = 977$ N/m

(b) $f = \dfrac{\omega}{2\pi} = \dfrac{1}{2\pi}\sqrt{\dfrac{k'}{m_r}} = \dfrac{1}{2\pi}\sqrt{\dfrac{\dfrac{m_H m_O}{m_H + m_O}}{k'}}$. Substituting the appropriate numbers gives us

$$f = \dfrac{1}{2\pi}\sqrt{\dfrac{\dfrac{(1.67\times10^{-27}\ \text{kg})(2.656\times10^{-26}\ \text{kg})}{1.67\times10^{-27}\ \text{kg} + 2.656\times10^{-26}\ \text{kg}}}{977\ \text{N/m}}} = 1.25\times10^{14}\ \text{Hz}$$

EVALUATE: The frequency is close to, but not quite in, the visible range.

42.37. **IDENTIFY** and **SET UP:** Eq. (21.14) gives the electric dipole moment as $p = qd$, where the dipole consists of charges $\pm q$ separated by distance d.

EXECUTE: **(a)** Point charges $+e$ and $-e$ separated by distance d, so
$p = ed = (1.602\times10^{-19}\ \text{C})(0.24\times10^{-9}\ \text{m}) = 3.8\times10^{-29}\ \text{C}\cdot\text{m}$

(b) $p = qd$ so $q = \dfrac{p}{d} = \dfrac{3.0\times10^{-29}\ \text{C}\cdot\text{m}}{0.24\times10^{-9}\ \text{m}} = 1.3\times10^{-19}\ \text{C}$

(c) $\dfrac{q}{e} = \dfrac{1.3\times10^{-19}\ \text{C}}{1.602\times10^{-19}\ \text{C}} = 0.81$

(d) $q = \dfrac{p}{d} = \dfrac{1.5\times10^{-30}\ \text{C}\cdot\text{m}}{0.16\times10^{-9}\ \text{m}} = 9.37\times10^{-21}\ \text{C}$

$\dfrac{q}{e} = \dfrac{9.37\times10^{-21}\ \text{C}}{1.602\times10^{-19}\ \text{C}} = 0.058$

EVALUATE: The fractional ionic character for the bond in HI is much less than the fractional ionic character for the bond in NaCl. The bond in HI is mostly covalent and not very ionic.

42.39. **(a) IDENTIFY:** $E(\text{Na}) + E(\text{Cl}) = E(\text{Na}^+) + E(\text{Cl}^-) + U(r)$. Solving for $U(r)$ gives

$U(r) = -[E(\text{Na}^+) - E(\text{Na})] + [E(\text{Cl}) - E(\text{Cl}^-)]$.

SET UP: $[E(\text{Na}^+) - E(\text{Na})]$ is the ionization energy of Na, the energy required to remove one electron, and is equal to 5.1 eV. $[E(\text{Cl}) - E(\text{Cl}^-)]$ is the electron affinity of Cl, the magnitude of the decrease in energy when an electron is attached to a neutral Cl atom, and is equal to 3.6 eV.

EXECUTE: $U = -5.1\ \text{eV} + 3.6\ \text{eV} = -1.5\ \text{eV} = -2.4\times10^{-19}\ \text{J}$, and $-\dfrac{1}{4\pi\epsilon_0}\dfrac{e^2}{r} = -2.4\times10^{-19}\ \text{J}$

$r = \left(\dfrac{1}{4\pi\epsilon_0}\right)\dfrac{e^2}{2.4\times10^{-19}\ \text{J}} = (8.988\times10^9\ \text{N}\cdot\text{m}^2/\text{C}^2)\dfrac{(1.602\times10^{-19}\ \text{C})^2}{2.4\times10^{-19}\ \text{J}}$

$r = 9.6\times10^{-10}\ \text{m} = 0.96\ \text{nm}$

(b) ionization energy of K = 4.3 eV; electron affinity of Br = 3.5 eV

Thus $U = -4.3\ \text{eV} + 3.5\ \text{eV} = -0.8\ \text{eV} = -1.28\times10^{-19}\ \text{J}$, and $-\dfrac{1}{4\pi\epsilon_0}\dfrac{e^2}{r} = -1.28\times10^{-19}\ \text{J}$

$r = \left(\dfrac{1}{4\pi\epsilon_0}\right)\dfrac{e^2}{1.28\times10^{-19}\ \text{J}} = (8.988\times10^9\ \text{N}\cdot\text{m}^2/\text{C}^2)\dfrac{(1.602\times10^{-19}\ \text{C})^2}{1.28\times10^{-19}\ \text{J}}$

$r = 1.8\times10^{-9}\ \text{m} = 1.8\ \text{nm}$

EVALUATE: K has a smaller ionization energy than Na and the electron affinities of Cl and Br are very similar, so it takes less energy to make $\text{K}^+ + \text{Br}^-$ from K + Br than to make $\text{Na}^+ + \text{Cl}^-$ from Na + Cl. Thus, the stabilization distance is larger for KBr than for NaCl.

42.45. **IDENTIFY** and **SET UP:** $E_l = l(l+1)\hbar^2/2I$, so E_l and the transition energy ΔE depend on I. Different isotopic molecules have different I.

EXECUTE: (a) Calculate I for $Na^{35}Cl$:

$$m_r = \frac{m_{Na} m_{Cl}}{m_{Na} + m_{Cl}} = \frac{(3.8176 \times 10^{-26} \text{ kg})(5.8068 \times 10^{-26} \text{ kg})}{3.8176 \times 10^{-26} \text{ kg} + 5.8068 \times 10^{-26} \text{ kg}} = 2.303 \times 10^{-26} \text{ kg}$$

$$I = m_r r^2 = (2.303 \times 10^{-26} \text{ kg})(0.2361 \times 10^{-9} \text{ m})^2 = 1.284 \times 10^{-45} \text{ kg} \cdot \text{m}^2$$

$l = 2 \rightarrow l = 1$ transition

$$\Delta E = E_2 - E_1 = (6-2)\left(\frac{\hbar^2}{2I}\right) = \frac{2\hbar^2}{I} = \frac{2(1.055 \times 10^{-34} \text{ J} \cdot \text{s})^2}{1.284 \times 10^{-45} \text{ kg} \cdot \text{m}^2} = 1.734 \times 10^{-23} \text{ J}$$

$$\Delta E = \frac{hc}{\lambda} \text{ so } \lambda = \frac{hc}{\Delta E} = \frac{(6.626 \times 10^{-34} \text{ J} \cdot \text{s})(2.998 \times 10^8 \text{ m/s})}{1.734 \times 10^{-23} \text{ J}} = 1.146 \times 10^{-2} \text{ m} = 1.146 \text{ cm}$$

$l = 1 \rightarrow l = 0$ transition

$$\Delta E = E_1 - E_0 = (2-0)\left(\frac{\hbar^2}{2I}\right) = \frac{\hbar^2}{I} = \frac{1}{2}(1.734 \times 10^{-23} \text{ J}) = 8.67 \times 10^{-24} \text{ J}$$

$$\lambda = \frac{hc}{\Delta E} = \frac{(6.626 \times 10^{-34} \text{ J} \cdot \text{s})(2.998 \times 10^8 \text{ m/s})}{8.67 \times 10^{-24} \text{ J}} = 2.291 \text{ cm}$$

(b) Calculate I for $Na^{37}Cl$: $m_r = \frac{m_{Na} m_{Cl}}{m_{Na} + m_{Cl}} = \frac{(3.8176 \times 10^{-26} \text{ kg})(6.1384 \times 10^{-26} \text{ kg})}{3.8176 \times 10^{-26} \text{ kg} + 6.1384 \times 10^{-26} \text{ kg}} = 2.354 \times 10^{-26} \text{ kg}$

$$I = m_r r^2 = (2.354 \times 10^{-26} \text{ kg})(0.2361 \times 10^{-9} \text{ m})^2 = 1.312 \times 10^{-45} \text{ kg} \cdot \text{m}^2$$

$l = 2 \rightarrow l = 1$ transition

$$\Delta E = \frac{2\hbar^2}{I} = \frac{2(1.055 \times 10^{-34} \text{ J} \cdot \text{s})^2}{1.312 \times 10^{-45} \text{ kg} \cdot \text{m}^2} = 1.697 \times 10^{-23} \text{ J}$$

$$\lambda = \frac{hc}{\Delta E} = \frac{(6.626 \times 10^{-34} \text{ J} \cdot \text{s})(2.998 \times 10^8 \text{ m/s})}{1.697 \times 10^{-23} \text{ J}} = 1.171 \times 10^{-2} \text{ m} = 1.171 \text{ cm}$$

$l = 1 \rightarrow l = 0$ transition

$$\Delta E = \frac{\hbar^2}{I} = \frac{1}{2}(1.697 \times 10^{-23} \text{ J}) = 8.485 \times 10^{-24} \text{ J}$$

$$\lambda = \frac{hc}{\Delta E} = \frac{(6.626 \times 10^{-34} \text{ J} \cdot \text{s})(2.998 \times 10^8 \text{ m/s})}{8.485 \times 10^{-24} \text{ J}} = 2.341 \text{ cm}$$

The differences in the wavelengths for the two isotopes are:
$l = 2 \rightarrow l = 1$ transition: $1.171 \text{ cm} - 1.146 \text{ cm} = 0.025 \text{ cm}$
$l = 1 \rightarrow l = 0$ transition: $2.341 \text{ cm} - 2.291 \text{ cm} = 0.050 \text{ cm}$

EVALUATE: Replacing ^{35}Cl by ^{37}Cl increases I, decreases ΔE and increases λ. The effect on λ is small but measurable.

42.47. **IDENTIFY:** The vibrational energy levels are given by $E_n = \left(n + \frac{1}{2}\right)\hbar \sqrt{\frac{k'}{m_r}}$. The zero-point energy is

$$E_0 = \frac{1}{2}\hbar \sqrt{\frac{2k'}{m_H}}.$$

SET UP: For H_2, $m_r = \frac{m_H}{2}$.

EXECUTE: $E_0 = \frac{1}{2}(1.054 \times 10^{-34} \text{ J} \cdot \text{s})\sqrt{\frac{2(576 \text{ N/m})}{1.67 \times 10^{-27} \text{ kg}}} = 4.38 \times 10^{-20} \text{ J} = 0.274 \text{ eV}.$

EVALUATE: This is much less than the magnitude of the H_2 bond energy.

42.49. **IDENTIFY and SET UP:** Use Eq. (42.6) to calculate I. The energy levels are given by Eq. (42.9). The transition energy ΔE is related to the photon wavelength by $\Delta E = hc/\lambda$.

EXECUTE: **(a)** $m_r = \dfrac{m_H m_I}{m_H + m_I} = \dfrac{(1.67 \times 10^{-27} \text{ kg})(2.11 \times 10^{-25} \text{ kg})}{1.67 \times 10^{-27} \text{ kg} + 2.11 \times 10^{-25} \text{ kg}} = 1.657 \times 10^{-27} \text{ kg}$

$I = m_r r^2 = (1.657 \times 10^{-27} \text{ kg})(0.160 \times 10^{-9} \text{ m})^2 = 4.24 \times 10^{-47} \text{ kg} \cdot \text{m}^2$

(b) The energy levels are $E_{nl} = l(l+1)\left(\dfrac{\hbar^2}{2I}\right) + \left(n + \tfrac{1}{2}\right)\hbar\sqrt{\dfrac{k'}{m_r}}$ (Eq. (42.9))

$\sqrt{\dfrac{k'}{m}} = \omega = 2\pi f$ so $E_{nl} = l(l+1)\left(\dfrac{\hbar^2}{2I}\right) + \left(n + \tfrac{1}{2}\right)hf$

(i) transition $n = 1 \rightarrow n = 0, l = 1 \rightarrow l = 0$

$\Delta E = (2 - 0)\left(\dfrac{\hbar^2}{2I}\right) + \left(1 + \tfrac{1}{2} - \tfrac{1}{2}\right)hf = \dfrac{\hbar^2}{I} + hf$

$\Delta E = \dfrac{hc}{\lambda}$ so $\lambda = \dfrac{hc}{\Delta E} = \dfrac{hc}{(\hbar^2/I) + hf} = \dfrac{c}{(\hbar/2\pi I) + f}$

$\dfrac{\hbar}{2\pi I} = \dfrac{1.055 \times 10^{-34} \text{ J} \cdot \text{s}}{2\pi(4.24 \times 10^{-47} \text{ kg} \cdot \text{m}^2)} = 3.960 \times 10^{11} \text{ Hz}$

$\lambda = \dfrac{c}{(\hbar/2\pi I) + f} = \dfrac{2.998 + 10^8 \text{ m/s}}{3.960 \times 10^{11} \text{ Hz} + 6.93 \times 10^{13} \text{ Hz}} = 4.30 \; \mu\text{m}$

(ii) transition $n = 1 \rightarrow n = 0, l = 2 \rightarrow l = 1$

$\Delta E = (6 - 2)\left(\dfrac{\hbar^2}{2I}\right) + hf = \dfrac{2\hbar^2}{I} + hf$

$\lambda = \dfrac{c}{2(\hbar/2\pi I) + f} = \dfrac{2.998 \times 10^8 \text{ m/s}}{2(3.960 \times 10^{11} \text{ Hz}) + 6.93 \times 10^{13} \text{ Hz}} = 4.28 \; \mu\text{m}$

(iii) transition $n = 2 \rightarrow n = 1, l = 2 \rightarrow l = 3$

$\Delta E = (6 - 12)\left(\dfrac{\hbar^2}{2I}\right) + hf = -\dfrac{3\hbar^2}{I} + hf$

$\lambda = \dfrac{c}{-3(\hbar/2\pi I) + f} = \dfrac{2.998 \times 10^8 \text{ m/s}}{-3(3.960 \times 10^{11} \text{ Hz}) + 6.93 \times 10^{13} \text{ Hz}} = 4.40 \; \mu\text{m}$

EVALUATE: The vibrational energy change for the $n = 1 \rightarrow n = 0$ transition is the same as for the $n = 2 \rightarrow n = 1$ transition. The rotational energies are much smaller than the vibrational energies, so the wavelengths for all three transitions don't differ much.

42.51.   **IDENTIFY:** E_{F0} is given by Eq. (42.20). Since potassium is a metal and E does not change much with T for metals, we approximate E_F by E_{F0}, so $E_F = \dfrac{3^{2/3} \pi^{4/3} \hbar^2 n^{2/3}}{2m}$.

SET UP: The number of atoms per m^3 is ρ/m. If each atom contributes one free electron, the electron concentration is $n = \dfrac{\rho}{m} = \dfrac{851 \text{ kg/m}^3}{6.49 \times 10^{-26} \text{ kg}} = 1.31 \times 10^{28}$ electrons/m^3.

EXECUTE: $E_F = \dfrac{3^{2/3} \pi^{4/3} (1.054 \times 10^{-34} \text{ J} \cdot \text{s})^2 (1.31 \times 10^{28} /\text{m}^3)^{2/3}}{2(9.11 \times 10^{-31} \text{ kg})} = 3.24 \times 10^{-19} \text{ J} = 2.03 \text{ eV}.$

EVALUATE: The E_F we calculated for potassium is about a factor of three smaller than the E_F for copper that was calculated in Example 42.7.

42.53.   **IDENTIFY** and **SET UP:**   Use the description of the bcc lattice in Fig.42.11c in the textbook to calculate the number of atoms per unit cell and then the number of atoms per unit volume.

EXECUTE: (a) Each unit cell has one atom at its center and 8 atoms at its corners that are each shared by 8 other unit cells. So there are $1 + 8/8 = 2$ atoms per unit cell.

$$\frac{n}{V} = \frac{2}{(0.35 \times 10^{-9}\,\text{m})^3} = 4.66 \times 10^{-8}\ \text{atoms/m}^3$$

(b) $E_{F0} = \dfrac{3^{2/3}\pi^{4/3}\hbar^2}{2m}\left(\dfrac{N}{V}\right)^{2/3}$

In this equation N/V is the number of free electrons per m^3. But the problem says to assume one free electron per atom, so this is the same as n/V calculated in part (a).

$m = 9.109 \times 10^{-31}$ kg (the electron mass), so $E_{F0} = 7.563 \times 10^{-19}$ J $= 4.7$ eV

EVALUATE: Our result for metallic lithium is similar to that calculated for copper in Example 42.7.

NUCLEAR PHYSICS

43.3. **IDENTIFY:** Calculate the spin magnetic energy shift for each spin state of the $1s$ level. Calculate the energy splitting between these states and relate this to the frequency of the photons.

SET UP: When the spin component is parallel to the field the interaction energy is $U = -\mu_z B$. When the spin component is antiparallel to the field the interaction energy is $U = +\mu_z B$. The transition energy for a transition between these two states is $\Delta E = 2\mu_z B$, where $\mu_z = 2.7928\mu_n$. The transition energy is related to the photon frequency by $\Delta E = hf$, so $2\mu_z B = hf$.

EXECUTE: $B = \dfrac{hf}{2\mu_z} = \dfrac{(6.626\times10^{-34}\text{ J}\cdot\text{s})(22.7\times10^{6}\text{ Hz})}{2(2.7928)(5.051\times10^{-27}\text{ J/T})} = 0.533\text{ T}$

EVALUATE: This magnetic field is easily achievable. Photons of this frequency have wavelength $\lambda = c/f = 13.2$ m. These are radio waves.

43.7. **IDENTIFY and SET UP:** The text calculates that the binding energy of the deuteron is 2.224 MeV. A photon that breaks the deuteron up into a proton and a neutron must have at least this much energy.

$E = \dfrac{hc}{\lambda}$ so $\lambda = \dfrac{hc}{E}$

EXECUTE: $\lambda = \dfrac{(4.136\times10^{-15}\text{ eV}\cdot\text{s})(2.998\times10^{8}\text{ m/s})}{2.224\times10^{6}\text{ eV}} = 5.575\times10^{-13}$ m $= 0.5575$ pm.

EVALUATE: This photon has gamma-ray wavelength.

43.15. **IDENTIFY:** Compare the mass of the original nucleus to the total mass of the decay products.

SET UP: Subtract the electron masses from the neutral atom mass to obtain the mass of each nucleus.

EXECUTE: If β^- decay of ^{14}C is possible, then we are considering the decay $^{14}_{6}\text{C} \rightarrow {}^{14}_{7}\text{N} + \beta^-$.

$\Delta m = M(^{14}_{6}\text{C}) - M(^{14}_{7}\text{N}) - m_e$

$\Delta m = (14.003242\text{ u} - 6(0.000549\text{ u})) - (14.003074\text{ u} - 7(0.000549\text{ u})) - 0.0005491\text{ u}$

$\Delta m = +1.68\times10^{-4}$ u. So $E = (1.68\times10^{-4}\text{ u})(931.5\text{ MeV/u}) = 0.156$ MeV $= 156$ keV

EVALUATE: In the decay the total charge and the nucleon number are conserved.

43.17. **IDENTIFY:** The energy released is the energy equivalent of the difference in the masses of the original atom and the final atom produced in the capture. Apply conservation of energy to the decay products.

SET UP: 1 u is equivalent to 931.5 MeV.

EXECUTE: **(a)** As in the example, $(0.000897\text{ u})(931.5\text{ MeV/u}) = 0.836$ MeV.

(b) $0.836\text{ MeV} - 0.122\text{ MeV} - 0.014\text{ MeV} = 0.700$ MeV.

EVALUATE: We have neglected the rest mass of the neutrino that is emitted.

43.19. **IDENTIFY and SET UP:** $T_{1/2} = \dfrac{\ln 2}{\lambda}$ The mass of a single nucleus is $124 m_p = 2.07\times10^{-25}$ kg.

$|dN/dt| = 0.350$ Ci $= 1.30\times10^{10}$ Bq, $|dN/dt| = \lambda N$.

EXECUTE: $N = \dfrac{6.13 \times 10^{-3} \text{ kg}}{2.07 \times 10^{-25} \text{ kg}} = 2.96 \times 10^{22}$; $\lambda = \dfrac{|dN/dt|}{N} = \dfrac{1.30 \times 10^{10} \text{ Bq}}{2.96 \times 10^{22}} = 4.39 \times 10^{-13} \text{ s}^{-1}$.

$T_{1/2} = \dfrac{\ln 2}{\lambda} = 1.58 \times 10^{12} \text{ s} = 5.01 \times 10^4 \text{ y}$.

EVALUATE: Since $T_{1/2}$ is very large, the activity changes very slowly.

43.21.   **IDENTIFY:**   From the known half-life, we can find the decay constant, the rate of decay, and the activity.

SET UP: $\lambda = \dfrac{\ln 2}{T_{1/2}}$. $T_{1/2} = 4.47 \times 10^9 \text{ yr} = 1.41 \times 10^{17} \text{ s}$. The activity is $\left|\dfrac{dN}{dt}\right| = \lambda N$. The mass of one ^{238}U

is approximately $238 m_p$. $1 \text{ Ci} = 3.70 \times 10^{10}$ decays/s.

EXECUTE: **(a)** $\lambda = \dfrac{\ln 2}{1.41 \times 10^{17} \text{ s}} = 4.92 \times 10^{-18} \text{ s}^{-1}$.

(b) $N = \dfrac{|dN/dt|}{\lambda} = \dfrac{3.70 \times 10^{10} \text{ Bq}}{4.92 \times 10^{-18} \text{ s}^{-1}} = 7.52 \times 10^{27}$ nuclei. The mass m of uranium is the number of nuclei

times the mass of each one. $m = (7.52 \times 10^{27})(238)(1.67 \times 10^{-27} \text{ kg}) = 2.99 \times 10^3 \text{ kg}$.

(c) $N = \dfrac{10.0 \times 10^{-3} \text{ kg}}{238 m_p} = \dfrac{10.0 \times 10^{-3} \text{ kg}}{238(1.67 \times 10^{-27} \text{ kg})} = 2.52 \times 10^{22}$ nuclei.

$\left|\dfrac{dN}{dt}\right| = \lambda N = (4.92 \times 10^{-18} \text{ s}^{-1})(2.52 \times 10^{22}) = 1.24 \times 10^5$ decays/s.

EVALUATE: Because ^{238}U has a very long half-life, it requires a large amount (about 3000 kg) to have an activity of a 1.0 Ci.

43.23.   **IDENTIFY and SET UP:**   As discussed in Section 43.4, the activity $A = |dN/dt|$ obeys the same decay

equation as Eq. (43.17): $A = A_0 e^{-\lambda t}$. For ^{14}C, $T_{1/2} = 5730$ y and $\lambda = \ln 2 / T_{1/2}$ so $A = A_0 e^{-(\ln 2) t / T_{1/2}}$; calculate

A at each t; $A_0 = 180.0$ decays/min.

EXECUTE: **(a)** $t = 1000$ y, $A = 159$ decays/min

(b) $t = 50{,}000$ y, $A = 0.43$ decays/min

EVALUATE: The time in part (b) is 8.73 half-lives, so the decay rate has decreased by a factor or $\left(\dfrac{1}{2}\right)^{8.73}$.

43.25.   **IDENTIFY and SET UP:**   Find λ from the half-life and the number N of nuclei from the mass of one nucleus and the mass of the sample. Then use Eq. (43.16) to calculate $|dN/dt|$, the number of decays per second.

EXECUTE: **(a)** $|dN/dt| = \lambda N$

$\lambda = \dfrac{0.693}{T_{1/2}} = \dfrac{0.693}{(1.28 \times 10^9 \text{ y})(3.156 \times 10^7 \text{ s/1 y})} = 1.715 \times 10^{-17} \text{ s}^{-1}$

The mass of one ^{40}K atom is approximately 40 u, so the number of ^{40}K nuclei in the sample is

$N = \dfrac{1.63 \times 10^{-9} \text{ kg}}{40 \text{ u}} = \dfrac{1.63 \times 10^{-9} \text{ kg}}{40(1.66054 \times 10^{-27} \text{ kg})} = 2.454 \times 10^{16}$.

Then $|dN/dt| = \lambda N = (1.715 \times 10^{-17} \text{ s}^{-1})(2.454 \times 10^{16}) = 0.421$ decays/s

(b) $|dN/dt| = (0.421 \text{ decays/s})(1 \text{ Ci}/(3.70 \times 10^{10} \text{ decays/s})) = 1.14 \times 10^{-11} \text{ Ci}$

EVALUATE: The very small sample still contains a very large number of nuclei. But the half-life is very large, so the decay rate is small.

43.31.   **IDENTIFY:**   Knowing the equivalent dose in Sv, we want to find the absorbed energy.

SET UP: equivalent dose (Sv, rem) = RBE × absorbed dose(Gy, rad); 100 rad = 1 Gy

EXECUTE: **(a)** RBE = 1, so 0.25 mSv corresponds to 0.25 mGy.

Energy = $(0.25 \times 10^{-3} \text{ J/kg})/(5.0 \text{ kg}) = 1.2 \times 10^{-3} \text{ J}$.

(b) RBE $= 1$ so 0.10 mGy $= 10$ mrad and 10 mrem. $(0.10 \times 10^{-3}$ J/kg$)(75$ kg$) = 7.5 \times 10^{-3}$ J.

EVALUATE: **(c)** $\dfrac{7.5 \times 10^{-3} \text{ J}}{1.2 \times 10^{-3} \text{ J}} = 6.2.$ Each chest x ray delivers only about $1/6$ of the yearly background radiation energy.

43.33. **IDENTIFY** and **SET UP:** The unit for absorbed dose is 1 rad $= 0.01$ J/kg $= 0.01$ Gy. Equivalent dose in rem is RBE times absorbed dose in rad.

EXECUTE: 1 rad $= 10^{-2}$ Gy, so 1 Gy $= 100$ rad and the dose was 500 rad.

rem $= $ (rad)(RBE) $= (500$ rad$)(4.0) = 2000$ rem. 1 Gy $= 1$ J/kg, so 5.0 J/kg.

EVALUATE: Gy, rad and J/kg are all units of absorbed dose. Rem is a unit of equivalent dose, which depends on the RBE of the radiation.

43.35. **IDENTIFY** and **SET UP:** For x rays RBE $= 1$ and the equivalent dose equals the absorbed dose.

EXECUTE: **(a)** 175 krad $= 175$ krem $= 1.75$ kGy $= 1.75$ kSv. $(1.75 \times 10^3$ J/kg$)(0.220$ kg$) = 385$ J.

(b) 175 krad $= 1.75$ kGy; $(1.50)(175$ krad$) = 262.5$ krem $= 2.625$ kSv. The energy deposited would be 385 J, the same as in (a).

EVALUATE: The energy required to raise the temperature of 0.150 kg of water 1 C° is 628 J, and 385 J is less than this. The energy deposited corresponds to a very small amount of heating.

43.37. **IDENTIFY:** Apply Eq. (43.16), with $\lambda = \ln 2 / T_{1/2}$, to find the number of tritium atoms that were ingested. Then use Eq. (43.17) to find the number of decays in one week.

SET UP: 1 rad $= 0.01$ J/kg. rem $= $ RBE $\times$ rad.

EXECUTE: **(a)** We need to know how many decays per second occur.

$\lambda = \dfrac{0.693}{T_{1/2}} = \dfrac{0.693}{(12.3 \text{ y})(3.156 \times 10^7 \text{ s/y})} = 1.785 \times 10^{-9} \text{ s}^{-1}.$ The number of tritium atoms is

$N_0 = \dfrac{1}{\lambda}\left|\dfrac{dN}{dt}\right| = \dfrac{(0.35 \text{ Ci})(3.70 \times 10^{10} \text{ Bq/Ci})}{1.79 \times 10^{-9} \text{ s}^{-1}} = 7.2540 \times 10^{18}$ nuclei. The number of remaining nuclei after

one week is $N = N_0 e^{-\lambda t} = (7.25 \times 10^{18})e^{-(1.79 \times 10^{-9} \text{ s}^{-1})(7)(24)(3600 \text{ s})} = 7.2462 \times 10^{18}$ nuclei.

$\Delta N = N_0 - N = 7.8 \times 10^{15}$ decays. So the energy absorbed is

$E_{\text{total}} = \Delta N \, E_\gamma = (7.8 \times 10^{15})(5000 \text{ eV})(1.60 \times 10^{-19} \text{ J/eV}) = 6.25$ J. The absorbed dose is

$\dfrac{6.25 \text{ J}}{67 \text{ kg}} = 0.0932$ J/kg $= 9.32$ rad. Since RBE $= 1$, then the equivalent dose is 9.32 rem.

EVALUATE: **(b)** In the decay, antineutrinos are also emitted. These are not absorbed by the body, and so some of the energy of the decay is lost.

43.39. **(a) IDENTIFY** and **SET UP:** Determine X by balancing the charge and nucleon number on the two sides of the reaction equation.

EXECUTE: X must have $A = 2 + 14 - 10 = 6$ and $Z = 1 + 7 - 5 = 3$. Thus X is ${}^{6}_{3}\text{Li}$ and the reaction is

${}^{2}_{1}\text{H} + {}^{14}_{7}\text{N} \rightarrow {}^{6}_{3}\text{Li} + {}^{10}_{5}\text{B}.$

(b) IDENTIFY and **SET UP:** Calculate the mass decrease and find its energy equivalent.

EXECUTE: The neutral atoms on each side of the reaction equation have a total of 8 electrons, so the electron masses cancel when neutral atom masses are used. The neutral atom masses are found in Table 43.2.

mass of ${}^{2}_{1}\text{H} + {}^{14}_{7}\text{N}$ is 2.014102 u $+ 14.003074$ u $= 16.017176$ u

mass of ${}^{6}_{3}\text{Li} + {}^{10}_{5}\text{B}$ is 6.015121 u $+ 10.012937$ u $= 16.028058$ u

The mass increases, so energy is absorbed by the reaction. The Q value is
$(16.017176 \text{ u} - 16.028058 \text{ u})(931.5 \text{ MeV/u}) = -10.14$ MeV

(c) IDENTIFY and **SET UP:** The available energy in the collision, the kinetic energy K_{cm} in the center of mass reference frame, is related to the kinetic energy K of the bombarding particle by Eq. (43.24).

EXECUTE: The kinetic energy that must be available to cause the reaction is 10.14 MeV. Thus $K_{cm} = 10.14$ MeV. The mass M of the stationary target ($^{14}_7$N) is $M = 14$ u. The mass m of the colliding particle ($^2_1$H) is 2 u. Then by Eq. (43.24) the minimum kinetic energy K that the $^2_1$H must have is

$$K = \left(\frac{M+m}{M}\right) K_{cm} = \left(\frac{14\text{ u} + 2\text{ u}}{14\text{ u}}\right)(10.14\text{ MeV}) = 11.59\text{ MeV}.$$

EVALUATE: The projectile ($^2_1$H) is much lighter than the target ($^{14}_7$N) so K is not much larger than K_{cm}. The K we have calculated is what is required to allow the mass increase. We would also need to check to see if at this energy the projectile can overcome the Coulomb repulsion to get sufficiently close to the target nucleus for the reaction to occur.

43.41.   **IDENTIFY** and **SET UP:**   Determine X by balancing the charge and the nucleon number on the two sides of the reaction equation.

EXECUTE: X must have $A = +2 + 9 - 4 = 7$ and $Z = +1 + 4 - 2 = 3$. Thus X is $^7_3$Li and the reaction is

$$^2_1\text{H} + ^9_4\text{Be} = ^7_3\text{Li} + ^4_2\text{He}$$

(b) IDENTIFY and **SET UP:**   Calculate the mass decrease and find its energy equivalent.

EXECUTE: If we use the neutral atom masses then there are the same number of electrons (five) in the reactants as in the products. Their masses cancel, so we get the same mass defect whether we use nuclear masses or neutral atom masses. The neutral atoms masses are given in Table 43.2.

$^2_1$H + $^9_4$Be has mass 2.014102 u + 9.012182 u = 11.26284 u

$^7_3$Li + $^4_2$He has mass 7.016003 u + 4.002603 u = 11.018606 u

The mass decrease is 11.026284 u − 11.018606 u = 0.007678 u.

This corresponds to an energy release of 0.007678 u(931.5 MeV/1 u) = 7.152 MeV.

(c) IDENTIFY and **SET UP:**   Estimate the threshold energy by calculating the Coulomb potential energy when the $^2_1$H and $^9_4$Be nuclei just touch. Obtain the nuclear radii from Eq. (43.1).

EXECUTE: The radius R_{Be} of the $^9_4$Be nucleus is $R_{Be} = (1.2 \times 10^{-15}$ m$)(9)^{1/3} = 2.5 \times 10^{-15}$ m.

The radius R_H of the $^2_1$H nucleus is $R_H = (1.2 \times 10^{-15}$ m$)(2)^{1/3} = 1.5 \times 10^{-15}$ m.

The nuclei touch when their center-to-center separation is

$R = R_{Be} + R_H = 4.0 \times 10^{-15}$ m.

The Coulomb potential energy of the two reactant nuclei at this separation is

$$U = \frac{1}{4\pi\epsilon_0}\frac{q_1 q_2}{r} = \frac{1}{4\pi\epsilon_0}\frac{e(4e)}{r}$$

$$U = (8.988 \times 10^9 \text{ N}\cdot\text{m}^2/\text{C}^2)\frac{4(1.602 \times 10^{-19} \text{ C})^2}{(4.0 \times 10^{-15} \text{ m})(1.602 \times 10^{-19} \text{ J/eV})} = 1.4 \text{ MeV}$$

This is an estimate of the threshold energy for this reaction.

EVALUATE: The reaction releases energy but the total initial kinetic energy of the reactants must be 1.4 MeV in order for the reacting nuclei to get close enough to each other for the reaction to occur. The nuclear force is strong but is very short-range.

43.43.   **IDENTIFY** and **SET UP:**   The energy released is the energy equivalent of the mass decrease. 1 u is equivalent to 931.5 MeV. The mass of one ^{235}U nucleus is $235 m_p$.

EXECUTE: **(a)** $^{235}_{92}\text{U} + ^1_0\text{n} \rightarrow ^{144}_{56}\text{Ba} + ^{89}_{36}\text{Kr} + 3^1_0\text{n}$. We can use atomic masses since the same number of electrons are included on each side of the reaction equation and the electron masses cancel. The mass decrease is $\Delta M = m(^{235}_{92}\text{U}) + m(^1_0\text{n}) - [m(^{144}_{56}\text{Ba}) + m(^{89}_{36}\text{Kr}) + 3m(^1_0\text{n})]$,

$\Delta M = 235.043930$ u + 1.0086649 u − 143.922953 u − 88.917630 u − 3(1.0086649 u), $\Delta M = 0.1860$ u. The energy released is (0.1860 u)(931.5 MeV/u) = 173.3 MeV.

(b) The number of ^{235}U nuclei in 1.00 g is $\dfrac{1.00 \times 10^{-3} \text{ kg}}{235 m_p} = 2.55 \times 10^{21}$. The energy released per gram is

$(173.3 \text{ MeV/nucleus})(2.55 \times 10^{21} \text{ nuclei/g}) = 4.42 \times 10^{23} \text{ MeV/g}$.

EVALUATE: The energy released is 7.1×10^{10} J/kg. This is much larger than typical heats of combustion, which are about 5×10^4 J/kg.

43.45. **IDENTIFY:** The energy released is the energy equivalent of the mass decrease that occurs in the reaction.
SET UP: 1 u is equivalent to 931.5 MeV.
EXECUTE: The energy liberated will be

$M(^3_2\text{He}) + M(^4_2\text{He}) - M(^7_4\text{Be}) = (3.016029 \text{ u} + 4.002603 \text{ u} - 7.016929 \text{ u})(931.5 \text{ MeV/u}) = 1.586 \text{ MeV}$.

EVALUATE: Using neutral atom masses includes four electrons on each side of the reaction equation and the result is the same as if nuclear masses had been used.

43.53. **IDENTIFY:** The minimum energy to remove a proton or a neutron from the nucleus is equal to the energy difference between the two states of the nucleus, before and after removal.
(a) SET UP: $^{17}_8\text{O} = {}^1_0\text{n} + {}^{16}_8\text{O}$. $\Delta m = m({}^1_0\text{n}) + m({}^{16}_8\text{O}) - m({}^{17}_8\text{O})$. The electron masses cancel when neutral atom masses are used.
EXECUTE: $\Delta m = 1.008665 \text{ u} + 15.994915 \text{ u} - 16.999132 \text{ u} = 0.004448 \text{ u}$. The energy equivalent of this mass increase is $(0.004448 \text{ u})(931.5 \text{ MeV/u}) = 4.14 \text{ MeV}$.

(b) SET UP and EXECUTE: Following the same procedure as in part (a) gives

$\Delta M = 8 M_\text{H} + 9 M_\text{n} - {}^{17}_8 M = 8(1.007825 \text{ u}) + 9(1.008665 \text{ u}) - 16.999132 \text{ u} = 0.1415 \text{ u}$.

$E_\text{B} = (0.1415 \text{ u})(931.5 \text{ MeV/u}) = 131.8 \text{ MeV}$. $\dfrac{E_\text{B}}{A} = 7.75 \text{ MeV/nucleon}$.

EVALUATE: The neutron removal energy is about half the binding energy per nucleon.

43.55. **IDENTIFY:** Use the decay scheme and half-life of ^{90}Sr to find out the product of its decay and the amount left after a given time.
SET UP: The particle emitted in β^- decay is an electron, $^0_{-1}\text{e}$. In a time of one half-life, the number of radioactive nuclei decreases by a factor of 2. $6.25\% = \dfrac{1}{16} = 2^{-4}$

EXECUTE: **(a)** $^{90}_{38}\text{Sr} \rightarrow {}^0_{-1}\text{e} + {}^{90}_{39}\text{Y}$. The daughter nucleus is $^{90}_{39}\text{Y}$.

(b) 56 y is $2 T_{1/2}$ so $N = N_0/2^2 = N_0/4$; 25% is left.

(c) $\dfrac{N}{N_0} = 2^{-n}$; $\dfrac{N}{N_0} = 6.25\% = \dfrac{1}{16} = 2^{-4}$ so $t = 4 T_{1/2} = 112$ y.

EVALUATE: After half a century, ¼ of the ^{90}Sr would still be left!

43.57. **(a) IDENTIFY and SET UP:** The heavier nucleus will decay into the lighter one.
EXECUTE: $^{25}_{13}\text{Al}$ will decay into $^{25}_{12}\text{Mg}$.

(b) IDENTIFY and SET UP: Determine the emitted particle by balancing A and Z in the decay reaction.
EXECUTE: This gives $^{25}_{13}\text{Al} \rightarrow {}^{25}_{12}\text{Mg} + {}^0_{+1}\text{e}$. The emitted particle must have charge $+e$ and its nucleon number must be zero. Therefore, it is a β^+ particle, a positron.

(c) IDENTIFY and SET UP: Calculate the energy defect ΔM for the reaction and find the energy equivalent of ΔM. Use the nuclear masses for $^{25}_{13}\text{Al}$ and $^{25}_{12}\text{Mg}$, to avoid confusion in including the correct number of electrons if neutral atom masses are used.
EXECUTE: The nuclear mass for $^{25}_{13}\text{Al}$ is $M_\text{nuc}({}^{25}_{13}\text{Al}) = 24.990429 \text{ u} - 13(0.000548580 \text{ u}) = 24.983297 \text{ u}$.
The nuclear mass for $^{25}_{12}\text{Mg}$ is $M_\text{nuc}({}^{25}_{12}\text{Mg}) = 24.985837 \text{ u} - 12(0.000548580 \text{ u}) = 24.979254 \text{ u}$.
The mass defect for the reaction is

$$\Delta M = M_{\text{nuc}}(^{25}_{13}\text{Al}) - M_{\text{nuc}}(^{25}_{12}\text{Mg}) - M(^{0}_{+1}\text{e}) = 24.983297 \text{ u} - 24.979254 \text{ u} - 0.00054858 \text{ u} = 0.003494 \text{ u}$$

$$Q = (\Delta M)c^2 = 0.003494 \text{ u}(931.5 \text{ MeV/1 u}) = 3.255 \text{ MeV}$$

EVALUATE: The mass decreases in the decay and energy is released. Note: $^{25}_{13}\text{Al}$ can also decay into $^{25}_{12}\text{Mg}$ by the electron capture.

$$^{25}_{13}\text{Al} + {}^{0}_{-1}\text{e} \rightarrow {}^{25}_{12}\text{Mg}$$

The $^{0}_{-1}$ electron in the reaction is an orbital electron in the neutral $^{25}_{13}\text{Al}$ atom. The mass defect can be calculated using the nuclear masses:

$$\Delta M = M_{\text{nuc}}(^{25}_{13}\text{Al}) + M(^{0}_{-1}\text{e}) - M_{\text{nuc}}(^{25}_{12}\text{Mg}) = 24.983287 \text{ u} + 0.00054858 \text{ u} - 24.979254 \text{ u} = 0.004592 \text{ u}.$$

$$Q = (\Delta M)c^2 = (0.004592 \text{ u})(931.5 \text{ MeV/1 u}) = 4.277 \text{ MeV}$$

The mass decreases in the decay and energy is released.

43.59.   **IDENTIFY and SET UP:** The amount of kinetic energy released is the energy equivalent of the mass change in the decay. $m_e = 0.0005486 \text{ u}$ and the atomic mass of $^{14}_{7}\text{N}$ is 14.003074 u. The energy equivalent of 1 u is 931.5 MeV. ^{14}C has a half-life of $T_{1/2} = 5730 \text{ yr} = 1.81 \times 10^{11} \text{ s}$. The RBE for an electron is 1.0.

EXECUTE: **(a)** $^{14}_{6}\text{C} \rightarrow \text{e}^- + {}^{14}_{7}\text{N} + \bar{v}_e$.

(b) The mass decrease is $\Delta M = m(^{14}_{6}\text{C}) - [m_e + m(^{14}_{7}\text{N})]$. Use nuclear masses, to avoid difficulty in accounting for atomic electrons. The nuclear mass of $^{14}_{6}\text{C}$ is $14.003242 \text{ u} - 6m_e = 13.999950 \text{ u}$. The nuclear mass of $^{14}_{7}\text{N}$ is $14.003074 \text{ u} - 7m_e = 13.999234 \text{ u}$.

$\Delta M = 13.999950 \text{ u} - 13.999234 \text{ u} - 0.000549 \text{ u} = 1.67 \times 10^{-4} \text{ u}$. The energy equivalent of ΔM is 0.156 MeV.

(c) The mass of carbon is $(0.18)(75 \text{ kg}) = 13.5 \text{ kg}$. From Example 43.9, the activity due to 1 g of carbon in a living organism is 0.255 Bq. The number of decay/s due to 13.5 kg of carbon is $(13.5 \times 10^3 \text{ g})(0.255 \text{ Bq/g}) = 3.4 \times 10^3 \text{ decays/s}$.

(d) Each decay releases 0.156 MeV so 3.4×10^3 decays/s releases $530 \text{ MeV/s} = 8.5 \times 10^{-11} \text{ J/s}$.

(e) The total energy absorbed in 1 year is $(8.5 \times 10^{-11} \text{ J/s})(3.156 \times 10^7 \text{ s}) = 2.7 \times 10^{-3} \text{ J}$. The absorbed dose is $\dfrac{2.7 \times 10^{-3} \text{ J}}{75 \text{ kg}} = 3.6 \times 10^{-5} \text{ J/kg} = 36 \text{ } \mu\text{Gy} = 3.6 \text{ mrad}$. With RBE = 1.0, the equivalent dose is $36 \text{ } \mu\text{Sv} = 3.6 \text{ mrem}$.

EVALUATE: Section 43.5 says that background radiation exposure is about 1.0 mSv per year. The radiation dose calculated in this problem is much less than this.

43.61.   **IDENTIFY and SET UP:** Find the energy equivalent of the mass decrease. Part of the released energy appears as the emitted photon and the rest as kinetic energy of the electron.

EXECUTE: $^{198}_{79}\text{Au} \rightarrow {}^{198}_{80}\text{Hg} + {}^{0}_{-1}\text{e}$

The mass change is $197.968225 \text{ u} - 197.966752 \text{ u} = 1.473 \times 10^{-3} \text{ u}$

(The neutral atom masses include 79 electrons before the decay and 80 electrons after the decay. This one additional electron in the product accounts correctly for the electron emitted by the nucleus.) The total energy released in the decay is $(1.473 \times 10^{-3} \text{ u})(931.5 \text{ MeV/u}) = 1.372 \text{ MeV}$. This energy is divided between the energy of the emitted photon and the kinetic energy of the β^- particle. Thus the β^- particle has kinetic energy equal to $1.372 \text{ MeV} - 0.412 \text{ MeV} = 0.960 \text{ MeV}$.

EVALUATE: The emitted electron is much lighter than the $^{198}_{80}\text{Hg}$ nucleus, so the electron has almost all the final kinetic energy. The final kinetic energy of the ^{198}Hg nucleus is very small.

43.63.   **IDENTIFY and SET UP:** The decay is energetically possible if the total mass decreases. Determine the nucleus produced by the decay by balancing A and Z on both sides of the equation. $^{13}_{7}\text{N} \rightarrow {}^{0}_{+1}\text{e} + {}^{13}_{6}\text{C}$. To avoid confusion in including the correct number of electrons with neutral atom masses, use nuclear masses, obtained by subtracting the mass of the atomic electrons from the neutral atom masses.

EXECUTE: The nuclear mass for $^{13}_{7}\text{N}$ is $M_{\text{nuc}}\,(^{13}_{7}\text{N}) = 13.005739\text{ u} - 7(0.00054858\text{ u}) = 13.001899\text{ u}$.

The nuclear mass for $^{13}_{6}\text{C}$ is $M_{\text{nuc}}\,(^{13}_{6}\text{C}) = 13.003355\text{ u} - 6(0.00054858\text{ u}) = 13.000064\text{ u}$.

The mass defect for the reaction is

$\Delta M = M_{\text{nuc}}\,(^{13}_{7}\text{N}) - M_{\text{nuc}}\,(^{13}_{6}\text{C}) - M(^{0}_{+1}\text{e})$. $\Delta M = 13.001899\text{ u} - 13.000064\text{ u} - 0.00054858\text{ u} = 0.001286\text{ u}$.

EVALUATE: The mass decreases in the decay, so energy is released. This decay is energetically possible.

43.65. **IDENTIFY:** Assume the activity is constant during the year and use the given value of the activity to find the number of decays that occur in one year. Absorbed dose is the energy absorbed per mass of tissue. Equivalent dose is RBE times absorbed dose.

SET UP: For α particles, RBE = 20 (from Table 43.3).

EXECUTE: $(0.63 \times 10^{-6}\text{ Ci})(3.7 \times 10^{10}\text{ Bq/Ci})(3.156 \times 10^{7}\text{ s}) = 7.357 \times 10^{11}\,\alpha$ particles. The absorbed

dose is $\dfrac{(7.357 \times 10^{11})(4.0 \times 10^{6}\text{ eV})(1.602 \times 10^{-19}\text{ J/eV})}{(0.50\text{ kg})} = 0.943\text{ Gy} = 94.3\text{ rad}$. The equivalent dose is (20)

(94.3 rad) = 1900 rem.

EVALUATE: The equivalent dose is 19 Sv. This is large enough for significant damage to the person.

43.69. **IDENTIFY:** Use Eq. (43.17) to relate the initial number of radioactive nuclei, N_0, to the number, N, left after time t.

SET UP: We have to be careful; after ^{87}Rb has undergone radioactive decay it is no longer a rubidium atom. Let N_{85} be the number of ^{85}Rb atoms; this number doesn't change. Let N_0 be the number of ^{87}Rb atoms on earth when the solar system was formed. Let N be the present number of ^{87}Rb atoms.

EXECUTE: The present measurements say that $0.2783 = N/(N + N_{85})$.

$(N + N_{85})(0.2783) = N$, so $N = 0.3856 N_{85}$. The percentage we are asked to calculate is $N_0/(N_0 + N_{85})$.

N and N_0 are related by $N = N_0 e^{-\lambda t}$ so $N_0 = e^{+\lambda t} N$.

Thus $\dfrac{N_0}{N_0 + N_{85}} = \dfrac{N e^{\lambda t}}{N e^{\lambda t} + N_{85}} = \dfrac{(0.3855 e^{\lambda t}) N_{85}}{(0.3856 e^{\lambda t}) N_{85} + N_{85}} = \dfrac{0.3856 e^{\lambda t}}{0.3856 e^{\lambda t} + 1}$.

$t = 4.6 \times 10^{9}\text{ y}$; $\lambda = \dfrac{0.693}{T_{1/2}} = \dfrac{0.693}{4.75 \times 10^{10}\text{ y}} = 1.459 \times 10^{-11}\text{ y}^{-1}$

$e^{\lambda t} = e^{(1.459 \times 10^{-11}\text{ y}^{-1})(4.6 \times 10^{9}\text{ y})} = e^{0.06711} = 1.0694$

Thus $\dfrac{N_0}{N_0 + N_{85}} = \dfrac{(0.3856)(1.0694)}{(0.3856)(1.0694) + 1} = 29.2\%$.

EVALUATE: The half-life for ^{87}Rb is a factor of 10 larger than the age of the solar system, so only a small fraction of the ^{87}Rb nuclei initially present have decayed; the percentage of rubidium atoms that are radioactive is only a bit less now than it was when the solar system was formed.

43.71. **IDENTIFY** and **SET UP:** Find the energy emitted and the energy absorbed each second. Convert the absorbed energy to absorbed dose and to equivalent dose.

EXECUTE: **(a)** First find the number of decays each second:

$2.6 \times 10^{-4}\text{ Ci}\left(\dfrac{3.70 \times 10^{10}\text{ decays/s}}{1\text{ Ci}}\right) = 9.6 \times 10^{6}\text{ decays/s}$. The average energy per decay is 1.25 MeV, and

one-half of this energy is deposited in the tumor. The energy delivered to the tumor per second then is

$\frac{1}{2}(9.6 \times 10^{6}\text{ decays/s})(1.25 \times 10^{6}\text{ eV/decay})(1.602 \times 10^{-19}\text{ J/eV}) = 9.6 \times 10^{-7}\text{ J/s}$.

(b) The absorbed dose is the energy absorbed divided by the mass of the tissue:

$\dfrac{9.6 \times 10^{-7}\text{ J/s}}{0.200\text{ kg}} = (4.8 \times 10^{-6}\text{ J/kg} \cdot \text{s})(1\text{ rad}/(0.01\text{ J/kg})) = 4.8 \times 10^{-4}\text{ rad/s}$.

(c) equivalent dose $(REM) = RBE \times$ absorbed dose (rad). In one second the equivalent dose is

$(0.70)(4.8 \times 10^{-4}$ rad$) = 3.4 \times 10^{-4}$ rem.

(d) $(200$ rem$)/(3.4 \times 10^{-4}$ rem/s$) = (5.9 \times 10^{5}$ s$)(1$ h$/3600$ s$) = 164$ h $= 6.9$ days.

EVALUATE: The activity of the source is small so that absorbed energy per second is small and it takes several days for an equivalent dose of 200 rem to be absorbed by the tumor. A 200-rem dose equals 2.00 Sv and this is large enough to damage the tissue of the tumor.

43.75. **(a)** **IDENTIFY** and **SET UP:** Use Eq. (43.1) to calculate the radius R of a $_{1}^{2}$H nucleus. Calculate the Coulomb potential energy (Eq. 23.9) of the two nuclei when they just touch.

EXECUTE: The radius of $_{1}^{2}$H is $R = (1.2 \times 10^{-15}$ m$)(2)^{1/3} = 1.51 \times 10^{-15}$ m. The barrier energy is the Coulomb potential energy of two $_{1}^{2}$H nuclei with their centers separated by twice this distance:

$$U = \frac{1}{4\pi\epsilon_0}\frac{e^2}{r} = (8.988 \times 10^{9} \text{ N} \cdot \text{m}^2/\text{C}^2)\frac{(1.602 \times 10^{-19} \text{ C})^2}{2(1.51 \times 10^{-15} \text{ m})} = 7.64 \times 10^{-14} \text{ J} = 0.48 \text{ MeV}$$

(b) **IDENTIFY** and **SET UP:** Find the energy equivalent of the mass decrease.

EXECUTE: $_{1}^{2}$H $+ _{1}^{2}$H $\rightarrow _{2}^{3}$He $+ _{0}^{1}$n

If we use neutral atom masses there are two electrons on each side of the reaction equation, so their masses cancel. The neutral atom masses are given in Table 43.2.

$_{1}^{2}$H $+ _{1}^{2}$H has mass $2(2.014102$ u$) = 4.028204$ u

$_{2}^{3}$He $+ _{0}^{1}$n has mass 3.016029 u $+ 1.008665$ u $= 4.024694$ u

The mass decrease is 4.028204 u $- 4.024694$ u $= 3.510 \times 10^{-3}$ u. This corresponds to a liberated energy of $(3.510 \times 10^{-3}$ u$)(931.5$ MeV/u$) = 3.270$ MeV, or $(3.270 \times 10^{6}$ eV$)(1.602 \times 10^{-19}$ J/eV$) = 5.239 \times 10^{-13}$ J.

(c) **IDENTIFY** and **SET UP:** We know the energy released when two $_{1}^{2}$H nuclei fuse. Find the number of reactions obtained with one mole of $_{1}^{2}$H.

EXECUTE: Each reaction takes two $_{1}^{2}$H nuclei. Each mole of D_2 has 6.022×10^{23} molecules, so 6.022×10^{23} pairs of atoms. The energy liberated when one mole of deuterium undergoes fusion is $(6.022 \times 10^{23})(5.239 \times 10^{-13}$ J$) = 3.155 \times 10^{11}$ J/mol.

EVALUATE: The energy liberated per mole is more than a million times larger than from chemical combustion of one mole of hydrogen gas.

PARTICLE PHYSICS AND COSMOLOGY

44.3. **IDENTIFY:** The energy released is the energy equivalent of the mass decrease that occurs in the decay.
SET UP: The mass of the pion is $m_{\pi^+} = 270 m_e$ and the mass of the muon is $m_{\mu^+} = 207 m_e$. The rest
energy of an electron is 0.511 MeV.
EXECUTE: (a) $\Delta m = m_{\pi+} - m_{\mu+} = 270 m_e - 207 m_e = 63 m_e \Rightarrow E = 63(0.511\,\text{MeV}) = 32\,\text{MeV}$.

EVALUATE: (b) A positive muon has less mass than a positive pion, so if the decay from muon to pion
was to happen, you could always find a frame where energy was not conserved. This cannot occur.

44.5. **IDENTIFY:** The kinetic energy of the alpha particle is due to the mass decrease.
SET UP and EXECUTE: ${}_0^1 n + {}_5^{10}B \to {}_3^7 Li + {}_2^4 He$. The mass decrease in the reaction is

$m({}_0^1 n) + m({}_5^{10}B) - m({}_3^7 Li) - m({}_2^4 He) = 1.008665\,\text{u} + 10.012937\,\text{u} - 7.016004\,\text{u} - 4.002603\,\text{u} = 0.002995\,\text{u}$

and the energy released is $E = (0.002995\,\text{u})(931.5\,\text{MeV/u}) = 2.79\,\text{MeV}$. Assuming the initial momentum

is zero, $m_{Li} v_{Li} = m_{He} v_{He}$ and $v_{Li} = \dfrac{m_{He}}{m_{Li}} v_{He}$. $\frac{1}{2} m_{Li} v_{Li}^2 + \frac{1}{2} m_{He} v_{He}^2 = E$ becomes

$\frac{1}{2} m_{Li} \left(\dfrac{m_{He}}{m_{Li}}\right)^2 v_{He}^2 + \frac{1}{2} m_{He} v_{He}^2 = E$ and $v_{He} = \sqrt{\dfrac{2E}{m_{He}}\left(\dfrac{m_{Li}}{m_{Li} + m_{He}}\right)}$. $E = 4.470 \times 10^{-13}\,\text{J}$.

$m_{He} = 4.002603\,\text{u} - 2(0.0005486\,\text{u}) = 4.0015\,\text{u} = 6.645 \times 10^{-27}\,\text{kg}$.

$m_{Li} = 7.016004\,\text{u} - 3(0.0005486\,\text{u}) = 7.0144\,\text{u}$. This gives $v_{He} = 9.26 \times 10^6\,\text{m/s}$.

EVALUATE: The speed of the alpha particle is considerably less than the speed of light, so it is not
necessary to use the more complicated relativistic formulas.

44.9. (a) **IDENTIFY** and **SET UP:** Eq. (44.7) says $\omega = |q| B/m$ so $B = m\omega / |q|$. And since $\omega = 2\pi f$, this becomes
$B = 2\pi m f / |q|$.

EXECUTE: A deuteron is a deuterium nucleus (${}_1^2 H$). Its charge is $q = +e$. Its mass is the mass of the

neutral ${}_1^2 H$ atom (Table 43.2) minus the mass of the one atomic electron:

$m = 2.014102\,\text{u} - 0.0005486\,\text{u} = 2.013553\,\text{u}(1.66054 \times 10^{-27}\,\text{kg/1 u}) = 3.344 \times 10^{-27}\,\text{kg}$

$B = \dfrac{2\pi m f}{|q|} = \dfrac{2\pi(3.344 \times 10^{-27}\,\text{kg})(9.00 \times 10^6\,\text{Hz})}{1.602 \times 10^{-19}\,\text{C}} = 1.18\,\text{T}$

(b) Eq. (44.8): $K = \dfrac{q^2 B^2 R^2}{2m} = \dfrac{[(1.602 \times 10^{-19}\,\text{C})(1.18\,\text{T})(0.320\,\text{m})]^2}{2(3.344 \times 10^{-27}\,\text{kg})}$.

$K = 5.471 \times 10^{-13}\,\text{J} = (5.471 \times 10^{-13}\,\text{J})(1\,\text{eV}/1.602 \times 10^{-19}\,\text{J}) = 3.42\,\text{MeV}$

$K = \frac{1}{2} m v^2$ so $v = \sqrt{\dfrac{2K}{m}} = \sqrt{\dfrac{2(5.471 \times 10^{-13}\,\text{J})}{3.344 \times 10^{-27}\,\text{kg}}} = 1.81 \times 10^7\,\text{m/s}$

EVALUATE: $v/c = 0.06$, so it is ok to use the nonrelativistic expression for kinetic energy.

44.11. **(a) IDENTIFY and SET UP:** The masses of the target and projectile particles are equal, so Eq. (44.10) can be used. $E_a^2 = 2mc^2(E_m + mc^2)$. E_a is specified; solve for the energy E_m of the beam particles.

EXECUTE: $E_m = \dfrac{E_a^2}{2mc^2} - mc^2$

The mass for the alpha particle can be calculated by subtracting two electron masses from the ^4_2He atomic mass:

$m = m_\alpha = 4.002603 \text{ u} - 2(0.0005486 \text{ u}) = 4.001506 \text{ u}$

Then $mc^2 = (4.001506 \text{ u})(931.5 \text{ MeV/u}) = 3.727 \text{ GeV}$.

$E_m = \dfrac{E_a^2}{2mc^2} - mc^2 = \dfrac{(16.0 \text{ GeV})^2}{2(3.727 \text{ GeV})} - 3.727 \text{ GeV} = 30.6 \text{ GeV}$.

(b) Each beam must have $\frac{1}{2}E_a = 8.0 \text{ GeV}$.

EVALUATE: For a stationary target the beam energy is nearly twice the available energy. In a colliding beam experiment all the energy is available and each beam needs to have just half the required available energy.

44.13. **(a) IDENTIFY and SET UP:** For a proton beam on a stationary proton target and since E_a is much larger than the proton rest energy we can use Eq. (44.11): $E_a^2 = 2mc^2 E_m$.

EXECUTE: $E_m = \dfrac{E_a^2}{2mc^2} = \dfrac{(77.4 \text{ GeV})^2}{2(0.938 \text{ GeV})} = 3200 \text{ GeV}$

(b) IDENTIFY and SET UP: For colliding beams the total momentum is zero and the available energy E_a is the total energy for the two colliding particles.

EXECUTE: For proton-proton collisions the colliding beams each have the same energy, so the total energy of each beam is $\frac{1}{2}E_a = 38.7 \text{ GeV}$.

EVALUATE: For a stationary target less than 3% of the beam energy is available for conversion into mass. The beam energy for a colliding beam experiment is a factor of (1/83) times smaller than the required energy for a stationary target experiment.

44.15. **IDENTIFY:** The kinetic energy comes from the mass decrease.

SET UP: Table 44.3 gives $m(K^+) = 493.7 \text{ MeV}/c^2$, $m(\pi^0) = 135.0 \text{ MeV}/c^2$, and $m(\pi^\pm) = 139.6 \text{ MeV}/c^2$.

EXECUTE: **(a)** Charge must be conserved, so $K^+ \to \pi^0 + \pi^+$ is the only possible decay.

(b) The mass decrease is
$m(K^+) - m(\pi^0) - m(\pi^+) = 493.7 \text{ MeV}/c^2 - 135.0 \text{ MeV}/c^2 - 139.6 \text{ MeV}/c^2 = 219.1 \text{ MeV}/c^2$. The energy released is 219.1 MeV.

EVALUATE: The π mesons do not share this energy equally since they do not have equal masses.

44.19. **IDENTIFY and SET UP:** Find the energy equivalent of the mass decrease.

EXECUTE: The mass decrease is $m(\Sigma^+) - m(p) - m(\pi^0)$ and the energy released is
$mc^2(\Sigma^+) - mc^2(p) - mc^2(\pi^0) = 1189 \text{ MeV} - 938.3 \text{ MeV} - 135.0 \text{ MeV} = 116 \text{ MeV}$. (The mc^2 values for each particle were taken from Table 44.3.)

EVALUATE: The mass of the decay products is less than the mass of the original particle, so the decay is energetically allowed and energy is released.

44.33. **(a) IDENTIFY and SET UP:** Use Eq. (44.14) to calculate v.

EXECUTE: $v = \left[\dfrac{(\lambda_0/\lambda_S)^2 - 1}{(\lambda_0/\lambda_S)^2 + 1}\right]c = \left[\dfrac{(658.5 \text{ nm}/590 \text{ nm})^2 - 1}{(658.5 \text{ nm}/590 \text{ nm})^2 + 1}\right]c = 0.1094c$

$v = (0.1094)(2.998 \times 10^8 \text{ m/s}) = 3.28 \times 10^7 \text{ m/s}$

(b) IDENTIFY and SET UP: Use Eq. (44.15) to calculate r.

EXECUTE: $r = \dfrac{v}{H_0} = \dfrac{3.28 \times 10^4 \text{ km/s}}{(71(\text{km/s})/\text{Mpc})(1 \text{ Mpc}/3.26 \text{ Mly})} = 1510 \text{ Mly}$

EVALUATE: The red shift $\lambda_0 / \lambda_S - 1$ for this galaxy is 0.116. It is therefore about twice as far from earth as the galaxy in Examples 44.8 and 44.9, that had a red shift of 0.053.

44.35. **(a) IDENTIFY and SET UP:** Hubble's law is Eq. (44.15), with $H_0 = 71$ (km/s)/(Mpc). 1 Mpc = 3.26 Mly.

EXECUTE: $r = 5210$ Mly so $v = H_0 r = ((71 \text{ km/s})/\text{Mpc})(1 \text{ Mpc}/3.26 \text{ Mly})(5210 \text{ Mly}) = 1.1 \times 10^5$ km/s

(b) IDENTIFY and SET UP: Use v from part (a) in Eq. (44.13).

EXECUTE: $\dfrac{\lambda_0}{\lambda_S} = \sqrt{\dfrac{c+v}{c-v}} = \sqrt{\dfrac{1+v/c}{1-v/c}}$

$\dfrac{v}{c} = \dfrac{1.1 \times 10^8 \text{ m/s}}{2.9980 \times 10^8 \text{ m/s}} = 0.367$ so $\dfrac{\lambda_0}{\lambda_S} = \sqrt{\dfrac{1+0.367}{1-0.367}} = 1.5$

EVALUATE: The galaxy in Examples 44.8 and 44.9 is 710 Mly away so has a smaller recession speed and redshift than the galaxy in this problem.

44.39. **IDENTIFY and SET UP:** Find the energy equivalent of the mass decrease.

EXECUTE: **(a)** $p + {}_1^2\text{H} \rightarrow {}_2^3\text{He}$ or can write as ${}_1^1\text{H} + {}_1^2\text{H} \rightarrow {}_2^3\text{He}$

If neutral atom masses are used then the masses of the two atomic electrons on each side of the reaction will cancel.

Taking the atomic masses from Table 43.2, the mass decrease is $m({}_1^1\text{H}) + m({}_1^2\text{H}) - m({}_2^3\text{He}) = 1.007825$ u + 2.014102 u - 3.016029 u = 0.005898 u. The energy released is the energy equivalent of this mass decrease: $(0.005898 \text{ u})(931.5 \text{ MeV/u}) = 5.494$ MeV.

(b) ${}_0^1\text{n} + {}_2^3\text{He} \rightarrow {}_2^4\text{He}$

If neutral helium masses are used then the masses of the two atomic electrons on each side of the reaction equation will cancel. The mass decrease is $m({}_0^1\text{n}) + m({}_2^3\text{He}) - m({}_2^4\text{He}) = 1.008665$ u + 3.016029 u - 4.002603 u = 0.022091 u. The energy released is the energy equivalent of this mass decrease: $(0.022091 \text{ u})(931.15 \text{ MeV/u}) = 20.58$ MeV.

EVALUATE: These are important nucleosynthesis reactions, discussed in Section 44.7.

44.41. **IDENTIFY:** The reaction energy Q is defined in Eq. (43.23) and is the energy equivalent of the mass change in the reaction. When Q is negative the reaction is endoergic. When Q is positive the reaction is exoergic.
SET UP: Use the particle masses given in Section 43.1. 1 u is equivalent to 931.5 MeV.
EXECUTE: $\Delta m = m_e + m_p - m_n - m_{\nu_e}$ so assuming $m_{\nu_e} \approx 0$,

$\Delta m = 0.0005486 \text{ u} + 1.007276 \text{ u} - 1.008665 \text{ u} = -8.40 \times 10^{-4}$ u

$\Rightarrow E = (\Delta m)c^2 = (-8.40 \times 10^{-4} \text{ u})(931.5 \text{ MeV/u}) = -0.783$ MeV and is endoergic.

EVALUATE: The energy consumed in the reaction would have to come from the initial kinetic energy of the reactants.

44.43. **IDENTIFY and SET UP:** The Wien displacement law (Eq. 39.21) sys $\lambda_m T$ equals a constant. Use this to relate $\lambda_{m,1}$ at T_1 to $\lambda_{m,2}$ at T_2.

EXECUTE: $\lambda_{m,1} T_1 = \lambda_{m,2} T_2$

$\lambda_{m,1} = \lambda_{m,2}\left(\dfrac{T_2}{T_1}\right) = 1.062 \times 10^{-3} \text{ m}\left(\dfrac{2.728 \text{ K}}{3000 \text{ K}}\right) = 966 \text{ nm}$

EVALUATE: The peak wavelength was much less when the temperature was much higher.

44.45. **IDENTIFY and SET UP:** For colliding beams the available energy is twice the beam energy. For a fixed-target experiment only a portion of the beam energy is available energy (Eqs. 44.9 and 44.10).

EXECUTE: **(a)** $E_a = 2(7.0 \text{ TeV}) = 14.0 \text{ TeV}$

(b) Need $E_a = 14.0 \text{ TeV} = 14.0 \times 10^6 \text{ MeV}$. Since the target and projectile particles are both protons Eq. (44.10) can be used: $E_a^2 = 2mc^2(E_m + mc^2)$

$$E_m = \frac{E_a^2}{2mc^2} - mc^2 = \frac{(14.0 \times 10^6 \text{ MeV})^2}{2(938.3 \text{ MeV})} - 938.3 \text{ MeV} = 1.0 \times 10^{11} \text{ MeV} = 1.0 \times 10^5 \text{ TeV}.$$

EVALUATE: This shows the great advantage of colliding beams at relativistic energies.

44.47. **IDENTIFY:** The energy comes from a mass decrease.
SET UP: A charged pion decays into a muon plus a neutrino. The muon in turn decays into an electron or positron plus two neutrinos.
EXECUTE: **(a)** $\pi^- \rightarrow \mu^- + \text{neutrino} \rightarrow e^- + \text{three neutrinos}$.

(b) If we neglect the mass of the neutrinos, the mass decrease is

$m(\pi^-) - m(e^-) = 273 m_e - m_e = 272 m_e = 2.480 \times 10^{-28} \text{ kg}.$

$E = mc^2 = 2.23 \times 10^{-11} \text{ J} = 139 \text{ MeV}.$

(c) The total energy delivered to the tissue is $(50.0 \text{ J/kg})(10.0 \times 10^{-3} \text{ kg}) = 0.500 \text{ J}$. The number of π^- mesons required is $\dfrac{0.500 \text{ J}}{2.23 \times 10^{-11} \text{ J}} = 2.24 \times 10^{10}$.

(d) The RBE for the electrons that are produced is 1.0, so the equivalent dose is

$1.0(50.0 \text{ Gy}) = 50.0 \text{ Sv} = 5.0 \times 10^3 \text{ rem}.$

EVALUATE: The π^- are heavier than electrons and therefore behave differently as they hit the tissue.

44.49. **IDENTIFY:** With a stationary target, only part of the initial kinetic energy of the moving proton is available. Momentum conservation tells us that there must be nonzero momentum after the collision, which means that there must also be leftover kinetic energy. Therefore not all of the initial energy is available.
SET UP: The available energy is given by $E_a^2 = 2mc^2(E_m + mc^2)$ for two particles of equal mass when one is initially stationary. The *minimum* available energy must be equal to the rest mass energies of the products, which in this case is two protons, a K^+ and a K^-. The available energy must be at least the sum of the final rest masses.
EXECUTE: The minimum amount of available energy must be

$E_a = 2m_p + m_{K^+} + m_{K^-} = 2(938.3 \text{ MeV}) + 493.7 \text{ MeV} + 493.7 \text{ MeV} = 2864 \text{ MeV} = 2.864 \text{ GeV}$

Solving the available energy formula for E_m gives $E_a^2 = 2mc^2(E_m + mc^2)$ and

$$E_m = \frac{E_a^2}{2mc^2} - mc^2 = \frac{(2864 \text{ MeV})^2}{2(938.3 \text{ MeV})} - 938.3 \text{ MeV} = 3432.6 \text{ MeV}$$

Recalling that E_m is the *total* energy of the proton, including its rest mass energy (RME), we have

$$K = E_m - \text{RME} = 3432.6 \text{ MeV} - 938.3 \text{ MeV} = 2494 \text{ MeV} = 2.494 \text{ GeV}$$

Therefore the threshold kinetic energy is $K = 2494 \text{ MeV} = 2.494 \text{ GeV}$.
EVALUATE: Considerably less energy would be needed if the experiment were done using colliding beams of protons.

44.53. **IDENTIFY** and **SET UP:** Apply the Heisenberg uncertainty principle in the form $\Delta E \Delta t \approx \hbar/2$. Let ΔE be the energy width and let Δt be the lifetime.

EXECUTE: $\dfrac{\hbar}{2\Delta E} = \dfrac{(1.054 \times 10^{-34} \text{ J} \cdot \text{s})}{2(4.4 \times 10^6 \text{ eV})(1.6 \times 10^{-19} \text{ J/eV})} = 7.5 \times 10^{-23} \text{ s}.$

EVALUATE: The shorter the lifetime, the greater the energy width.

44.55. **IDENTIFY:** Apply $\left| \dfrac{dN}{dt} \right| = \lambda N$ to find the number of decays in one year.

SET UP: Water has a molecular mass of $18.0 \times 10^{-3} \text{ kg/mol}$.

EXECUTE: **(a)** The number of protons in a kilogram is

$$(1.00 \text{ kg})\left(\frac{6.022\times10^{23} \text{ molecules/mol}}{18.0\times10^{-3} \text{ kg/mol}}\right)(2 \text{ protons/molecule}) = 6.7\times10^{25}.$$ Note that only the protons in the

hydrogen atoms are considered as possible sources of proton decay. The energy per decay is

$m_p c^2 = 938.3 \text{ MeV} = 1.503\times10^{-10} \text{ J}$, and so the energy deposited in a year, per kilogram, is

$$(6.7\times10^{25})\left(\frac{\ln(2)}{1.0\times10^{18} \text{ y}}\right)(1 \text{ y})(1.50\times10^{-10} \text{ J}) = 7.0\times10^{-3} \text{ Gy} = 0.70 \text{ rad}.$$

(b) For an RBE of unity, the equivalent dose is $(1)(0.70 \text{ rad}) = 0.70 \text{ rem}$.

EVALUATE: The equivalent dose is much larger than that due to the natural background. It is not feasible
for the proton lifetime to be as short as 1.0×10^{18} y.

44.59.   **IDENTIFY:** The matter density is proportional to $1/R^3$.
  SET UP and **EXECUTE:** **(a)** When the matter density was large enough compared to the dark energy density,
  the slowing due to gravitational attraction would have dominated over the cosmic repulsion due to dark energy.

  (b) Matter density is proportional to $1/R^3$, so $R \propto \dfrac{1}{\rho^{1/3}}$. Therefore $\dfrac{R}{R_0} = \left(\dfrac{1/\rho_{past}}{1/\rho_{now}}\right)^{1/3} = \left(\dfrac{\rho_{now}}{\rho_{past}}\right)^{1/3}$. If ρ_m

  and ρ_{DE} are the present-day densities of matter of all kinds and of dark energy, we have $\rho_{DE} = 0.726\rho_{crit}$

  and $\rho_m = 0.274\rho_{crit}$ at the present time. Putting this into the above equation for R/R_0 gives

  $$\frac{R}{R_0} = \left(\frac{\dfrac{0.274}{0.726}\rho_{DE}}{2\rho_{DE}}\right)^{1/3} = 0.574.$$

  EVALUATE: **(c)** 300 My: speeding up $(R/R_0 = 0.98)$; 10.2 Gy: slowing down $(R/R_0 = 0.35)$.

44.61.   **IDENTIFY:** The kinetic energy comes from the mass difference.
  SET UP and **EXECUTE:** $K_\Sigma = 180 \text{ MeV}$. $m_\Sigma c^2 = 1197 \text{ MeV}$. $m_n c^2 = 939.6 \text{ MeV}$. $m_\pi c^2 = 139.6 \text{ MeV}$.

  $E_\Sigma = K_\Sigma + m_\Sigma c^2 = 180 \text{ MeV} + 1197 \text{ MeV} = 1377 \text{ MeV}$. Conservation of the x-component of momentum

  gives $p_\Sigma = p_{nx}$. Then $p_{nx}^2 c^2 = p_\Sigma^2 c^2 = E_\Sigma^2 - (m_\Sigma c)^2 = (1377 \text{ MeV})^2 - (1197 \text{ MeV})^2 = 4.633\times10^5 \text{ (MeV)}^2$.

  Conservation of energy gives $E_\Sigma = E_\pi + E_n$. $E_\Sigma = \sqrt{m_\pi^2 c^4 + p_\pi^2 c^2} + \sqrt{m_n^2 c^4 + p_n^2 c^2}$.

  $E_\Sigma - \sqrt{m_n^2 c^4 + p_n^2 c^2} = \sqrt{m_\pi^2 c^4 + p_\pi^2 c^2}$. Square both sides:

  $E_\Sigma^2 + m_n^2 c^4 + p_{nx}^2 c^2 + p_{ny}^2 c^2 - 2E_\Sigma E_n = m_\pi^2 c^4 + p_\pi^2 c^2$. $p_\pi = p_{ny}$ so

  $E_\Sigma^2 + m_n^2 c^4 + p_{nx}^2 c^2 - 2E_\Sigma E_n = m_\pi^2 c^4$ and $E_n = \dfrac{E_\Sigma^2 + m_n^2 c^4 - m_\pi^2 c^4 + p_{nx}^2 c^2}{2E_\Sigma}$.

  $E_n = \dfrac{(1377 \text{ MeV})^2 + (939.6 \text{ MeV})^2 - (139.6 \text{ MeV})^2 + 4.633\times10^5 \text{ (MeV)}^2}{2(1377 \text{ MeV})} = 1170 \text{ MeV}$.

  $K_n = E_n - m_n c^2 = 1170 \text{ MeV} - 939.6 \text{ MeV} = 230 \text{ MeV}$.

  $E_\pi = E_\Sigma - E_n = 1377 \text{ MeV} - 1170 \text{ MeV} = 207 \text{ MeV}$.

  $K_\pi = E_\pi - m_\pi c^2 = 207 \text{ MeV} - 139.6 \text{ MeV} = 67 \text{ MeV}$.

  $p_n^2 c^2 = E_n^2 - m_n^2 c^2 = (1170 \text{ MeV})^2 - (939.6 \text{ MeV})^2 = 4.861\times10^5 \text{ (MeV)}^2$. The angle θ the velocity of the

  neutron makes with the $+x$-axis is given by $\cos\theta = \dfrac{p_{nx}}{p_n} = \sqrt{\dfrac{4.633\times10^5}{4.861\times10^5}}$ and $\theta = 12.5°$ below the $+x$-axis.

  EVALUATE: The decay particles do not have equal energy because they have different masses.